Marine Wastewater Outfalls
and Treatment Systems

Marine Wastewater Outfalls and Treatment Systems

Philip J. W. Roberts
Georgia Institute of Technology
Atlanta, Georgia, USA

Henry J. Salas
Pan American Health Organization
Lima, Peru

Fred M. Reiff
Reiff Engineering
Chevy Chase, Maryland, USA

Menahem Libhaber
The World Bank
Washington, DC, USA

Alejandro Labbe
Halcrow Group Limited-Chile
Santiago, Chile

James C. Thomson
Consulting Engineer
Geneva, Switzerland

Publishing
London · New York

Published by   IWA Publishing
Alliance House
12 Caxton Street
London SW1H 0QS, UK
Telephone: +44 (0)20 7654 5500
Fax: +44 (0)20 7654 5555
Email: publications@iwap.co.uk
Web: www.iwapublishing.com

First published 2010
Reprinted 2011
© 2010 IWA Publishing

Typeset in India by OKS Prepress Services.
Printed in Great Britain by the MPG Books Group, Bodmin and King's Lynn

Apart from any fair dealing for the purposes of research or private study, or criticism or review, as permitted under the UK Copyright, Designs and Patents Act (1998), no part of this publication may be reproduced, stored or transmitted in any form or by any means, without the prior permission in writing of the publisher, or, in the case of photographic reproduction, in accordance with the terms of licences issued by the Copyright Licensing Agency in the UK, or in accordance with the terms of licenses issued by the appropriate reproduction rights organization outside the UK. Enquiries concerning reproduction outside the terms stated here should be sent to IWA Publishing at the address printed above.

The publisher makes no representation, express or implied, with regard to the accuracy of the information contained in this book and cannot accept any legal responsibility or liability for errors or omissions that may be made.

Disclaimer
The information provided and the opinions given in this publication are not necessarily those of IWA Publishing and should not be acted upon without independent consideration and professional advice. IWA Publishing and the Author will not accept responsibility for any loss or damage suffered by any person acting or refraining from acting upon any material contained in this publication.

British Library Cataloguing in Publication Data
A CIP catalogue record for this book is available from the British Library

Library of Congress Cataloging-in-Publication Data
A catalog record for this book is available from the Library of Congress

ISBN 10: 1843391899
ISBN 13: 9781843391890

Contents

Preface	xv
Acknowledgments	xix
About the Authors	xxi

1 INTRODUCTION 1
  1.1 Objectives and Philosophy of Marine Wastewater Disposal 1
  1.2 Water Quality Aspects 4
  1.3 Appropriate Treatment for Ocean Outfall Discharges 5
  1.4 Wastewater Disposal in Latin America and the Caribbean 8
  1.5 Book Outline 11

2 WATER QUALITY ASPECTS 15
  2.1 Introduction 15
  2.2 General Water Quality Issues 16
  2.3 Concept of a Mixing Zone 18

© 2010 IWA Publishing. *Marine Wastewater Outfalls and Treatment Systems.* By Philip JW Roberts, Henry J Salas, Fred M Reiff, Menahem Libhaber, Alejandro Labbe, and James C Thomson. ISBN: 9781843391890. Published by IWA Publishing, London, UK.

2.4	The California Ocean Plan	20
	2.4.1 General provisions	21
	2.4.2 Bacterial characteristics	21
	2.4.3 Physical characteristics	22
	2.4.4 Chemical characteristics	22
	2.4.5 Biological characteristics	24
	2.4.6 Radioactivity	24
	2.4.7 Discussion of the plan	24
2.5	Standards for Toxics	25
2.6	Bacterial Contaminants	26
	2.6.1 Introduction	26
	2.6.2 Historical review	27
	2.6.2.1 The United States of America (USA)	27
	2.6.2.2 International organizations	33
	2.6.3 Standards for human health protection	35
	2.6.3.1 WHO guidelines	35
	The 95th percentile approach	38
	Guidelines for seawater	42
	2.6.3.2 Other standards	42
	2.6.4 Standards for shellfish	45
	2.6.5 Standards for indigenous organisms	47
2.7	Recommendations	47

3 WASTEWATER MIXING AND DISPERSION 51

3.1	Introduction	51
3.2	Features of Coastal Waters	54
	3.2.1 Introduction	54
	3.2.2 Water motions	55
	3.2.3 Density stratification	56
	3.2.4 Winds	57
3.3	Mechanisms and Prediction of Wastefield Fate and Transport	59
	3.3.1 Introduction	59
	3.3.2 Near field mixing	59
	3.3.2.1 Introduction	59
	3.3.2.2 Horizontal buoyant jet in stationary, homogeneous environment	60
	3.3.2.3 Multiple buoyant jets in stationary, homogeneous environment	66

		3.3.2.4	Effect of flowing currents: single plume	68
		3.3.2.5	Merging plumes in a flowing current	70
		3.3.2.6	Single plume into stationary, stratified flow	73
		3.3.2.7	Merging plumes into stationary, stratified environment	75
		3.3.2.8	Merging plumes into a flowing, stratified current	76
		3.3.2.9	Discussion and summary	80
	3.3.3	Far field mixing		81
		3.3.3.1	Introduction	81
		3.3.3.2	Oceanic turbulent mixing	82
		3.3.3.3	Statistical far field model	84
	3.3.4	Long-term flushing		91
3.4	Conclusions			93

4 MODELING TRANSPORT PROCESSES AND WATER QUALITY ... 95

4.1	Introduction		95
4.2	Near Field and Mixing Zone Models		96
	4.2.1	Length-scale models	96
	4.2.2	Entrainment models	96
	4.2.3	CFD Models	101
4.3	Some Near Field and Mixing Zone Models		103
	4.3.1	Introduction	103
	4.3.2	Visual Plumes	103
	4.3.3	NRFIELD	107
	4.3.4	Cormix	108
	4.3.5	VISJET	109
4.4	Far Field Models		109
	4.4.1	Hydrodynamic models	109
	4.4.2	Nesting technique	113
4.5	Water Quality Models		115
	4.5.1	Introduction	115
	4.5.2	Eulerian models	115
	4.5.3	Lagrangian models	116
4.6	Model Coupling		119
4.7	Numerical Modeling Case Studies		123
	4.7.1	Introduction	123
	4.7.2	Mamala Bay	123

viii   Marine Wastewater Outfalls and Treatment Systems

		4.7.3	Rio de Janiero	127
		4.7.4	Cartagena	128
		4.7.5	Boston	132
	4.8	Physical Models		133
	4.9	Discussion		134

5 FIELD SURVEYS AND DATA REQUIREMENTS 137

	5.1	Introduction		137
	5.2	Physical Oceanography and Meteorology		138
		5.2.1	Introduction	138
		5.2.2	Currents	138
		5.2.3	Density stratification	145
		5.2.4	Waves and tides	149
		5.2.5	Meteorology	149
	5.3	Bathymetry and Geophysical Studies		151
		5.3.1	Bathymetry	151
		5.3.2	Geophysical	153
	5.4	Water Quality		155
	5.5	Bacterial Decay (T_{90})		156
		5.5.1	On-site measurement in artificial plume	156
		5.5.2	*In-situ* measurement in existing wastewater discharge	157
		5.5.3	Bottle method	157

6 WASTEWATER MANAGEMENT AND TREATMENT ... 159

	6.1	Introduction		159
	6.2	Global Water Supply and Wastewater Disposal		162
	6.3	Proposed Guidelines for Good Wastewater Management Practice in Developing Countries		164
		6.3.1	Overview	164
		6.3.2	The utility perspective	165
		6.3.3	The government perspective	166
	6.4	Effluent Standards		167
		6.4.1	Standards in industrialized countries	167
		6.4.2	Standards in developing countries	168
	6.5	Key Pollutants in Wastewater		169
	6.6	Principals and Processes of Wastewater Treatment and Disposal		171
		6.6.1	Introduction	171

	6.6.2	Treatment processes	172
	6.6.3	Sludge treatment and disposal	177
	6.6.4	Summary and costs	179
6.7	Appropriate Treatment Technology		180
	6.7.1	Introduction	180
	6.7.2	Preliminary treatment	183
	6.7.3	Physico-chemical treatment	190
		6.7.3.1 The case for physico-chemical treatment	190
		6.7.3.2 Chemically Enhanced Primary Treatment (CEPT)	192
		6.7.3.3 CEPT followed by filtration and disinfection	193
		6.7.3.4 Chemically enhanced solids separation by rotating fine screens (CERFS)	195
		6.7.3.5 CERFS followed by filtration and disinfection	196
	6.7.4	Management and disposal of solid wastes	197
		6.7.4.1 Solid wastes from preliminary treatment processes	197
		6.7.4.2 Sludge management	199
	6.7.5	Treatment costs	200
6.8	Wastewater Treatment and Global Warming		202
6.9	Conclusions		203

7 MATERIALS FOR SMALL AND MEDIUM DIAMETER OUTFALLS ... **205**

7.1	General Considerations	205
7.2	Specific Considerations	208
	7.2.1.1 Pipe connections	208
	7.2.2 Internal pressure	209
	7.2.3 Resistance to externally imposed forces	209
	7.2.4 Flexibility	210
7.3	Types of Pipe Used in Submarine Outfalls	210
	7.3.1 Concrete pipe	210
	7.3.2 Steel pipe	213
	7.3.3 Ductile iron pipe	214
	7.3.4 GRP pipe	216
	7.3.5 PVC Pipe	217
	7.3.6 Polyethylene (HDPE) pipe	218
7.4	Corrosion Protection	219

8 OCEANIC FORCES ON SUBMARINE OUTFALLS 223

- 8.1 Introduction 223
- 8.2 Currents 224
- 8.3 Waves 226
  - 8.3.1 Design wave 226
  - 8.3.2 Prediction of wind-driven waves 228
  - 8.3.3 Goda's simplified procedure for wind wave prediction 229
  - 8.3.4 Transformation of deep water waves approaching shore 231
  - 8.3.5 Velocity and acceleration under a passing wave 233
- 8.4 Forces due to a Steady Current 236
- 8.5 Wave-Induced Forces 240
  - 8.5.1 Horizontal forces 241
  - 8.5.2 Vertical forces 244
- 8.6 Hydrostatic Pressure Forces 245

9 DESIGN OF POLYETHYLENE OUTFALLS 247

- 9.1 Introduction 247
- 9.2 Reference Standards for Polyethylene Materials and Pipe 248
- 9.3 Selecting the Pipe Diameter 253
- 9.4 Stabilizing and Protecting the Outfall 256
  - 9.4.1 Stabilization with concrete ballast weights 257
  - 9.4.2 Stabilization with mechanical anchors 262
  - 9.4.3 Protection by entrenchment 263
  - 9.4.4 Articulated concrete block mats (ACBM) 266
- 9.5 Stresses on an HDPE Outfall 268
  - 9.5.1 Pipe stress due to currents and waves 269
  - 9.5.2 Stress during flotation 270
  - 9.5.3 Bending stresses 271
    - 9.5.3.1 Minimum bending radius 272
    - 9.5.3.2 Stress due to bending 273
- 9.6 Hydraulic Considerations 274
  - 9.6.1 Criteria 274
    - 9.6.1.1 Self-cleaning velocities 275
    - 9.6.1.2 Elimination of entrained or trapped air 276
- 9.7 Designing the Diffuser 279
  - 9.7.1 Introduction 279
  - 9.7.2 Diffuser configuration 279

		9.7.3	Diffuser design objectives	281
		9.7.4	Port configurations	284
		9.7.5	Hydraulic calculations	286
		9.7.6	Check valves	290
	9.8	Extremely Deep Outfalls		293
	9.9	Very Shallow Outfalls		296

10 OCEAN OUTFALL CONSTRUCTION — 299

10.1	Introduction		299
10.2	Important Preliminary Considerations for Construction		301
	10.2.1 Permits and compliance with regulations		301
	10.2.2 Seabed conditions		301
	10.2.3 Sea conditions, wind, waves, tides, and currents		302
	10.2.4 Surf zone conditions		303
	10.2.5 Construction offices		303
	10.2.6 Supply of pipe and fittings		304
	10.2.7 Storage of materials and equipment		306
	10.2.8 Public relations		307
10.3	Outfall Installation Methods		307
	10.3.1 Flotation and submersion		308
		10.3.1.1 Temporary working platform and launch facility	309
		10.3.1.2 Butt fusion welding HDPE pipe	312
		10.3.1.3 Concrete ballasts	315
		10.3.1.4 Testing the outfall	319
		10.3.1.5 Launching the outfall	319
		10.3.1.6 Towing and positioning the outfall	321
		10.3.1.7 Submerging the outfall	322
		10.3.1.8 Post submergence activities	327
	10.3.2 Bottom pull method		327
	10.3.3 Mobile jack-up platforms		330
	10.3.4 Outfall installation from a lay barge		332
	10.3.5 Installation from a floating crane barge		333
10.4	Temporary Trestles		334
10.5	Subsea Excavation		336
	10.5.1 Excavation in an unconsolidated seabed		336
	10.5.2 Subsea excavation in rock		342
10.6	Installation of Mechanical Anchors		344
10.7	Diffuser Installation		344

xii   Marine Wastewater Outfalls and Treatment Systems

11 OUTFALL INSTALLATION BY TRENCHLESS TECHNIQUES **347**

- 11.1 Developments in the Underground Installation of Pipelines 347
- 11.2 Advantages of Installation by Trenchless Methods 348
- 11.3 The Site and Geotechnical Investigation 350
- 11.4 Marine Outfalls Installed by Tunneling 351
  - 11.4.1 Background 351
  - 11.4.2 The tunneling process 351
  - 11.4.3 Case studies 356
- 11.5 Marine Outfalls Installed by Microtunneling 364
  - 11.5.1 Background 364
  - 11.5.2 The microtunneling process 366
  - 11.5.3 Case studies 369
- 11.6 Marine Outfalls Installed by Horizontal Directional Drilling 376
  - 11.6.1 Background 376
  - 11.6.2 Horizontal directional drilling 376
  - 11.6.3 Marine outfall and intake installations 379
  - 11.6.4 Case studies 380
- 11.7 New Developments 384
- 11.8 Summary of Applications 388
- 11.9 Bibliography 389

12 OPERATION AND MAINTENANCE **391**

- 12.1 Introduction 391
- 12.2 Maintaining a Clean Pipe Bore 392
- 12.3 Corrosion Protection Systems 394
- 12.4 Diffuser Maintenance 396
- 12.5 Routine Visual Inspection 397
- 12.6 Recording Discharge Flow 398
- 12.7 Recording Total Head 399
- 12.8 Miscellaneous Maintenance and Repair 399
- 12.9 Costs 400
- 12.10 Summary 400

13 MONITORING **403**

- 13.1 Introduction 403
- 13.2 Monitoring Parameters 404
- 13.3 Outfall-Related Monitoring 405
  - 13.3.1 Introduction 405

Contents   xiii

      13.3.2 Pre-discharge monitoring 405
      13.3.3 Post-discharge monitoring 406
            13.3.3.1 Physical and chemical sampling 406
            13.3.3.2 Biological sampling 407
            13.3.3.3 Wastewater characteristics 408
      13.3.4 Sample Monitoring Program 408
  13.4 Routine Beach Monitoring 410
  13.5 Summary 411

14 CASE STUDIES **413**

  14.1 Introduction 413
  14.2 Cartagena, Colombia 414
      14.2.1 Background 414
      14.2.2 Institutional aspects of the Cartagena water and
            sanitation sector 416
      14.2.3 The World Bank's support to Cartagena 418
      14.2.4 The proposed wastewater disposal scheme 421
      14.2.5 Social aspects of the outfall 427
  14.3 Concepcion, Chile 428
  14.4 Santa Marta, Colombia 434

15 GAINING PUBLIC ACCEPTANCE **439**

  15.1 Introduction 439
  15.2 Cartagena Case Study 442

APPENDIX A: OUTFALL COSTS **447**

  1. Introduction 447
  2. Outfall Costs 448
  3. Cost Issues 456
      3.1 Overview 456
      3.2 Direct costs 458
      3.3 Indirect costs 458
      3.4 Example calculation for typical HDPE outfall 459
  4. Summary 460

APPENDIX B: ABBREVIATIONS **461**

References **465**

Index **479**

Preface

Extensive experience around the world has shown that disposing of properly treated domestic wastewater through effective ocean outfalls is an economical and reliable strategy for wastewater disposal with minimal environmental effects.

Unfortunately, this is not well known. Because of extensive publicity about "dying" oceans, polluted beaches, and failures of poorly or inadequately designed outfalls, outfalls are often perceived poorly in the eyes of the public and political officials. This, combined with lack of reliable information, has resulted in outfalls often not being considered, or disposal schemes chosen that are poorly suited to the task. This is unnecessary. The technology of marine wastewater disposal has advanced rapidly in recent years, with improved materials becoming available for oceanic pipelines, better marine construction techniques, and advances in oceanographic instrumentation, mathematical modeling, and understanding of the mechanisms that work to assimilate and render harmless the discharged pollutants. Nevertheless, ocean outfall design is a somewhat arcane endeavor that depends on many disciplines including, to list but a few, oceanography, civil and environmental engineering, construction,

© 2010 IWA Publishing. *Marine Wastewater Outfalls and Treatment Systems.* By Philip JW Roberts, Henry J Salas, Fred M Reiff, Menahem Libhaber, Alejandro Labbe, and James C Thomson. ISBN: 9781843391890. Published by IWA Publishing, London, UK.

economics, and public relations. These diverse disciplines are rarely brought together in a single book, and this is the aim here.

With increasing population and economic growth, treatment and safe disposal of wastewater is an essential objective to preserve public health and reduce intolerable levels of environmental degradation. Effective wastewater management is well established in developed countries, but is still limited in developing countries. Developing countries must first expand coverage of water supply and sanitation services and only then provide safe wastewater treatment and disposal. The governments in Europe, the USA and Japan have been quite generous in providing large capital subsidies to finance substantial portions of wastewater management investments, but this does not usually happen in developing countries. Population growth forecasts indicate that most of the world's population growth will occur in developing countries. This will reduce the potential to expand coverage of wastewater treatment, unless innovative affordable treatment technologies are made available. The key to expanding wastewater treatment is appropriate technologies based on simple processes that are less expensive than conventional ones in capital costs and operation and maintenance, simple to operate, reliable, and can yield an effluent of any specified quality. This book advocates appropriate treatment combined with effective outfalls as the best viable technology for wastewater management in coastal communities.

More than half of the World's population lives within 60 km of the coastline. The natural receiving body of the wastewater of much of this population is the sea. In developing countries, where wastewater is often discharged as raw sewage to urban ditches and creeks, there is a strong incentive to explore the safe discharge to the sea using an appropriate combination of pretreatment and an effective outfall. As explained in this book, this option is an affordable, effective, and simple to operate solution with minimal environmental impacts and a proven track record in many coastal cities worldwide, and is therefore often a good solution. The technology presented herein is not specific for developing countries, and it is hoped that this book will contribute to sound wastewater management in all coastal cities and communities.

This book concerns the design of marine wastewater disposal systems: that is an ocean outfall plus treatment plant. The emphasis is on the outfall, and discussions of wastewater treatment are limited to those issues most relevant to marine disposal. The outfall design is particularly concerned with the use of High Density Polyethylene (HDPE) pipe, whose use is increasing rapidly around the world. In the past, this would have restricted interest to "small" diameter outfalls serving small communities. Advances in construction technology, however, now means that HDPE pipes can be constructed in diameters up two meters or more and delivered almost anywhere in the world. The emphasis on HDPE is therefore not overly restrictive, and much of the material in this book is relevant to outfalls built of other materials such as concrete or steel, and even tunneled outfalls.

Our initial goal was to write a "design manual" for ocean outfalls. This proved to be impossible, however, due to the wide range of disciplines involved, rapid advances in all of them, the unique nature of many outfalls, and particularly the complexity of the tasks involved. To cover them all adequately would require making each chapter into a book. Indeed, an entire book has been recently published (Grace, 2009) that covers marine outfall construction, the topic discussed in Chapter 10. Similar comments could be made about modeling, etc. Therefore, we elected to cover essentially all of the issues and topics related to marine wastewater disposal, albeit at a less detailed level, with references to where the reader can go for more information.

In addition to scientific articles, the extensive, and rapidly growing, literature on marine wastewater disposal includes publications such as books and manuals. These include text books that cover all issues, such as Grace (1978), Gunnerson and French (1996), Wood et al. (1993); books that specialize on policy issues, such as NRC (1993); engineering design manuals such as WRC (1990), Pipelife (2002), Neville-Jones and Chitty (1996), and Quetin and de Rouville (1986); and collected papers such as Myers and Harding (1983), and the proceedings of specialized conferences such as MWWD (International Conferences on Marine Waste Water Discharges). Since publication of these works, however, almost all facets of marine outfalls have advanced rapidly. Changes include materials and construction methods such as large diameter HDPE pipe, trenchless construction, and dredging; instrumentation for oceanographic measurements such as recording profiling current meters and thermistor and conductivity strings, low cost GPS drifters, and surface current radar; numerical modeling, especially of three-dimensional coastal water dynamics and near field processes; and knowledge of near field mixing gained by new laboratory techniques such as laser-induced fluorescence. There is a strong need therefore for a book that covers these new topics and presents advances. We attempt to do so, at least partially, in this book.

The authors hope that the book will provide information that will be of value to all who are involved in any way with wastewater disposal through marine outfalls. This includes those involved in the decision-making process for outfalls, those involved in their design, at least to a preliminary level, and those concerned with their environmental impacts. We hope that the book contributes to rational choices for sanitary schemes based on sound technical and scientific information about the options available.

Acknowledgments

Much of this book was written while the first author (PJWR) was a Distinguished Scholar in the Oceans and Human Health Initiative (OHHI) of the National Oceanic and Atmospheric Administration (NOAA). PJWR is especially grateful to the Director of OHHI, Ms. Juli Trtanj, for her support which helped make this book possible. Many people contributed to the writing of this book. The authors are particularly indebted to Dr. Walter Frick of the U.S. EPA for supplying material on Visual Plumes in Chapter 4, to Steve Bartlett of Ocean Surveys, Inc. for supplying photographs of oceanographic instrumentation, to Mr. Trygve Blomster of PipeLife for photographs of HDPE pipelines, to Mr. Hans G. Huber of Huber Technology for ideas, information and photographs of preliminary wastewater treatment equipment, David Velasco of Sontek/YSI for illustrations of current meters, Mike Duer of Tideflex Technologies, and many others who supplied illustrations.

    The authors also recognize the efforts of the Pan American Health Organization through its Regional Program coordinated by Henry Salas to disseminate the marine outfall option throughout Latin America and the Caribbean as a viable alternative for wastewater disposal and the protection of

© 2010 IWA Publishing. *Marine Wastewater Outfalls and Treatment Systems.* By Philip JW Roberts, Henry J Salas, Fred M Reiff, Menahem Libhaber, Alejandro Labbe, and James C Thomson. ISBN: 9781843391890. Published by IWA Publishing, London, UK.

human health through documents and numerous courses and workshops for which the authors prepared material that subsequently made its way into this book.

Similarly, the authors acknowledge the support of the World Bank in promoting and financing projects which incorporate effluent disposal through marine outfalls, supporting the acceptance of the technology of outfalls, and helping to prepare and publish this book. Finally, the authors are very appreciative of the financial support from Halcrow Group Limited, Huber Technology, and Pipelife Norge AS, which allowed us to publish this book in color.

About the Authors

Philip Roberts, PhD, PE, Consulting Engineer and Professor, Georgia Institute of Technology, Atlanta, USA. Dr. Roberts' professional interests include the mixing and dynamics of natural water bodies, the engineering design of intakes and ocean outfalls, mathematical modeling of water quality, field studies, and laboratory studies using sophisticated instrumentation of mixing in stratified fluids. He is an authority on the fluid mechanics of outfall diffuser mixing and the development and application of mathematical models of wastewater fate and transport. He has extensive international experience in marine waste disposal including the design of ocean outfalls, review of schemes, numerical modeling, and oceanographic field work program design and data interpretation His mathematical models and methods have been adopted by the US EPA and are widely used. He is a regular lecturer at the EPA Mixing Zone Workshops on the use of mathematical models and on outfall design for the Pan American Health Organization. He conducts research on diffuser mixing processes and has published extensively in this area.

© 2010 IWA Publishing. *Marine Wastewater Outfalls and Treatment Systems.* By Philip JW Roberts, Henry J Salas, Fred M Reiff, Menahem Libhaber, Alejandro Labbe, and James C Thomson. ISBN: 9781843391890. Published by IWA Publishing, London, UK.

For this research he was awarded the Collingwood Prize of ASCE in 1980, and was UPS Foundation Visiting Professor at Stanford University in 1993–94. Dr. Roberts has lectured widely on outfall design around the world and is presently Co-Chairman of the Specialist Group on Marine Wastewater Disposal, International Association on Water Quality, London. He is presently one of only two Distinguished Scholars in the National Ocean and Atmospheric Administration (NOAA) Oceans and Human Health Initiative (OHHI) in which he is conducting research on the hydrodynamic aspects of bacterial and pathogen transport in coastal waters. E-mail: proberts@ce.gatech.edu

Henry Salas, BCE, MSc, PE, Consulting Engineer. Henry Salas has a BS in Civil Engineering and MSc in Environmental Engineering from Manhattan College and is a registered professional engineer. He worked for Hydroscience, Inc, in the USA (1970–79) developing and applying water quality mathematical models to waste load allocation and eutrophication in rivers, estuaries, lakes and coastal waters. He worked as an environmental engineering consultant to the Environmental Quality Board of Puerto Rico (1979–1982). Then, during his 25 years with the Pan American Health Organization (PAHO), he held various posts at the Pan American Center for Sanitary Engineering and Environmental Sciences (CEPIS), in Lima, Peru as Regional Advisor in Water Pollution Control; Environmental Impact Assessment and Health; and Water Resources for Public Health as well as coordinator of the Environmental Impact Department and subsequently the Environmental Risk Unit. He was also Regional Advisor in Environmental Protection of PAHO in Washington, DC. He has coordinated various PAHO regional programs, projects and investigations including preparation of seventeen training courses and workshops concerning marine outfalls and microbiological water quality standards. He participated in the development of the World Health Organization (WHO) Guidelines for Safe Recreational Water Environments and protocols for epidemiological investigation in bathing beaches. Other PAHO regional programs over which he had responsible charge were methodologies for evaluating eutrophication in warm-water tropical lakes, toxic substances in surface waters, groundwater contamination, minimization of hazardous wastes and industrial effluents, coordination of the Global Environmental Monitoring System (GEMS), the development of water quality models, and environmental and health impact

About the Authors

assessment (EHIA). Presently he is a private consulting engineer. He has provided consulting services to almost all the counties of Latin America and the Caribbean for PAHO and post-PAHO consulting to the World Meteorological Organization, ProInversión (Peru), and private consulting firms including Andrade Gutierrez (Brazil/Peru), OIST, SA (Peru), Halcrow (England/Chile), Odebrecht (Brazil/Peru) and ACS (Spain). E-mail: hsalas130@hotmail.com

Fred M. Reiff, PE, Reiff Engineering, Chevy Chase, Maryland, USA. Fred Reiff is a registered professional engineer with more than 50 years of experience in environmental, sanitary, civil engineering and public health administration in both the private and government sectors at the local, national, and international levels. His experience encompasses a number of areas including management, planning, design, construction, research and development, studies and investigations, derivation of technical standards and criteria, teaching, organization of international and national technical symposiums, emergency preparedness and relief as well as development of government programs. This work has been carried out for wide ranging geographical and climatic conditions including arctic, tropical, temperate, arid deserts, mountainous terrain, flood plains, and oceanic environments and for both municipal and rural settings. Mr. Reiff has planned, designed, constructed, supervised and/or inspected more than 20 submarine outfalls and pipelines of small diameters. He has participated as a lecturer in PAHO submarine outfall courses in Barbados, Chile, Colombia, Costa Rica, Ecuador, Honduras, México, Panama, Peru, and the Dominican Republic. E-mail: fm7reiff@rcn.com

Menahem Libhaber, PhD, World Bank. Dr. Libhaber received an MSc in Chemical Engineering and a PhD in Water Resources and Environmental Engineering from the Technion, Israel Institute of Technology, Haifa, Israel. Prior to joining the World Bank in 1991, he worked for 18 years for Tahal, Consulting Engineers as a water and sanitation engineer in Israel and many other countries including Brazil, Costa Rica, Peru, El Salvador, Chile, Mexico, Honduras, Turkey, Spain, Yugoslavia, and Nigeria. He served for three years as a consultant to UNEP – United Nations Environmental Program, the Mediterranean Action Plan. He joined the World Bank in 1991 as a Lead Water and Sanitation

Engineer in the Latin America Region. He has served as task manager of water and sewerage projects in Colombia, Bolivia, Argentina, Costa Rica, the Dominican Republic, Peru, and Trinidad and Tobago. He has also worked on other projects in Brazil, Paraguay, St. Lucia, Guyana, Venezuela, Jamaica, Honduras, Panama, Haiti, India, and China. Many of these projects involved marine outfalls for coastal cities. Dr. Libhaber retired from the World Bank in July 2009. He is presently a private consulting engineer to the World Bank and others. E-mail: mlibhaber@worldbank.org

Alejandro Labbe Fluhmann. BCE, Halcrow, Director Chile, Peru & Colombia. Alejandro Labbe is a Civil Engineer who graduated from the University of Chile. He has 25 years of experience in the design, construction, supervision, and inspection of more than 20 submarine outfalls. He is presently the Director of the Halcrow Group Limited-Chile and Global Chief Engineer of Sea Outfalls. He has designed marine outfalls and preliminary treatment systems for various cities in Chile and other countries. Labbe has supervised technical inspection and assessment during construction and initial operation of outfalls in ten Chilean and four other cities in Latin America. He has conducted studies to evaluate the construction and operating costs of outfalls to ascertain the tariffs for wastewater treatment services. He has provided consulting services for outfall projects to firms in Chile, Columbia, Ecuador, Costa Rica, the Dominican Republic, and the UAE. He is a consultant to the World Bank, Inter-American Development Bank (IDB), and PAHO. He has lectured on the construction, operation, and cost of outfalls, including numerous PAHO courses, in more than ten Latin American countries. E-mail: LabbeAA@halcrow.com

James C. Thomson C. Eng Eur. Ing. Consulting Engineer, Geneva Switzerland. Mr. Thomson has extensive experience in the design and construction of a wide range of international civil engineering projects. In his early years he pioneered and developed pipe jacking and microtunnelling techniques. After a period working for major European contractors in senior management he founded Jason Consultants in 1979 which became a leading group in the field of underground infrastructure.

In 2004 he sold the company to his colleagues but has continued to work with them as an independent consultant. He is an acknowledged expert in pipe jacking microtunnelling and soft ground tunneling for utilities including outfalls and intakes. He has been the Principal Investigator on a number of research projects for Water Environment Research Foundation and USEPA on condition assessment and inspection of water and wastewater networks. He is the author of several books and nearly 100 technical papers. E-mail: jamescthomson@gmail.com

1
Introduction

1.1 OBJECTIVES AND PHILOSOPHY OF MARINE WASTEWATER DISPOSAL

The oceans of the world, which cover 70% of the earth surface, have always been the recipient of human wastes. They have changed little, however, as evidenced by the fact that the chemical composition of the sea has remained essentially the same for over a million years (Alder, 1973). When compared to the enormous quantities of organics and sediments carried to the oceans by rivers, man's contribution is quite small. And Dr. John Isaacs of the Scripps Institute of Oceanography has pointed out that the fecal discharge into Southern California coastal waters of anchovy alone is equivalent in organic content to the sewage discharge of about 90 million persons, and this is only one of hundreds of species of marine life (Ludwig, 1983). Dr. Peter Franks, also of Scripps, has calculated that one upwelling event contributes more nutrients to coastal waters than all of the Southern California outfalls combined for one year. These

© 2010 IWA Publishing. *Marine Wastewater Outfalls and Treatment Systems.* By Philip JW Roberts, Henry J Salas, Fred M Reiff, Menahem Libhaber, Alejandro Labbe, and James C Thomson. ISBN: 9781843391890. Published by IWA Publishing, London, UK.

observations would seem to refute a prominent point of view, advocated by some environmentalists and supported by the policy decisions of some developed countries, which would eliminate ocean discharges.

Nevertheless, problems can arise when waste products are poorly disposed of and are concentrated at one location. This may occur with large population centers, and the rapidly increasing numbers of people who live near coasts can exacerbate the problem. About fifty percent of the world's population, more than 3 billion people, presently lives within sixty kilometers of the coast.

The wastewaters that these coastal populations generate are usually discharged into the adjacent ocean. But whether or not this creates a health risk to these same populations depends on how the sewage is discharged. This issue was addressed by the World Health Organization (WHO, 2003) in their guidelines for recreational water quality. Their findings are summarized in Table 1.1 for the major types of treatment and disposal practiced by coastal communities around the world.

Table 1.1 Risk to human health from exposure to sewage (including stormwater runoff and combined sewer overflows) (WHO, 2003)

Treatment process	Human health risk		
	Discharge directly on beach	Discharge from short outfall[a]	Discharge from effective outfall[b]
None[c]	Very high	High	NA[d]
Preliminary	Very high	High	Low
Primary (including septic tanks)	Very high	High	Low
Secondary	High	High	Low
Secondary plus disinfection[e]	–	–	–
Tertiary	Moderate	Moderate	Very low
Tertiary plus disinfection[e]	–	–	–
Lagoons	High	High	Low

[a] The relative risk is modified by population size. Relative risk is increased for discharges from large populations and decreased for discharges from small populations.
[b] Assumes the design capacity has not been exceeded and that climatic and oceanic extreme conditions are considered in the design objective (i.e., no sewage on the beach zone).
[c] Includes combined sewer overflows if active during the bathing season (a positive history of total non-discharge during the bathing season can be treated as "Low").
[d] NA = not applicable.
[e] Additional investigations recommended to account for the likely lack of prediction with faecal index organisms.

Introduction 3

Coastal wastewater disposal is often posed as a choice between treatment or outfall, but Table 1.1 shows this to be a false choice. Even treated effluent must ultimately be disposed of through an ocean outlet or to rivers that eventually flow to the ocean. Table 1.1 clearly demonstrates that an effective outfall is suprisk to human health is very low, and more advanced treatment does not significantly lower this risk. The underlying principal is that effective outfalls physically separate people from sewage. The reliability of an outfall, which is a civil engineering structure with minimal operation and maintenance requirements, is also much higher than treatment plants, which require high operation and maintenance and are subject to upsets, especially in developing countries.

The three principal types of discharge in Table 1.1 are:

(1) Directly onto the beach;
(2) A "short" outfall, with likely contamination of recreational waters;
(3) An "effective" outfall, designed so that the sewage is efficiently diluted and dispersed and does not pollute recreational areas.

Although the terms "short" and "long" are often used, outfall length is generally less important than proper location and effective diffusion. An effective outfall has sufficient length and depth to ensure high initial dilution and to prevent sewage from reaching areas of human usage.

A typical disposal scheme consists of a treatment plant and an outfall, as shown schematically in Figure 1.1. The outfall is a pipeline or tunnel, or combination of the two, which terminates in a diffuser that efficiently mixes the effluent in the receiving water. Outfalls typically range from 1 to 4 km long and discharge into waters 20 to 70 m deep. Some may lie outside these ranges, for example lengths of 500 m or less and discharge depths of 150 m or more when the seabed slope is very steep, or lengths of more than 5 km when the slope is very gradual. The disposal system can be thought of as the treatment plant, outfall, diffuser, and also the region round the diffuser (known as the near field) where rapid mixing and dilution occurs.

The objectives of the system are to dispose of wastewater in a safe, economical, and reliable way with minimal impacts to the receiving water. This means that the local ecosystem and the health of the public that are swimming and using nearby beaches are protected, and that the outfall functions reliably with minimal maintenance. How to accomplish these objectives is the subject of this book.

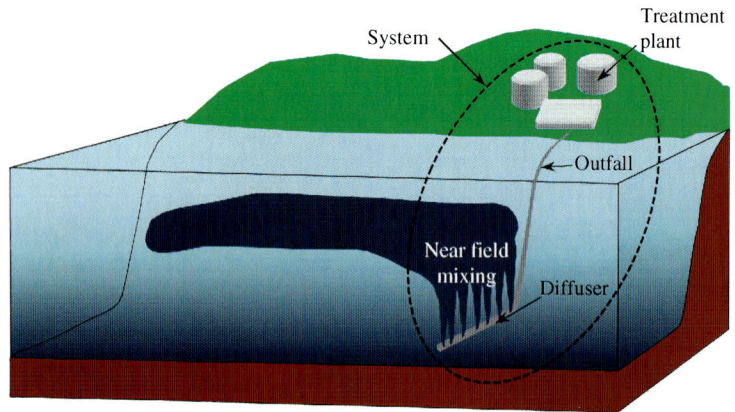

Figure 1.1 A marine wastewater disposal system: Treatment plant, outfall pipe, diffuser, and near field

1.2 WATER QUALITY ASPECTS

Receiving water quality considerations for ocean outfalls are discussed in Chapter 4, but here we give some general comments and observations.

To ensure that water quality effects are minimal requires that:

- Concentrations of bacteria, toxics, and other contaminants are reduced to safe levels;
- Ecosystem products of the effluent (organic carbon, nutrients, etc.) and dissolved oxygen concentrations are kept within allowable limits;
- Local particulate deposition is not excessive;
- The wastefield is not visible on the water surface.

These can readily be accomplished by a suitable combination of outfall and diffuser location, effective dispersion and dilution of the effluent, and treatment. Discharges near environmentally sensitive areas such as coral reefs or shell fishing beds should be avoided. If the diffuser causes rapid dilution and dispersion of the effluent and is positioned so that wastefield transport to critical areas (especially the shoreline) is minimized, only preliminary wastewater treatment such as milliscreening may be needed. To achieve these objectives, it is necessary for the designer to understand how wastewaters mix in coastal waters, to design the outfall and diffuser to promote efficient mixing, and to match the treatment level accordingly.

Efficient mixing with dilutions consistently exceeding 100:1 can be readily achieved within the first few minutes after discharge. This reduces concentrations of organics and nutrients (which are characteristic of sewage) to levels that have no adverse ecological effects. In some situations, the introduction of such substances to a nutrient deficient ocean environment can be beneficial. For a discharge that contains toxic substances such as PCBs (polychlorinated biphenyls), pesticides, mercury and other potentially toxic substances, however, source control may be needed.

Public health is protected by reducing levels of pathogenic organisms to meet established bathing beach water quality criteria. The orders-of-magnitude reductions required for this are achieved through physical dilution and mortality in the ocean environment. As demonstrated by numerous studies, properly designed ocean outfalls discharging domestic wastewaters do not cause significant ecological impacts.

1.3 APPROPRIATE TREATMENT FOR OCEAN OUTFALL DISCHARGES

Wastewater treatment for ocean discharges is a contentious issue that is rife with misinformation. It is discussed in detail in Chapter 6.

According to Table 1.1, the level of treatment has little bearing on the human health risk of discharge from an effective outfall. The risk from any effluent discharged through an effective outfall is low, even if only treated to preliminary or primary levels. Preliminary treatment is therefore often sufficient for domestic sewage discharges to the sea and can be undertaken with minimal water quality impacts.

This is illustrated by studies (see Chapter 14) around two outfalls near Concepción, Chile, that discharge wastewater with only preliminary treatment. It was found that, except for fecal and total coliforms, the concentrations of all water quality parameters at a distance of 100 m from the discharge are comparable to background levels. Although the concentrations of coliforms in the effluents is extremely high, their concentrations are markedly reduced 100 m from the outfall to levels that meet the most stringent standards. And with mortality and diffusion, concentrations fall to background levels within a farther short distance. It was concluded that: (i) The impact of the outfalls on ocean water quality is minor; (ii) Heavy metals and micropollutants were below detection limits; and (iii) There were no negative effects on the benthic communities. These findings are typical of effective outfalls.

Conversely, if a short (ineffective) outfall is used, Table 1.1 shows that even secondary treatment cannot reduce the human health risk to acceptable levels.

Effluents from stabilization lagoons discharged directly on the beach or from a short outfall also constitute a high health risk. Unfortunately, this method of disposal is common in developing countries.

A central thesis of this book is that preliminary treatment alone will usually suffice with an effective outfall. For domestic sewage this treatment (see extended discussion in Chapter 6) consists of milliscreens with apertures that usually range from 0.5 to 1.5 mm. This eliminates larger particles and significant quantities of floatable material, including grease and oil, which could reach bathing beaches.

To understand why advanced treatment is usually unnecessary, consider an outfall with a diffuser that effects an initial dilution of 100:1 (which can usually be easily accomplished). This corresponds to a 99% reduction in contaminant concentrations in the receiving water, which is far beyond the capabilities of conventional secondary treatment. Diffuser mixing is therefore usually much more important than treatment in mitigating environmental impacts, in fact the diffuser and near field could be considered as a treatment plant. In addition, oceanic mixing and mortality can reduce bacteria at nearby recreational waters to levels comparable to or better than is achievable by secondary treatment and chlorination alone.

For coastal cities in developing countries, the strategy of wastewater disposal through an effective outfall with preliminary treatment is an affordable, effective, and reliable solution that is simple to operate and with minimal health and environmental impacts. Many such systems are successfully functioning with a proven track record around the world. It is shown in Chapter 6 that the costs of preliminary treatment are about one-tenth that of secondary treatment. And mandating more advanced levels of treatment that are unaffordable often results in "no action," with continued contamination of water bodies and their associated health risks. Finally, the allocation of scarce resources must be made in the face of shortages in hospitals, schools, safe water supplies, or even the food necessary for survival.

In spite of these observations, the WHO findings, and operating experience from many outfalls around the world, many countries require a high level of treatment, usually secondary, prior to outfall discharge. This is often for political, rather than scientific or technical reasons. Developing countries tend to copy the standards of industrialized countries, so many of their discharge requirements are also quite stringent and require secondary treatment. These policies are not recommended for developing countries unless there is a clear justification for them. Some countries, such as Colombia and Costa Rica, are now taking more enlightened approaches (see Chapter 6). They require a level of treatment that, in combination with the processes of initial dilution, dispersion, and decay, guarantees compliance with the receiving water quality objectives. The legislation recognizes that the outfall is part of a system, so that the *combination* of treatment and outfall must comply with the quality standards, not just the treatment alone.

In this book we advocate preliminary treatment as often being the only treatment needed for outfall discharges. Some countries require more than preliminary treatment, however, and secondary treatment may be needed if the discharge could cause significant dissolved oxygen depletion, such as in enclosed bays or embayments with poor circulation and flushing. Therefore, more advanced treatment options that are especially suited to ocean discharges are also discussed. These options can obtain a higher quality effluent at much lower cost than conventional secondary treatment.

The negative consequences of secondary treatment must also be considered. These include much higher capital and operation and maintenance costs, the problems of sludge disposal, and reliability. Secondary treatment processes are often subject to upsets, and if the discharge is through a short outfall, this could result in direct onshore or nearshore discharge of raw wastes. This does not occur with effective outfalls (discounting structural outfall failure which is rare with modern designs). Furthermore, biological secondary treatment systems do not have the flexibility to accommodate large seasonal variations in flows resulting from population fluctuations such as occur in tourist areas, but ocean outfalls can readily be designed to handle large flow variations.

Secondary treatment separates the effluent at great expense into two waste streams: treated effluent, which is often chlorinated, and sludge. If both find their way into the ocean environment via separate outfalls, such separation is pointless. Such a practice was prevalent in California in the USA, and, although no significant impacts were observed, the discharge of sludge was subsequently prohibited by federal legislation. Sludge disposal then becomes a difficult and costly problem with potential transfer of contaminants to other media. Many toxic substances can be essentially untouched in the effluent streams.

It is highly unlikely that tertiary treatment would be needed for an ocean discharge. It is mainly for nutrient removal, and would only be needed for discharges to water bodies that have very limited assimilative capacity with the potential for eutrophication. Discharges near coral reefs (which should be avoided), may require nutrient removal to minimize eutrophication stimulation that could cause small changes in sunlight penetration that could have significant impacts on coral reef formations.

Ludwig (1988) showed that the lifetime cost of a typical urban wastewater scheme with secondary treatment is much higher than one with primary treatment combined with an effective long ocean outfall. This is because effective outfalls will almost always meet bacterial bathing water standards at beaches. If the treatment is limited to removal of floatables and grease and oil, the economic comparison is even more favorable for the outfall. Also, increasing use of more economical plastics (HDPE) makes outfalls even more attractive, especially for small to intermediate communities.

Local needs must, of course, be considered. For example, in arid coastal areas, reuse of treated sewage can be better than disposal through an outfall.

1.4 WASTEWATER DISPOSAL IN LATIN AMERICA AND THE CARIBBEAN

To illustrate the widespread use of ocean outfalls (and also deficiencies in wastewater management), we discuss here wastewater practice in Latin America and the Caribbean (LAC).

The total population in LAC was 523 million in 2000 and is growing at an annual rate of 1.4% (United Nations, 2003). As shown in Table 1.2, there are 503 cities with more than 100,000 inhabitants, containing about 57% (296 million) of the region's population. More than a quarter of these cities, which contain more than a third of the total urban population of 97 million, and near coast or estuaries. This population could be serviced by outfall systems for wastewater disposal. The total number of cities increases to 1,654 (CEPAL, 2005) when urban centers of 10,000 to 100,000 are considered (this estimate does not include the following countries where data were not readily available: Anguilla, Aruba, British Virgin Islands, Cayman Islands, Dominica, Grenada, Haiti, Montserrat, Saint Kitts and Nevis, Turks and Caicos Islands, Guyana). There are hundreds of coastal towns with populations less than 10,000.

Table 1.2 Distribution of urban centers in Latin America and the Caribbean in 2000 (United Nations, 2003)

Population greater than	Regional total — Number of cities	Regional total — Total population (millions)	Coastal or estuarine areas — Number of cities	Coastal or estuarine areas — Total population (millions)
100,000	503	295.8	115	96.7
500,000	121	215.4	42	80.9
1,000,000	53	166.9	20	64.6
3,000,000	13	99.6	5	37.6

If the present rapid urban growth rate continues, the coastal population that could be serviced by marine outfalls will increase from 97 million to almost 120 million by the year 2015 with a total wastewater flow of about 280 m^3/s. Proper disposal of these wastewaters is critical to the future sustainable development and environmental well-being of the region.

Common practice in many coastal cities has been to discharge untreated wastewaters to the nearest or most convenient water body. Due to lack of economic resources minimal considerations are given to the environmental consequences of this practice. Raw sewage discharges often occurred on or very near to popular bathing beaches. This causes gross beach contamination, with bacterial levels on the beaches at times approaching levels of raw sewage. In addition to potential health hazards, this can cause aesthetic problems and ecological impacts. These can often bring severe economic consequences due to curtailed tourism.

According to a survey conducted by CEPIS in 1983 and updated in 2006, there are 138 constructed or planned sewage outfalls in the LAC region that are longer than 500 m. As discussed in this book, in order to meet common recreational beach microbiological standards, modern designs require an appropriate combination of outfall length and discharge depth. The 500 m outfall length is arbitrary, and longer outfalls would usually be required for major discharges (although shorter outfalls discharging in very deep waters nearshore may suffice).

Puerto Rico, with a total population of about 3.9 million, has eleven operating outfalls, with all coastal areas discharging through regional disposal systems. This is the highest per capita use of outfalls in the LAC. At least primary treatment is utilized. The final discharge permits are granted by the Environmental Quality Board (EQB) which conducts extensive and detailed reviews of final designs according to the procedures, models, and criteria of the U.S. Environmental Protection Agency. Thus, modern criteria are applied and post-discharge water quality monitoring is carried out to ascertain performance and compliance. In 1998, the new outfall of Ponce was able to obtain a waiver to allow primary treatment instead of secondary treatment.

Three of the five most populated coastal cities of Brazil (Rio de Janeiro, Salvador and Fortaleza) are served by outfalls. Many have no treatment, but modern criteria are usually applied to the diffuser design to ensure high dilution. Brazil has about 22 major outfalls (six for industrial discharges).

Mexico has about nine major outfalls (three for industrial discharges), most designed according to modern design criteria. Primary treatment is usually applied.

Venezuela has about 39 outfalls. Two were constructed in 1949, and are the oldest in South America. Only 17 of these outfalls are longer than 1000 m. Twelve shorter outfalls serve small towns and recreational facilities in the Federal District, where the public beaches are frequented by up to two million persons during national holidays. Based on bacteriological surveys, 75% of these public beaches were found to have acceptable coliform levels (DMSA, 1971). Poor water quality conditions were usually limited to the vicinity of raw discharges on or nearshore, or tributaries heavily contaminated by animal wastes. Beaches in areas serviced by outfalls were generally classified as acceptable. Therefore, despite their relatively short lengths (less than 1000 m), because of favorable oceanographic conditions the outfalls apparently performed well. However, structural deterioration has been reported with leaks along some outfalls so water quality has probably degraded.

Chile has about 39 operating outfalls, all constructed of modern plastics. There are many other shorter outfalls but these are generally simply extensions of the sewerage systems. Preliminary treatment is common.

Other countries have fewer outfalls. The Punta Carretas outfall in Montevideo, Uruguay was constructed in 1990 and an additional outfall is currently under consideration. An outfall in Limon, Costa Rica was constructed in 2005. Colombia has an outfall in Santa Marta (2000), and an outfall for Cartagena has been designed and financed and is under construction. Two sewage outfalls in Lima and one combined sewage and fish meal industrial wastewater outfall in Chimbote, Peru, have been designed. There are numerous outfalls discharging fishmeal plant wastewaters along the Peruvian coast, but only a few were designed to modern standards.

In the Greater Caribbean Sea, there are outfalls in Barbados, Bermuda, and Jamaica. Anguilla, Dominica and Martinique (Fort-de-France) also have outfalls. An outfall at Sosua in the Dominican Republic was constructed in 2005.

In addition to estuarine and coastal areas, outfalls may also be used for discharging sewage into large fresh water lakes or rivers. An example is Manaus, Brazil, where sewage is discharged into the Black River, a tributary of the Amazon River, through a one meter diameter outfall 3600 m long. Most of the outfall is parallel to shore, so the actual discharge occurs only 300 m offshore. Paraguay has two outfalls discharging to the Paraguay River. This inland use of outfalls increases the population that could be served by outfalls above the 97 million cited in Table 1.2 and further emphasizes the importance of this technology.

Although there are about 138 existing and planned outfalls in the LAC, the population served is relatively small. Only 43 (including Manaus) of these outfalls, serve cities with populations more than 100,000, and most of these

Introduction

cities are only partially served. Therefore, the greater part of the wastewaters generated by the estuarine and coastal population continues to be discharged on or near shore without any treatment, often resulting in the aesthetic, public health, ecological and economic problems previously mentioned.

Great improvements in water quality, especially on beaches, are often observed when outfalls are commissioned. This is exemplified by the water quality conditions observations on the beaches of Ipanema and Leblon in Rio de Janeiro, Brazil. The Ipanema Outfall was inaugurated in September 1975 serving a population of about 2.5 million of the southern zone of Rio de Janeiro with a design flow of 12 m^3/s. Continuous water quality monitoring conducted by the local water and sewage authority at the stations shown in Figure 1.2 (CEDAEH, 1982), demonstrates significantly improved conditions. Except for coarse screening to protect pumps, no other wastewater treatment or chlorination was practiced on the Ipanema outfall effluent during this period primarily due to cost and the unavailability of land. Many similar "success" stories have been observed elsewhere around the world.

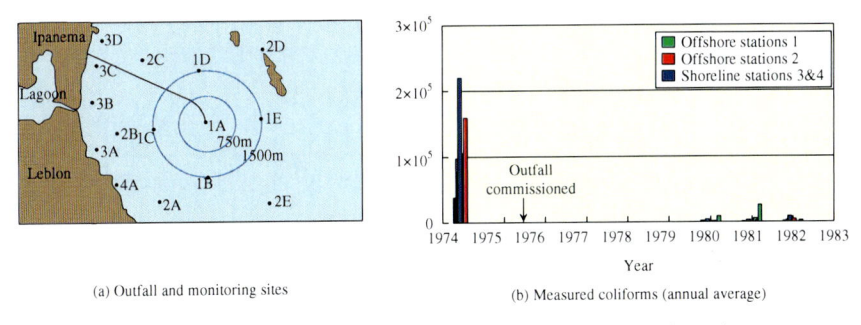

Figure 1.2 Total coliforms before and after construction of the Ipanema outfall (CEDAEH, 1982).

Chile has also done extensive monitoring over the last decade that has documented the excellent performance of two submarine outfalls in Penco and Tome. The results of this monitoring are discussed in Chapter 14.

1.5 BOOK OUTLINE

When properly designed, marine outfalls are a reliable and economical technology for wastewater disposal with minimal adverse environmental, ecological, and public health impacts. This has been proven by extensive experience with many outfalls operating around the world.

This book considers the "system" of treatment plant and outfall shown in Figure 1.1. It does not cover the onshore sewerage piping required to bring the sewage to the treatment plant. We do not cover treatment plant design in detail, just those aspects that are particularly relevant to marine discharges. The intent is to cover most issues involved in the design of ocean outfalls and to provide information useful to the engineers, politicians, and other stakeholders involved in the planning and decision making process.

Designing an outfall system requires an unusual and complex mix of skills including oceanography, environmental fluid mechanics, hydraulics, treatment plant design, ocean engineering, environmental engineering, coastal construction techniques, and public relations, among others. Improper design can result in structural failure of the outfall due to wave effects, resources wasted on inappropriate and expensive unnecessary treatment, and unnecessary impacts on the environment or public health. And poor planning and inadequate building of support for the project can lead to unnecessary objections and delays. These diverse aspects of outfall design and planning are rarely brought together in one publication, and this is the aim of this book.

The emphasis is on outfalls constructed from high density polyethylene (HDPE). This is not a severe restriction. The use of HDPE for marine outfalls has increased rapidly in recent years, and advances in technology now permits pipes of up to 2 m diameter or more that can be delivered to most parts of the world. HDPE pipes therefore have wide applicability, and are no longer limited to "small diameter" outfalls. A single HDPE pipe can serve communities of several hundred thousand, and multiple pipelines can serve communities of up to a million and more. They can also economically serve very small coastal communities with populations less than 10,000. Although HDPE is emphasized, much of the material in this book, such as water quality aspects, wave forces, etc., are also applicable to the design of outfalls using other materials, such as concrete or steel or even tunneled. Because of the increasing use of tunneling, sometimes in conjunction with HDPE pipelines, a chapter on trenchless construction is also included.

The book outline is as follows. We first address environmental issues, particularly receiving water quality. The main issues pertinent to coastal wastewater discharges are discussed in Chapter 2, along with a summary of the various local and international regulations that they must meet. Water quality impacts depend on the dilution and dispersion capability of the diffuser and the level of wastewater treatment. Of these, dilution and dispersion is usually more important in mitigating environmental impacts, so they are discussed in detail in two chapters. First, in Chapter 3 the processes that govern the fate and transport of coastal discharges are described and simple methods to predict them are presented. Next, in Chapter 4, the mathematical models that are now widely used are discussed. These models (and also the structural outfall design) require extensive oceanographic data that are gathered in field surveys. The planning and gathering of these data to ensure their relevance and usefulness is discussed in Chapter 5. Wastewater treatment systems with an emphasis on disposal through marine outfalls are discussed in Chapter 6.

Structural design aspects are then addressed. Many materials have been used for outfalls, and the considerations involved in choosing the best pipeline material are discussed in Chapter 7. Forces on marine outfalls are discussed in Chapter 8. Details of HDPE pipe design are given in Chapter 9, including design for structural stability, internal hydraulics, and other considerations. Marine pipeline construction is a specialized discipline that requires special techniques; the extensive experience that has been gathered over the years is discussed in Chapter 10. Installation by tunneling, microtunneling and horizontal directional drilling is discussed in Chapter 11. Operation and maintenance is essential for reliable operation over many years, and is discussed in Chapter 12. Before and after discharge commences, monitoring is essential to ensure proper operation and protection of the environment and public health as presented in Chapter 13. Some case studies are presented in Chapter 14. Chapter 15 deals with public acceptance issues. Appendix A discusses cost issues and presents data from which the construction cost of an outfall can be estimated. Appendix B contains a summary of the many abbreviations and acronyms used in the book.

2
Water quality aspects

2.1 INTRODUCTION

In this chapter, we discuss receiving water quality considerations for marine outfalls and the regulations that they must meet. Section 2.2 discusses the main issues, which are protection of public health, the environment and ecosystems, and aesthetics. Environmental impacts depend strongly on the dilution and dispersion capability of the outfall. And central to understanding these impacts is the concept of a mixing zone (Section 2.3), which in turn has a major effect on how the discharge is regulated. One of the first major attempts to regulate ocean discharges that acknowledges their unique characteristics and addresses their myriad water quality issues was the California Ocean Plan, so we discuss this in some detail in Section 2.4. Water quality standards for toxics and bacteria (for protection of public health) are discussed in Sections 2.5 and 2.6. Because bacterial standards are often the primary driver of outfall design we discuss them extensively in Section 2.6. This section includes a review of the history of the

© 2010 IWA Publishing. *Marine Wastewater Outfalls and Treatment Systems.* By Philip JW Roberts, Henry J Salas, Fred M Reiff, Menahem Libhaber, Alejandro Labbe, and James C Thomson. ISBN: 9781843391890. Published by IWA Publishing, London, UK.

formulation of bacterial standards around the world and a summary of the major international standards for human health protection.

2.2 GENERAL WATER QUALITY ISSUES

According to the National Research Council (NRC, 1993), a wastewater constituent may be considered to be of high concern if it:

> "...poses a significant risk to human health and ecosystems (e.g. if it contaminates fish, shellfish and wildlife, causes eutrophication, or otherwise damages marine plant and animal communities) well beyond points of discharge and is not under demonstrable control. A wastewater constituent may be generally considered to be of lower concern if it causes only local impact or is under demonstrable control."

Using these criteria, the NRC developed a list of anticipated priorities for wastewater constituents in coastal urban areas as shown in Table 2.1. Some general comments on this table are given below, and details of how these constituents are controlled by outfall systems are discussed in later chapters.

Table 2.1 Wastewater constituents of potential concern in coastal waters (After NRC, 1993)

Priority	Pollutant Groups	Examples
High	Nutrients	Nitrogen, Phosphorus
	Pathogens	Enteric viruses
	Toxic organic chemicals	PAHs
Intermediate	Selected trace metals	Lead
	Other hazardous materials	Oil, chlorine
	Plastics and floatables	Beach trash, oil, and grease
Low	Biochemical oxygen demand (BOD)	
	Solids	

The high priority pollutants can be readily controlled by an effective outfall combined with appropriate wastewater treatment. Although nutrients are listed as high priority, they are not a concern for discharges to open coastal waters with good flushing. They are more of a problem in enclosed water bodies with poor flushing such as lakes, bays, or estuaries, where eutrophication may occur. Pathogens are microorganisms that can cause disease in humans. They are assumed to be controlled if the level of an indicator organism (an organism that indicates the presence of sewage) is below some specified standard or guideline. They can be controlled by a combination of initial dilution, oceanic diffusion, and mortality, as discussed in Chapters 3 and 4. Toxic organic chemicals and trace metals can cause adverse effects in aquatic organisms and humans. They can be addressed by, for example, applying the limitations prescribed in the California Ocean Plan, as discussed in Section 2.4. This can usually be accomplished by initial dilution alone for regular domestic sewage, but source control may be needed for industrial discharges.

Intermediate level constituents can also be controlled by treatment and dilution. Plastics and other particulate floatables should be removed by treatment such as screening (see Chapter 6). Other floatables, especially grease and oil, are of more concern since they may contain pathogens and may be blown onshore by winds (see Figure 1.1). As discussed in Chapter 6, milliscreening, especially when combined with other forms of treatment such as flotation, will remove substantial quantities of grease and oil and other floatables.

It may seem surprising that biochemical oxygen demand (BOD) is low priority. This is because high initial dilution and the large surface area available for re-aeration generally results in negligible depletion of dissolved oxygen. Solids are also ranked low priority due to the ability to control them with treatment and high dilution. The potential for accumulation on the seabed, and their possible association with toxic organic chemicals, metals, and pathogens should be addressed, however.

Mathematical modeling and monitoring of operating outfalls generally show that their effects are limited to a small area, typically a few hundred meters around the discharges. This is true even for substantial discharges of essentially raw sewage, for example the Ipanema outfall in Rio de Janeiro discussed in Section 1.4.

The key parameters in the design of municipal wastewater systems are generally bacteria, floatables, and grease and oil. Toxics are readily controlled by dilution. Bacteria are best controlled by locating the outfall so that transport of wastewater to beaches or other water contact areas is virtually eliminated. The outfall should be designed, however, that, in the unlikely event that transport to beaches does occur, the combination of initial dilution, oceanic diffusion, and bacterial mortality reduces the bacteria to very low levels. Chlorination of the effluent is then unnecessary. As discussed above, other parameters such as nutrients, BOD, and dissolved oxygen will not usually be a concern unless the sewage is discharged to a shallow, poorly flushed coastline, or embayment.

2.3 CONCEPT OF A MIXING ZONE

Central to understanding the environmental impacts of an ocean outfall and how they are regulated is the concept of a mixing zone. This is discussed in detail in Chapter 3, but a brief summary is given here.

The mixing zone is a region of non-compliance and limited water use around the diffuser. Water quality criteria must be met at the edge of the mixing zone. Within this zone the discharge undergoes energetic mixing (see Figure 1.1) that rapidly reduces the concentrations of most contaminants to safe levels. The mixing is caused by the turbulence generated by the high velocity of the jets issuing from the diffuser ports and by the effluent buoyancy that causes it to rise through the water column. These mechanisms entrain substantial quantities of ocean water that readily dilutes the effluent to at least 100:1 within a few minutes after discharge and within a few hundred meters from the diffuser. Another way of looking at this is that the concentrations of contaminants are reduced by more than 99% within the mixing zone.

This rapid and very substantial contaminant reduction is recognized by the concept of a regulatory mixing zone. For example, the US EPA regulations for toxics (USEPA, 1991), defines a mixing zone as:

> "An area where an effluent discharge undergoes initial dilution and is extended to cover the secondary mixing in the ambient water body. A mixing zone is an allocated impact zone where water quality criteria can be exceeded as long as acutely toxic conditions are prevented." (Water quality criteria must be met at the edge of a mixing zone.)

Thus, water quality requirements are specified at the edge of the mixing zone rather than by end-of-pipe requirements for conventional and toxic discharges.

There is much terminology associated with wastewater mixing processes and the regulations that cover them. Unfortunately, there do not appear to be universal definitions of these terms and they are often used interchangeably and imprecisely. As summarized in Table 2.2, they include zone of initial dilution, regulatory and hydrodynamic mixing zones, and near and far field mixing. In this book, we will use the definitions given in Table 2.2.

The mixing zone may not correspond to actual physical mixing processes. It may fully encompass the near field and extend some distance into the far field, or it may not even fully contain the near field. Mixing zones can be defined as lengths, areas, or water volumes. An example is contained in the guidelines for the US National Pollutant Discharge Elimination System (NPDES) permits for the discharge of pollutants from a point source into the oceans at 40 CFR

Water quality aspects

125.121(c) U.S. Federal Water Quality that defines a mixing zone for federal waters as:

> "...the zone extending from the sea's surface to seabed and extending laterally to a distance of 100 meters in all directions from the discharge point(s) or to the boundary of the zone of initial dilution as calculated by a plume model approved by the director, whichever is greater, unless the director determines that the more restrictive mixing zone or another definition of the mixing zone is more appropriate for a specific discharge."

Table 2.2 Outfall mixing and mixing zone terminology

Term	Definition	Comments
Mixing zone	A limited area where rapid mixing takes place and where numeric water quality criteria can be exceeded but acutely toxic conditions must be prevented. Specified dilution factors and water quality requirements must be met at the edge of the mixing zone	
Allocated impact zone (AIZ)	Same as a mixing zone	
Regulatory mixing zone	As defined by the appropriate regulatory authority	Can be a length, an area, or a volume of the water body
Legal mixing zone (LMZ)	Same as a regulatory mixing zone	
Near field	Region where mixing is caused by turbulence and other processes generated by the discharge itself	Near field processes are intimately linked to the discharge parameters and are under the control of the designer. For further discussion, see Chapter 3, and Doneker and Jirka (1999), and Roberts (1999c)
Hydrodynamic mixing zone	Same as near field	Near field and hydrodynamic mixing zone are synonymous with these definitions
Far field	Region where mixing is due to ambient oceanic turbulence	Far field processes are not under control of the designer

(continued)

Table 2.2 (*continued*)

Term	Definition	Comments
Toxic dilution zone (TDZ)	A more restrictive mixing zone within the usual mixing zone	
Initial dilution	No specific definition	A general term for the rapid dilution that occurs near the diffuser
Zone of initial dilution (ZID)	A region extending over the water column and extending up to one water depth around the diffuser.	A regulatory mixing zone, as defined in the U.S. EPA's 301(h) regulations (USEPA, 1994)

The California Ocean Plan (discussed below) defines initial dilution (which is therefore a regulatory mixing zone) as:

> "...the process which results in the rapid and irreversible turbulent mixing of wastewater with ocean water around the point of discharge. For a submerged buoyant discharge, characteristic of most municipal and industrial wastes that are released from the submarine outfalls, the momentum of the discharge and its initial buoyancy act together to produce turbulent mixing. Initial dilution in this case is completed when the diluting wastewater ceases to rise in the water column and first begins to spread horizontally."

Clearly, application of these regulations require much judgment, such as which oceanographic conditions, currents, density stratification, flow rates, and averaging times are used. These must be carefully chosen and explicitly specified in the outfall design documentation.

Mixing zone water quality standards are usually limited to parameters for acute toxicity protection (sometimes determined by bioassays) and to minimize visual impacts. They are not usually applied to BOD, dissolved oxygen, or nutrients. Bacterial standards are also not normally imposed within or at the boundary of mixing zones unless the diffuser is located near areas of shellfish harvesting or recreational uses.

2.4 THE CALIFORNIA OCEAN PLAN

Because of the unique behavior of wastewaters discharged from an outfall into a coastal environment and the many water quality and other issues involved, specification of regulations and guidelines is quite difficult. Probably the first major attempt to do so was the California Ocean Plan, more formally *"California Ocean Plan: Water Quality Control Plan, Ocean Waters of California,"* (SWRCB, 2005, hereafter referred to as *The Plan*). It was first published in 1972 and has been updated several times since; the latest version was adopted February 14, 2006. *The Plan* specifies beneficial uses of the ocean and requirements for water quality and point-source discharges to protect them. These include physical, chemical, biological, radioactive, aesthetic, and particulate deposition. *The Plan*, or parts of it, has been adopted by many environmental agencies around the world. Because of its importance we discuss it in some detail here, especially the parts most relevant to planning and design of ocean outfalls.

2.4.1 General provisions

The general provisions of *The Plan* specify:

- ... limits or levels of water quality characteristics for ocean waters to ensure the reasonable protection of beneficial uses and the prevention of nuisance. The discharge of waste shall not cause violation of these objectives.
- The Water Quality Objectives and Effluent Limitations are defined by a statistical distribution when appropriate. This method recognizes the normally occurring variations in treatment efficiency and sampling and analytical techniques...
- Compliance with the water quality objectives...shall be determined from samples collected at stations representative of the area within the waste field where initial dilution is completed.

2.4.2 Bacterial characteristics

Bacterial standards are specified for areas used for water contact recreation and shellfish. The most recent standards add enterococcus as an indicator organism to total and fecal coliforms and replace percentage exceedance limits with geometric means.

The water contact standards are applied *"Within a zone bounded by the shoreline and a distance of 1,000 feet from the shoreline or the 30-foot depth contour, whichever is further from the shoreline, and in areas outside this zone used for water contact sport...the following bacterial objectives shall be maintained throughout the water column:"*

The 30-day geometric means of:

- Total coliform density shall not exceed 1,000 per 100 ml;
- Fecal coliform density shall not exceed 200 per 100 ml;
- Enterococcus density shall not exceed 35 per 100 ml.

Single sample maximum:

- Total coliform density shall not exceed 10,000 per 100 ml;
- Fecal coliform density shall not exceed 400 per 100 ml;
- Enterococcus density shall not exceed 104 per 100 ml; and

The standards for waters where shellfish may be harvested for human consumption state that, throughout the water column:

- The median total coliform density shall not exceed 70 per 100 ml,
- Not more than 10 percent of the samples shall exceed 230 per 100 ml.

The Plan specifies that samples should be obtained weekly and the five most recent samples are to be used in computing the geometric mean. If any single-sample standards are exceeded, the sampling becomes more frequent. Other international bacterial standards that are frequently used are discussed in Section 2.5.

2.4.3 Physical characteristics

Physical characteristics specify that there be no visible floating particulates and oil and grease, no aesthetically undesirable discoloration of the ocean surface, and natural light should not be significantly reduced outside the initial dilution zone.

2.4.4 Chemical characteristics

The standards for control of chemical substances specify that:

- The dissolved oxygen concentration shall not at any time be depressed more than 10 percent from that which occurs naturally;
- The pH shall not be changed at any time more than 0.2 units from that which occurs naturally;
- The dissolved sulfide concentration of waters in and near sediments shall not be significantly increased above natural conditions;
- The concentration of substances set forth in Table B, in marine sediments shall not be increased to levels which would degrade indigenous biota;

Water quality aspects

- The concentration of organic materials in marine sediments shall not be increased to levels that would degrade marine life;
- Nutrient materials shall not cause objectionable aquatic growths or degrade indigenous biota.

Possibly the most important of these requirements are the numerical water quality objectives as set forth in "Table B." The portion of this table that applies to protection of aquatic life is reproduced as Table 2.3. It specifies three different limiting concentrations: a 6-month median, a daily maximum, and an instantaneous maximum.

Table 2.3 Table B of the California Ocean Plan: Water Quality Objectives for Protection of Marine Aquatic Life

Compound	Units	6-month median	Daily maximum	Instantaneous maximum
Arsenic	µg/l	8.0	32.0	80.0
Cadmium	µg/l	1.0	4.0	10.0
Chromium (Hexavalent)	µg/l	2.0	8.0	20.0
Copper	µg/l	3.0	12.0	30.0
Lead	µg/l	2.0	8.0	20.0
Mercury	µg/l	0.04	0.16	0.4
Nickel	µg/l	5.0	20.0	50.0
Selenium	µg/l	15.0	60.0	150.0
Silver	µg/l	0.7	2.8	7.0
Zinc	µg/l	20.0	80.0	200.0
Cyanide	µg/l	1.0	4.0	10.0
Total Chlorine Residual	µg/l	2.0	8.0	60.0
Ammonia (as nitrogen)	µg/l	600.0	2400.0	6000.0
Acute* Toxicity	TUa	N/A	0.3	N/A
Chronic* Toxicity	TUc	N/A	1.0	N/A
Phenolic Compounds (non-chlorinated)	µg/l	30.0	120.0	300.0
Chlorinated Phenolics	µg/l	1.0	4.0	10.0
Endosulfan	µg/l	0.009	0.018	0.027
Endrin	µg/l	0.002	0.004	0.006
HCH*	µg/l	0.004	0.008	0.012
Radioactivity	Not to exceed limits specified in Title 17, Division 1, Chapter 5, Subchapter 4, Group 3, Article 3, Section 30253 of the California Code of Regulations. Reference to Section 30253 is prospective, including future changes to any incorporated provisions of federal law, as the changes take effect.		

It is important to note that these limits are specified *at the completion of initial dilution* according to the following equation:

$$C_e = C_o + D_m(C_o - C_s) \qquad (2.1)$$

where C_e is the effluent concentration limit, C_o is the concentration to be met at the completion of initial dilution (the water quality objective), D_m is the minimum probable initial dilution, and C_s is the background seawater concentration. Minimum initial dilution is defined as:

> "...the lowest average initial dilution within any single month of the year. Dilution estimates shall be based on observed waste flow characteristics, observed receiving water density structure, and the assumption that no currents, of sufficient strength to influence the initial dilution process, flow across the discharge structure."

Background levels for arsenic, copper, mercury, silver, and zinc are specified in *The Plan*; the background levels of all other constituents are assumed to be zero.

Concentrations of other constituents for protection of human health are also specified in "Table B". They include non-carcinogens such as antimony, and carcinogens, such as benzene. These will not normally affect outfall designs, however.

2.4.5 Biological characteristics

The requirements for protection of marine organisms specify that:

(1) Marine communities, including vertebrate, invertebrate, and plant species, shall not be degraded;
(2) The natural taste, odor, and color of fish, shellfish, or other marine resources used for human consumption shall not be altered;
(3) The concentration of organic materials in fish, shellfish or other marine resources used for human consumption shall not bioaccumulate to levels that are harmful to human health.

2.4.6 Radioactivity

Discharge of radioactive waste shall not degrade marine life.

2.4.7 Discussion of the plan

The California Ocean Plan contains many other details such as how to perform computations and how to sample, and defines terms such as significant, shellfish,

Water quality aspects

etc. These, and other appropriate local regulations, should be consulted in an actual outfall design. We have attempted to summarize the most important features above from which we make the following observations.

The Plan does not generally specify effluent limitations or treatment levels. Instead, it specifies effluent limits that will achieve water quality objectives *after initial dilution*, i.e. at the edge of the mixing zone. To emphasize this, it states that:

> "Waste effluents shall be discharged in a manner which provides sufficient initial dilution to minimize the concentrations of substances not removed in the treatment."

It does set some limits in the waste stream for grease and oil, suspended solids, settleable solids, turbidity, pH, and acute toxicity, but other limits are determined as above.

It is important to note that bacterial levels are *not* specified at the edge of the mixing zone. Rather, they are specified at actual water contact areas, whose extents are defined. Because of the wide variations that always occur in bacterial sampling, the standards are generally expressed in statistical terms rather than instantaneous values. Indeed, *The Plan* states that:

> "It is state policy that the geometric mean bacterial objectives are strongly preferred for use in water body assessment decisions... because the geometric mean objectives are a more reliable measure of long-term water body conditions. In making assessment decisions on bacterial quality, single sample maximum data must be considered together with any available geometric mean data. The use of only single sample maximum bacterial data is generally inappropriate unless there is a limited data set, the water is subject to short-term spikes in bacterial concentrations, or other circumstances justify the use of only single sample maximum data."

2.5 STANDARDS FOR TOXICS

The issue of toxics frequently arises in discussion of marine wastewater discharges. They can be readily controlled by satisfying the requirements of Table B of the California Ocean Plan (Table 2.3). For regular domestic sewage, this will usually be accomplished if an initial dilution of 100:1 is maintained.

The USEPA maintains two water quality criteria for toxic substances (USEPA, 1991). The CMC (Criteria Maximum Concentration) is for protection of the aquatic ecosystem from acute or lethal effects; the CCC (Criteria Continuous Concentration) is for protection from chronic effects. The CCC is like a regular water quality standard and must be met at the edge of the mixing zone. It is *"...intended to be the highest concentration that could be maintained indefinitely in receiving water without causing an unacceptable effect on the aquatic community or its uses"*. The CCC limits may be sometimes exceeded, as organisms can tolerate higher concentrations for short periods so long as peak concentrations are limited. In other words, the CCC relates to average concentrations, which are in turn related to time-averaged dilutions. It is assumed that the CCC are the appropriate water quality criteria to apply in order to protect the aquatic ecosystem from chronic effects.

2.6 BACTERIAL CONTAMINANTS

2.6.1 Introduction

Prevention of microbial contamination is an essential part of outfall design; indeed, the main reason for an outfall project is often to solve a microbial contamination problem. Because of the importance of bacterial contamination, and ensuring that public health is protected at beaches and bathing waters, we deal with it in some detail in this section.

To provide context and background for the many bacterial standards have been adopted around the world (summarized in Section 2.6.3), we first present a review of the history and application of microbiological water quality standards for primary contact recreation and shellfish harvesting. Special note is taken of the first investigations conducted in the United States which concluded that enterococcus, as an indicator organism, provided the best correlation with gastrointestinal symptoms attributed to swimming in contaminated waters. The linear relationship developed between mean enterococcus density per 100 ml and swimming associated rate for gastrointestinal symptoms per 1000 persons is presented along with the US Environmental Protection Agency (USEPA) adaptation of enterococcus as the primary indicator organism in lieu of total and fecal coliforms.

After an extensive review of 33 epidemiological investigations conducted in marine waters worldwide, the World Health Organization (WHO) presented Guidelines for Safe Recreational Water Environments in 2002 using intestinal enterococcus as the indicator organism. These guidelines are based primarily on the controlled randomized epidemiological trial studies conducted in the coastal waters of the United Kingdom.

Existing international, national and local microbiological guidelines and standards in the marine environment are presented to provide a range for the water quality planner. Simply adapting a particular set of standards is inappropriate without a thorough review of local circumstances and local/national economic factors. Also, caution should be exercised in directly applying quantitative relationships between health risk and indicator organism in other areas where the general health and immunity of the local population may be different.

A recreational water quality criterion is defined as a quantifiable exposure-effect relationship based on scientific evidence between the level of some indicator of the quality of the water concerned and the potential human health risks associated with the recreational use of that water. A water quality guideline derived from such a criterion is a suggested maximum density of the indicator in the water which is associated with unacceptable health risks. The concept of acceptability implies that social, cultural, economic and political, as well as medical factors are involved. A water quality standard obtained from the criterion is a guideline fixed by law.

2.6.2 Historical review

2.6.2.1 The United States of America (USA)

The first evaluations of water contact recreation and disease incidence were conducted in the USA by the American Public Health Association with studies in the early 1920s to ascertain the prevalence of infectious diseases, which may be conveyed by swimming pools and other bathing places (Simons et al., 1922). As reported by Moore (1975), the application of bacterial guidelines to seawater can be traced back to the cautious suggestion by Winslow and Moxon (1928), in a pollution study of New Haven Harbor, where typhoid fever was attributed to bathing in grossly polluted water, that the coliform count of samples from bathing waters should not exceed 100/100 ml. No logical basis for this figure was provided, however. Coburn (1930) suggested a maximum permitted coliform count of 10,000 per 100 ml and quoted a bathing area where counts were consistently higher than this without apparently causing ill-health in bathers. Ludwig (1983) notes that the California coliform standard of 1,000 MPN/100 ml, which has been adopted in many other areas, was developed during the 1940s based entirely on aesthetic considerations, in that investigators found that when total coliform counts remained consistently (more than 80% of the time) below 1,000 MPN/100 ml, the beaches remained aesthetically satisfactory with no visual evidence of sewage pollution.

Cabelli et al. (1983) reports that the U.S. total coliform limit of 1,000/100 ml "apparently developed from two sources: the predicted risk of salmonellosis as obtained from calculations made by Streeter (1951) on the incidence of Salmonella species in bathing waters and attainability as determined by Scott (1951) from microbiological surveys conducted at Connecticut bathing beaches." This Connecticut standard was then adopted by many other USA State agencies.

As noted by the Committee on Bathing Beach Contamination of the Public Health Laboratory Service (1959), Garber (1956) reported on an inquest of different public health and control agencies in the USA concerning *"how bacteriological standards were determined and why they were decided upon."* The most frequent reply was that there was no analytical background for the limits set other than the fact that epidemiological experience under the given standards had been satisfactory. This argument was used for standards ranging from a median coliform count of less than 2,400/100 ml down to a requirement that no coliform organisms should be present.

Major epidemiological studies aimed directly at assessing the health risk of bathing in polluted waters were conducted during the years 1948–1950 by the United States Public Health Service. The findings (Stevenson, 1953) were that statistically significant epidemiologically detectable health effects at levels of 2,300 and 2,700 coliforms/100 ml were demonstrated by the studies on Lake Michigan at Chicago, Illinois in 1948 and on the Ohio River at Dayton, Kentucky in 1949, respectively. The third study conducted in 1950 in the saline tidal waters of Long Island Sound at New Rochelle and Mamaroneck, New York showed no relationship between total coliform levels and swimming related diseases. Subsequent work in the same stretch of the Ohio River indicated that the fecal coliforms represented 18% of the total coliforms (Cabelli *et al.*, 1983) and therefore would indicate that detectable health effects would occur at a fecal coliform level of about 400 MPN/100 ml. Applying a factor of safety, in that water quality should be better than that which would cause a health effect, the National Technical Advisory Committee (NTAC, 1968) to the USA Federal Water Pollution Control Administration developed in 1968 a national fecal coliform guideline of 200 MPN/100 ml for fresh and marine waters which was based primarily on the two fresh water studies of Stevenson (1953).

However, in 1972, the Committee on Water Quality Criteria of the National Academy of Sciences of the USA (1972), in a USEPA funded project, came to the following conclusion: *"No specific recommendation is made concerning the presence or concentration of microorganisms in bathing water because of the paucity of valid epidemiological data."* Subsequently in 1976, the USEPA (1976) presented fecal coliform guidelines which were essentially those presented in the NTAC (1968) document. Notwithstanding, the primary rationale was based on the relationship of fecal coliform densities to the frequency of Salmonella isolations in surface waters and the findings of the Stevenson (1953) studies were essentially abandoned as a rationale. The final guideline proposed by the USEPA (1976) was as follows: *"based on a minimum of not less than five samples taken over not more than a 30-day period, the fecal coliform content of primary contact recreational waters shall not exceed a log mean of 200/100 ml, nor shall more than 10% of total samples during any 30-day period exceed 400/100 ml."*

Based on a three-year (1973–1975) study conducted at New York City beaches, Cabelli *et al.* (1983) concluded that Enterococci (analytical procedures presented in USEPA, 1985), as an indicator organism, provided the best correlation with gastrointestinal (vomiting, diarrhea, nausea or stomachache) symptoms attributed to swimming in contaminated waters. Other indicators evaluated included total coliforms and their component genera (Escherichia, Klebsiella, Citrobacter-Enterobacter), fecal coliforms, Escherichia coli (E. Coli), Pseudomonas aeruginosa, Clostridium penfringens, Aeromonas hydrophila, Vibrio parahaemolyticus and Salmonella. Subsequent USA studies confirmed the superiority of enterococci as an indicator organism and Cabelli (1983) developed a linear relationship between mean enterococcus density/100 ml and swimming associated rate for gastrointestinal symptoms per 1000 persons as presented in Figure 2.1. Cabelli concluded that enterococci better mimicked the survival characteristics of the etiological agent which Cabelli (undated) concluded to be the human rotavirus with regard to gastroenteritis. A critique of this work is presented by Fleisher (1991).

It must be recognized that the pathogen to indicator organism ratio is variable due to its dependence on the overall health of the discharging population. As noted by Cabelli (1983), the swimming associated outbreak of Shigellosis on the Mississippi River below Dubuque, Iowa, USA (Rosenburg, 1976), appears to represent an instance where, although the 200/100 ml fecal coliform guideline was probably exceeded for some time, the outbreak did not occur until there was a large enough number of ill individuals and carriers in the discharging population. Also, comparisons made by Cabelli (1983, undated) of epidemiological studies conducted in Egypt with those conducted in the USA suggest the important role of population immunity in that gastrointestinal illness rates in the USA studies were associated with bathing in waters with relatively much lower enterococci densities. These studies also demonstrated that swimming-associated gastrointestinal symptoms were much more prevalent among children (10 years of age and under) with lesser developed immune systems than adults. This further suggests the importance of immunity in the epidemiology of the observed swimming-associated gastroenteritis. These factors imply that caution should be exercised in directly applying the relationships developed to other areas.

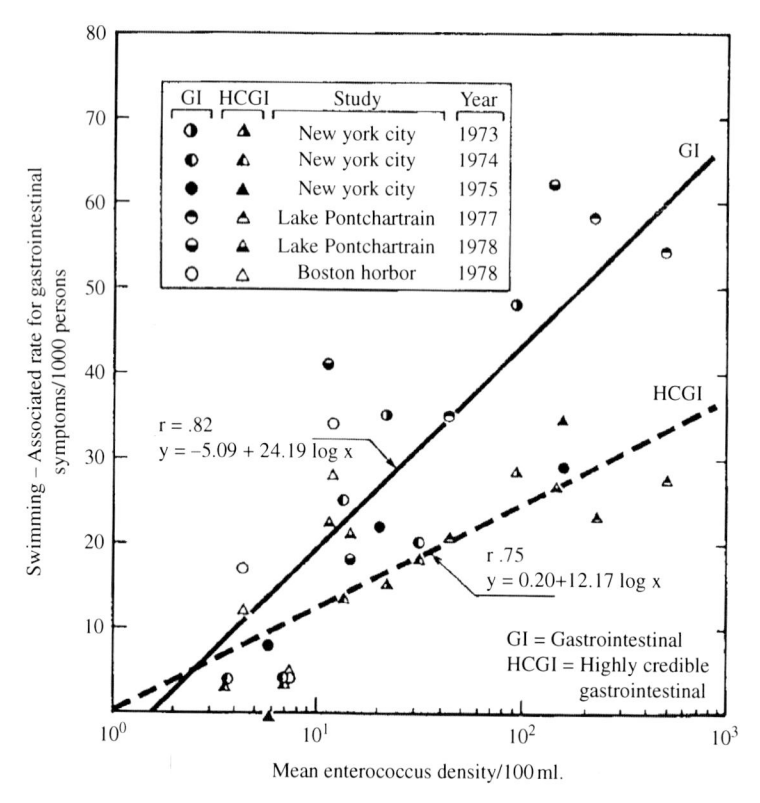

Figure 2.1 Effect of enterococcus on gastrointestinal symptoms for swimmers in marine waters (Cabelli, 1983)

The USEPA (1984) first presented the recommendation that enterococci be adapted by the States of the USA as the primary indicator organism for primary contact recreations in lieu of the indicators applied at that time (primarily total and fecal coliforms). After a review and recalculation of the data the USEPA (1986) adopted the following criteria:

Freshwater: E. Coli: not to exceed 126/100 ml, or
              Enterococci: not to exceed 33/100 ml
Marine water: Enterococci: not to exceed 35/100 ml

These guidelines were based on studies conducted by Dufour (1984) and Cabelli (1983) applying the empirical equations developed for highly credible gastrointestinal symptoms (HCGI) associated with swimming in fresh and marine waters, respectively. These guidelines were based on risk levels of 8 and 19 gastrointestinal illnesses per 1000 swimmers at fresh water and marine beaches (see Figure 2.1), respectively, which were estimated to be equivalent to the risk levels for 200/100 ml fecal coliform criteria. The USEPA (1984) premise was that by using the existing criterion of 200 fecal coliform bacteria per 100 ml the health risks mentioned above for fresh water and marine beaches have been unknowingly accepted.

These criteria are calculated as the geometric mean of a statistically sufficient number of samples, generally not less than five samples equally spaced over a thirty-day period. Single sample maximum allowable densities were also promulgated based on beach use and are presented in Table 2.4 taken from Dufour and Ballentine (1986).

As per footnote 5 in Table 2.4, the Single Sample Maximum Allowable Density is based on the observed log standard deviations during the USEPA studies: 0.4 for freshwater E. coli and enterococci; and 0.7 for marine water enterococci. It is also stated that each jurisdiction should establish its own standard deviation for its conditions which would then vary the single sample limit.

Based on the above log normal distribution, different confidence levels (CL), i.e., 75, 82, 90 and 95 percentiles, were assigned to different designated beach uses (see Table 2.4 and Figure 2.2). The geometric mean (CL = 50 percentile) remains at 35 enterococci/100 ml and all the CLs specified in Table 2.4 form part of the same log normal distribution with a standard deviation of 0.7, which would theoretically be equivalent 19 gastrointestinal illnesses per 1000 swimmers for marine waters.

For the category, Designated Beach Area (75% CL) in Table 2.4, there would be a 75% probability that the specified value of 104 enterococci/100 ml would not be exceeded in a statistically sufficient number of data. Consequently there would be a probability that 25% of the measurements would indeed exceed the specified value.

The Single Sample Maximum Allowable Density approach adopted in the USEPA guidelines essentially applies a safety factor under the apparent premise that bathers should not be exposed to the higher indicator levels of the log normal distribution, which is applied based on beach use ranging from the 75 percentile for a designated beach area to the 95 percentile for an infrequently used full body contact recreation beach area. This single sample maximum criterion would inherently drive down the geometric mean substantially for most data sets and consequently the gastrointestinal illness rate would be lower. Compliance would also be that more difficult to achieve.

Table 2.4 Indicator criteria for bacteriological densities
(from Dufour and Ballentine, 1986)

	Acceptable swimming associated gastro-enteritis rates per 1000 swimmers	Simple sample maximum allowable density[4,5]				
		Steady-state geometric mean indicator density	Designed beach area (Upper 75% CL)	Moderate full body contact recreation (Upper 82% CL)	Lightly used full body contact recreation (Upper 90% CL)	Infrequently used full body contact recreation (Upper 95% CL)
Freshwater:						
Enterococci	8	33[1]	61	89	108	151
E. Coli	8	126[2]	235	298	406	576
Marine waters:						
Enterococci	19	35[3]	104	158	275	500

(1) Calculated to nearest whole number using equation:

$$(\text{mean enterococci density}) = \text{antilog}_{10} \frac{\text{illness rate}/1000 \text{ people} + 6.28}{9.40}$$

(2) Calculated to nearest whole number using equation:

$$(\text{mean } E.\ Coli \text{ density}) = \text{antilog}_{10} \frac{\text{illness rate}/1000 \text{ people} + 11.74}{9.40}$$

(3) Calculated to nearest whole number using equation:

$$(\text{mean enterococci density}) = \text{antilog}_{10} \frac{\text{illness rate}/1000 \text{ people} - 0.20}{12.17}$$

(4) Single sample limit = $\text{antilog}_{10}\left\{\left(\begin{array}{c}\text{log indicator geometric}\\ \text{mean density}/100\text{ ml}\end{array}\right) + \left(\begin{array}{c}\text{factor determined from areas under}\\ \text{the normal probality curve for the}\\ \text{assumed level of probality (see below)}\end{array}\right) \times (\log_{10} \text{ standard deviation})\right\}$

The appropriate factors for the indicated one sided confidence levels are:
75% C.L. −.675
82% C.L. −.935
90% C.L. −1.28
95% C.L. −1.65

(5) Based on the observed log standard deviations during the USEPA studies: 0.4 for freshwater *E. Coli* and enterococci; and 0.7 for marine water enterococci. Each jurisdiction should establish its own standard deviation for its conditions which would then vary the single sample limit.

C.L. Confidence level

Dufour and Ballentine (1986)

Water quality aspects

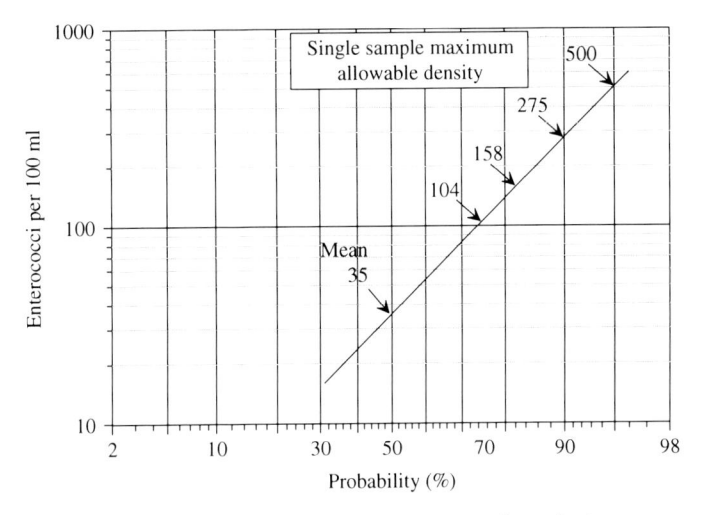

Figure 2.2 USEPA ambient water quality criteria

In 1997, the EPA embarked on the Beaches Environmental Assessment, Closure, and Health (BEACH) Program which included a program of epidemiological studies conducted in the USA in 2003–6 to substantiate the USEPA Guidelines (results are pending).

2.6.2.2 International organizations

In 1974, the World Health Organization (WHO) convened a Working Group of European Experts on Guides and Criteria for Recreational Quality of Beaches and Coastal Waters (WHO, 1975) in Bilthoven, Netherlands, which *"agreed that the recommended upper limits for indicator organisms should be expressed in broad terms of orders of magnitude rather than rigidly stated specific numbers. Highly satisfactory bathing areas should, however, show E. Coli counts consistently less than 100 per 100 ml and to be considered acceptable, bathing waters should not give counts consistently greater than 1000 E. Coli per 100 ml"*. Subsequently, in 1977 a group of experts jointly convened by WHO and UNEP in Athens, (WHO, 1977) concluded that there was no basis for recommending changes in the conclusions reached by the WHO Working Group in Bilthoven in 1974.

Saliba and Helmer (1990) reported that, in 1983, UNEP/WHO (1983) proposed that the Mediterranean governments adopt interim criteria for coastal recreational waters using, as a basis, the 1974 Bilthoven conclusions extrapolated to

Mediterranean conditions based on the results of the WHO organized pilot project on coastal water quality control within the framework of the UNEP-sponsored MED POL program (carried out between 1976 and 1981). These criteria were based on concentrations of both fecal coliforms and fecal streptococci. The part of the proposal concerning fecal coliform concentration limits, that is, 100 per 100 ml in at least 50% of the samples and 1,000 per 100 ml in at least 90%; minimum of 10 samples, were subsequently adopted in 1985 by the Mediterranean States on a joint basis as an interim measure (UNEP, 1985).

The European Economic Community (EEC 1976) published Quality Requirements (microbiological) for Bathing Waters as presented in Table 2.5. The latest European Union (EU) guidelines for coastal and transitional waters that were published on 15 February 2006 (EU 2006) are presented in Table 2.6.

Table 2.5 EEC microbiological requirements for bathing waters (EEC 1976)

Microbiological parameter	Guide[a]	Mandatory[b]	Minimum sampling frequency
1. Total coliforms/100 ml	500	10,000	Fortnightly
2. Fecal coliforms/100 ml	100	2,000	Fortnightly
3. Streptococci fecal/100 ml	100[c]	–	*
4. Salmonella/l	–	0	*
5. Enteroviruses PFU/10 l	–	0	*

[a] 80% of samples less than.
[b] 95% of samples less than.
[c] 90% of samples less than.
* Concentration to be checked by the competent authorities when an inspection in the bathing area shows that the substance may be present or that the quality of the water has deteriorated.

Table 2.6 Coastal and transitional waters (EU, 2006)

Parameter	Excellent quality	Good quality	Sufficient	Reference methods of analysis
Intestinal Enterococci (I.E.) (cfu/100 mL)	100*	200*	185**	ISO 7899-1 or ISO 7899-2
Escherischia coli (E.C.) (cfu/100 mL)	250*	500*	500**	ISO 9308-3 or ISO 9308-1

* Based upon a 95-percentile evaluation.
** Based upon a 90-percentile evaluation.

The Caribbean Environment Programme (CEPPOL) held Regional meetings in 1991 and 1993 on Monitoring and Control of Sanitary Quality Bathing and Shellfish-Growing Marine Waters in the Wider Caribbean. Due to the economic dependence of the Caribbean on tourism, both the bacteriological and aesthetic water quality are very important. It was concluded at these meetings that Member Countries should adopt EEC, WHO or USEPA (prior to 1986) standards and guidelines for bacteriological quality of bathing waters until sufficient information is available, based on future epidemiological studies conducted in the Caribbean, to modify the current standards (CEPPOL and UNEP, 1991).

As reported by Saliba and Helmer (1990), prospective cohort epidemiological studies similar to that conducted by Cabelli (generally referred to as "Cabelli style studies") were carried out in a number of countries between 1982 and 1989. Saliba and Helmer (1990) state that *"...practically all studies showed higher morbidity among bathers as compared to non-bathers, but correlation between specific symptoms and bacterial indicator concentrations varied considerably"*. Furthermore, they conclude that although difficult to quantify *"...the evidence clearly indicates that health risks do exist and are most pronounced in areas directly exposed to pollution by untreated sewage."*

2.6.3 Standards for human health protection

2.6.3.1 WHO guidelines

In 1994, the World Health Organization (WHO) embarked on the development of guidelines concerning recreational use of the water environment. Guidelines of this type are primarily a consensus view amongst experts on the risks to health represented by various media and activities and are based on critical review of the available evidence. WHO convened several experts' meetings:

- Bad Elster, Germany, June 1996
- St. Helier, Jersey, Channel Island, United Kingdom, May 1997
- Farnham United Kingdom April 1998 for the development of a Code of Good Practice for Monitoring

The *WHO Guidelines for Safe Recreational Water Environments*, which resulted from this process, were released in two volumes. Volume 1, released at the AIDIS (Spanish acronym for the Inter-American Association of Sanitary and Environmental Engineering) Congress in Cancun, Mexico in 2002 (hard copy released by WHO, 2003), addresses coastal and fresh waters. Volume 2, released in 2006 addresses swimming pools, spas, and similar recreational water environments.

The preliminary publication of the guidelines was made in 1998 (WHO, 1998). As part of this process, Prüss (1998) summarized the epidemiological studies conducted worldwide. Of the 37 studies evaluated, 22 qualified for inclusion in the evaluation. Figure 2.3 presents the relation between indicator organism density and illness risk for marine waters. Of the 22 studies selected, 18 were prospective cohort studies, two retrospective cohort studies and two were randomized controlled trials, as summarized in Table 2.7. In 19 of the 22 epidemiological studies examined in Prüss' (1998) review, the rate of certain symptoms or symptom groups was significantly related to the count of fecal indicator bacteria in recreational water. Hence, there was a consistency across the various studies, and gastrointestinal symptoms were the most frequent health outcome for which significant dose-related associations were reported. The overwhelming evidence provided by most of the epidemiological studies conducted worldwide over the past 30 years and reviewed by the WHO (WHO, 2003) has shown that the indicator organisms which correlate best with health outcome were enterococci/fecal streptococci for marine waters. Other indicators showing correlation were fecal coliforms and staphylococci.

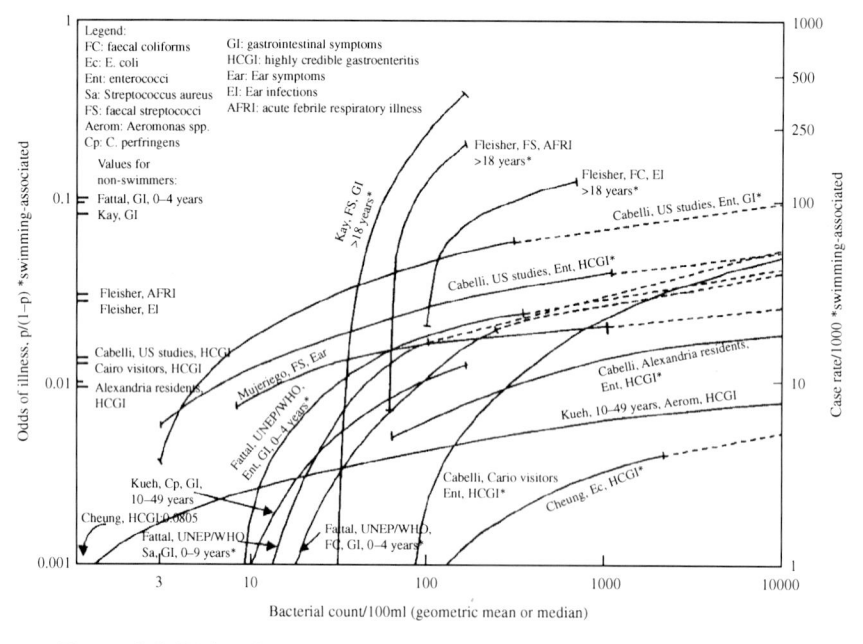

Figure 2.3 Risks of illness for swimmers in marine waters (WHO, 2003)

In marine bathing waters, the United Kingdom randomized controlled trials (Kay *et al.*, 1994; Fleisher *et al.*, 1996) probably contained the least amount of bias. These studies gave the most accurate measure of exposure, water quality, and illness compared with observational studies where an artificially low

Water quality aspects

threshold and flattened dose–response curve (due to misclassification bias) were likely to have been determined.

Table 2.7 Epidemiological studies evaluated for WHO guidelines

First author	Year	Country	Study design	Water	Notes
Fleisher	1996	UK	Randomized controlled trial	marine	d
Haile	1996	US	Prospective cohort	marine	
Van Dijk	1996	UK	Prospective cohort	marine	c
Bandaranayake	1995	New Zealand	Prospective cohort	marine	d
Kueh	1995	Hong Kong	Prospective cohort	marine	b
Medical Research Council	1995	South Africa	Prospective cohort	marine	a, c
Kay	1994	UK	Randomized controlled trial	marine	d
Pike	1994	UK	Prospective cohort**	marine	a, b, c
Corbett	1993	Australia	Prospective cohort	marine	a, d
Fewtrell*	1992	UK	Prospective cohort	fresh	d
UNEP/WHO no. 46	1991	Israel	Prospective cohort	marine	b, d
UNEP/WHO no. 53	1991	Spain	Prospective cohort	marine	a, b, d
Cheung	1989	Hong Kong	Prospective cohort	marine	a, b
Ferley	1989	France	Retrospective cohort	fresh	a, b, c
Lightfoot	1989	Canada	Prospective cohort	fresh	
Fattal, UNEP/WHO no. 20	1987	Israel	Prospective cohort	marine	b, d
Seyfried	1985	Canada	Prospective cohort	fresh	
Dufour	1984	US	Prospective cohort	fresh	a, b
Cabelli	1983	Egypt	Prospective cohort	marine	a, b, c
Cabelli	1982	US	Prospective cohort	fresh & marine	a, b
Mujeriego	1982	Spain	Retrospective cohort**	marine	b, a
Stevenson, 3-day study	1953	US	Prospective cohort	fresh	b, c, d

a. Only use of seasonal mean for analysis of association with outcome reported.
b. Control for less than three confounders reported, or no reporting at all.
c. Exposure not defined as head immersion/head splashing/water ingestion.
d. <1700 bathers and 1700 non-bathers participating in the study.
* Exposure is whitewater canoeing; similar to swimming, water intake is likely, while turnover or through ingestion or inhalation of droplets.
** Cross-sectional study.
Remark: Two studies analyze the same data sets 5, 10 but come to different conclusions.
Source: WHO (1998), "Guidelines for safe recreational-water environments."

The United Kingdom randomized controlled trials were therefore the key studies used to derive the guideline values for recreational waters. However, it should be emphasized that they are primarily indicative for adult populations in marine waters in temperate climates. Studies that reported higher thresholds and case rate values (for adult populations or populations of countries with higher endemicities) may suggest increased immunity, which is a plausible hypothesis. Most studies reviewed by Prüss (1998) suggested that symptom rates were higher in lower age groups, and the United Kingdom studies may therefore systematically underestimate risks to children.

WHO concluded that the controlled randomized trial studies were the most accurate and the WHO Experts' Committee based the new guidelines for marine waters on the only study of this type for enteric illness, reported by Kay and Fleisher *et al.* (1994), in the United Kingdom. It is noted that these are temperate waters and not characteristic of tropical waters.

The guideline values for microbial water quality given in Table 2.8 are derived from the key studies described above. The values are expressed in terms of the 95th percentile of numbers of intestinal enterococci per 100 ml and represent readily understood levels of risk based on the exposure conditions of the key studies. WHO defined a 1% risk for illness occurrence due to bathing as "an excess illness of one incidence in every 100 exposures," compared to non-bathers. The values may need to be adapted to account for different local conditions.

Figure 2.4, on which the WHO guideline values of Table 2.8 are derived, shows the dose-response relation between health risk and the 95th percentile value of the intestinal enterococci indicator for contracting gastro-enteritis and acute febrile respiratory illness (AFRI) (Fleisher 1996) by bathing in microbiologically contaminated water.

The 95th percentile approach
WHO and many agencies have chosen to base criteria for recreational-water compliance upon either 95% compliance levels (i.e., 95% of the samples must lie below a specific value in order to meet the standard) or geometric mean values of water quality data collected in the bathing zone. Both have significant drawbacks. The geometric mean is statistically a more stable measure, but this is because the inherent variability in the distribution of the water quality data is not characterized in the geometric mean. However, it is this variability that produces the high values at the top end of the distribution that are of greatest public health concern.

The 95% compliance system, on the other hand, does reflect much of the top-end variability in the distribution of water quality data and has the merit of being more easily understood. However, it is affected by greater statistical uncertainty and hence is a less reliable measure of water quality, thus requiring careful application to regulation.

Water quality aspects 39

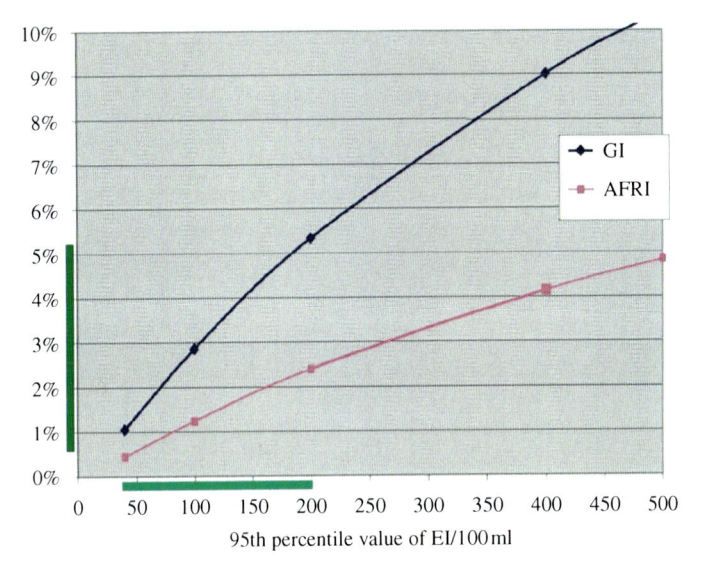

Figure 2.4 Risks for GI and AFRI due to EI exposure

Other options include the percentile approach, in which a specified percentile, most commonly the 80th, 90th, or 95th, is calculated. A limit can then be set for making judgments about the water quality, depending on whether the specified percentile value exceeds it or not. A simple ranking method by which a specified percentile may be calculated from the sample series being evaluated is given in Bartram and Rees (2000). Other methods for calculating sample series percentiles are given by Ellis (1989). Ninety-fifth percentile values calculated in this manner suffer from some of the same drawbacks described above for the 95% compliance system.

A more appropriate method of calculating the 95th percentile, which makes better use of all the data in the sample set, is to generate a probability density function (PDF) based on the distribution of indicator organisms over a defined bathing area and then to use the properties of this PDF to estimate the 95th percentile value of this distribution. In practice, the full procedure is rarely carried out, and 95th percentiles are calculated using the lognormal distribution method given in Bartram and Rees (2000). This is called a parametric method, since it requires the estimation of the population parameters known as the mean and standard deviation of the lognormal distribution. One limitation of the method is that if the samples are not lognormally distributed, it will yield erroneous estimates of the 95th percentile. Also, if there are data below the limit of detection, these data must be assigned an arbitrary value based on the limit of detection.

Table 2.8 WHO Guidelines for microbial water quality of recreational waters

95th percentile intestinal enterococci per 100 ml	Basis of derivation	Estimated risk per exposure
≤40 A	This range is below the NOAEL in most epidemiological studies.	<1% GI illness risk <0.3% AFRI risk The upper 95th percentile value of 40/100 ml relates to an average probability of less than one case of gastroenteritis in every 100 exposures. The AFRI burden would be negligible.
41–200 B	The 200/100 ml value is above the threshold of illness transmission reported in most epidemiological studies that have attempted to define a NOAEL or LOAEL for GI illness and AFRI.	1–<5% GI illness risk 0.3–<1.9% AFRI risk The upper 95th percentile value of 200/100 ml relates to an average probability of one case of gastroenteritis in 20 exposures. The AFRI illness rate at this upper value would be less than 19 per 1000 exposures, or less than approximately 1 in 50 exposures.
201–500 C	This range represents a substantial elevation in the probability of all adverse health outcomes for which dose–response data are available.	5–10% GI illness risk 1.9–3.9% AFRI risk This range of 95th percentiles represents a probability of 1 in 10 to 1 in 20 of gastroenteritis for a single exposure. Exposures in this category also suggest a risk of AFRI in the range of 19–39 per 1000 exposures, or a range of approximately 1 in 50 to 1 in 25 exposures.
>500 D	Above this level, there may be a significant risk of high levels of minor illness transmission.	>10% GI illness risk >3.9% AFRI risk There is a greater than 10% chance of gastroenteritis per single exposure. The AFRI illness rate at the 95th percentile point of >500/100 ml would be greater than 39 per 1000 exposures, or greater than approximately 1 in 25 exposures.

Notes:
1. Abbreviations used: A–D are the corresponding microbial water quality assessment categories (see section 4.6 of WHO, 2003) used as part of the classification procedure (Table 4.12 of WHO, 2003); AFRI = acute febrile

respiratory illness; GI = gastrointestinal; LOAEL = lowest-observed-adverse-effect level; NOAEL = no-observed-adverse-effect level.

2. The "exposure" in the key studies was a minimum of 10 min of swimming involving three head immersions. It is envisaged that this is equivalent to many immersion activities of similar duration, but it may underestimate risk for longer periods of water contact or for activities involving higher risks of water ingestion (see also note 8).
3. The "estimated risk" refers to the excess risk of illness (relative to a group of non-bathers) among a group of bathers who have been exposed to faecally contaminated recreational water under conditions similar to those in the key studies.
4. The functional form used in the dose–response curve assumes no further illness outside the range of the data (i.e., at concentrations above 158 intestinal enterococci/100 ml; see Box 4.3 of WHO, 2003). Thus, the estimates of illness rate reported above this value are likely to be underestimates of the actual disease incidence attributable to recreational water exposure.
5. The estimated risks were derived from sewage-impacted marine waters. Different sources of pollution and more or less aggressive environments may modify the risks.
6. This table relates to risk to "healthy adult bathers" exposed to marine waters in temperate north European waters.
7. This table may not relate to children, the elderly, or the immunocompromised, who could have lower immunity and might require a greater degree of protection. There are no adequate data with which to quantify this, and no correction factors are therefore applied.
8. Epidemiological data on fresh waters or exposures other than swimming (e.g., high-exposure activities such as surfing, dinghy boat sailing or whitewater canoeing) are currently inadequate to present a parallel analysis for defined reference risks. Thus, a single series of microbial values is proposed, for all recreational uses of water, because insufficient evidence exists at present to do otherwise. However, it is recommended that the length and frequency of exposure encountered by special interest groups (such as bodysurfers, board riders, windsurfers, sub-aqua divers, canoeists and dinghy sailors) be taken into account (Chapter 1 of WHO, 2003).
9. Where disinfection is used to reduce the density of index organisms in effluents and discharges, the presumed relationship between intestinal enterococci (as an index of faecal contamination) and pathogen presence may be altered. This alteration is, at present, poorly understood. In water receiving such effluents and discharges, intestinal enterococci counts may not provide an accurate estimate of the risk of suffering from gastrointestinal symptoms or AFRI.
10. Risk attributable to exposure to recreational water is calculated after the method given by Wyer et al. (1999), in which a log10 standard deviation of 0.8103 for faecal streptococci was assumed. If the true standard deviation for a beach is less than 0.8103, then reliance on this approach would tend to overestimate the health risk for people exposed above the threshold level, and vice versa.
11. Note that the values presented in this table do not take account of health outcomes other than gastroenteritis and AFRI. Where other outcomes are of public health concern, then the risks should also be assessed and appropriate action taken.
12. Guideline values should be applied to water used recreationally and at the times of recreational use. This implies care in the design of monitoring programs to ensure that representative samples are obtained.

Guidelines for seawater
The guideline values for microbiological quality given in Table 2.8 are derived from the key studies described above. The cutoff or bounding guideline values (40, 200, 500) are expressed in terms of the 95th percentile of numbers of intestinal enterococci per 100 ml and represent readily understood levels of risk based on the exposure conditions of the key studies. The values may need to be adapted to account for different local conditions and are recommended for use in the recreational-water environment classification scheme discussed in the WHO Guidelines document.

For the purposes of water quality monitoring, the terms fecal streptococci, intestinal enterococci and enterococci are considered to be synonymous (Figueras *et al.*, 2000). Intestinal enterococci are used in the *WHO Guidelines for Safe Recreational Environments*. Exposure to recreational waters with these measured indicators refers to body contact that is likely to involve head immersion, such as swimming, surfing, white-water canoeing, scuba diving, and dinghy boat sailing.

Available evidence suggests that the guideline values presented in Table 2.8 provide a lesser degree of health protection than that considered tolerable in other areas of environmental quality regulation. However, the central "200" cutoff or upper bounding value represents a stricter standard than is encountered in many areas at present. Measures to discourage water use at times or in locations of greater risk may provide cost-effective means to improve health protection and water quality classification.

2.6.3.2 Other standards

Many international, national, and local agencies have established guidelines and standards for water quality indicators to protect human health that can provide a reference point for planning. They are summarized in Table 2.9. The standards vary widely, reflecting different philosophies, levels of risk, and/or levels of water use protection. The principal factor responsible for the range of standards is the origin of the supporting criteria, be they epidemiological, aesthetic, or ecological.

It is noted that, except for Brazil and Peru, most of the countries in Latin America that have promulgated national standards have adopted them directly, with minor modifications, from those applied in the USA prior to 1986 with perhaps minimal considerations given to economic realities and development priorities. Developing nations, such as in Latin America, differ from industrialized nations, where most of the research is conducted, in that the developing country must allocate limited financial resources to a greater number of basic public works and economic development projects. It is important that the planner conduct a thorough review of the prevailing local water quality guidelines/standards (if any exist) to insure that local economic development priorities are reasonably accounted for. Control systems, such as ocean outfalls, are among the most capital intensive means of wastewater disposal although life time costs will be considerably less in comparison to secondary wastewater treatment with onshore disposal. Consequently, the decision to design the system for other than minimum water quality standards should be supported by demonstrated need, or a stated local/national policy decision.

We have not found any epidemiological investigations which were used as a basis for the Brazilian standard for primary contact recreation, which is essentially five times that of the USEPA guideline for fecal coliforms which was applied until 1986 (see Table 2.9). It is reasonable to conclude that the Peruvian standard, which is identical to that of Brazil, was highly influenced by the latter. In 1987, the "Companhia de Tecnologia de Saneamento Ambiental of Sao Paulo (CETESB)" (García Agudo, 1991) embarked on an epidemiological study to ascertain the relationship between swimming associated illnesses and microbiological indicators for Brazil (publication pending). CETESB (1999) has subsequently conducted additional prospective cohort studies. A similar epidemiological study was conducted during 1991–1993 in Trinidad Tobago (CEPPOL/UNEP, 1991). The studies in Brazil and Trinidad and Tobago demonstrated that the indicator organism which correlated best with health outcomes was enterococci.

As can be seen in Table 2.9, the microbiological standards are frequently expressed as a permissible mean concentration and a maximum value that should not be exceeded a given percent (90% is common) of the time. However, the relationship between these two criteria should be evaluated. For example, Kay et al. (1990), show that the 1976 EEC mandatory criteria that 95% of the samples be less than 2000 fecal coliforms/100 ml was more strict than the geometric mean of 200/100 ml used by the USEPA prior to 1986. This analysis assumes a log normal distribution with a standard deviation of 0.7 ($\log_{10}$) that implies that for a mean of 200/100 ml, 95% of the samples would have to be less than 2834/100 ml. However, it is noted that the pre-1986 USEPA guidelines also specified that 90% of the samples be less than 400/100 ml which translates to a geometric mean of about 50 fecal coliforms/100 ml utilizing the same assumptions of Kay et al. (1990). The present USEPA guidelines and the implications of the Single Sample Maximum Allowable Density approach adopted in these guidelines have already been discussed in Section 2.6.2.1.

The establishment of water quality objectives (standards) is dependent on existing or planned water uses in an area, and, as such, are site specific. The discussions in this document have been and will continue to be limited to the presentation of past and present guidelines and standards and there adaptation and application to protect different water uses.

Table 2.9 Guidelines and standards for Microbiological Water Quality for primary contact recreation (per 100 ml) (Salas, 2002)

Country	Coliforms Total	Coliforms Fecal	Other	Reference
USA (EPA)			Enterococci 35[a] (see Table 2.3)	USEPA (1986), Dufour and Ballentine (1986)
USA (California)	1000[a,h] 10000[m,h]	200[a,h] 400[m,h]	Enterococci 35[a,h] 100[m,h]	SWRCB (2005)
EEC[b], Europe			Intestinal Enterococci 100[c1,d1], 200[c2,d1], 185[c3,d2] E. Coli 250[c1,d1], 500[c2,d1], 500[c3,d2]	EC (2006)
UNEP/WHO		50% < 100[k] 90% < 1000[k]		WHO and UNEP (1978)
Brazil	80% < 5000[j]	80% < 1000[j]		Ministerio del Interior (1976)
Colombia	1000	200		Ministerio de Salud (1979)
Cuba	1000[a]	200[a] 90% < 400		Ministerio de Salud (1986)
Ecuador	1000	200		Ministerio de Salud Pública (1987)
Mexico	80% < 1000[e] 100% < 10,000[j]			SEDUE (1983)
Peru	80% < 5000[e]	80% < 1000[e]		Ministerio de Salud (1983)
Puerto Rico		200[g] 80% < 400		JCA (1983)
Venezuela	< 5000 90% < 1000	90% < 200 100% < 400		Venezuela (1978)
France	< 2000	< 500	Fecal streptococci < 100	WHO (1977)
Israel	80% < 1000[f]			Argentina INCYTH (1984)
Japan	1000			Japan Env. Agency (1981)
Poland			E. Coli < 1000	WHO (1975)
USSR			E. Coli < 100	WHO (1977)
Yugoslavia	2000			Argentina, INCYTH (1984)
People's Republic of China		< 200[l]	Coli index < 1000[l]	SEPA (1998)

a. Logarithmic average for a period of 30 days of at least 5 samples
b. Minimum sampling frequency – fortnightly
c. Excellent quality (c1); Good quality (c2); sufficient (c3)
d. Based upon 95-%tile (d1); Based upon 90-%tile (d2)
e. At least 5 samples per month
f. Minimum 10 samples per month
g. At least 5 samples taken sequentially from the waters in a given instance
h. Within a zone bounded by the shoreline and a distance of 1000 feet from the shoreline or the 30 foot depth contour, whichever is further from the shoreline

i. No sample taken during the verification period of 48 hours should exceed 10,000
j. "Satisfactory" waters, samples obtained in each of the preceding 5 weeks
k. Maximum permitted value in unit item determination
l. Sampling frequency no less than once a month. More than 95% of the samples in a year should accord with the standard.
m. Single sample maximum

2.6.4 Standards for shellfish

An ancient reference concerning the danger of eating certain marine animals is found in the Bible, where Moses in Deutoronomy alerts his people that *"Of all creatures living in the water, you may eat any that have fins and scales. But anything that does not have fins and scales you may not eat; for you it is unclean"* (Deut. 14:9-10).

The most stringent bacterial criteria are associated with shellfish harvesting areas. Various shellfish standards are summarized in Table 2.10. Certain shellfish such as oysters, clams, mussels and scallops, etc. feed by filtering water, and consequently, tend to concentrate contaminants, providing a favorable environment for the continued growth of harmful organisms. It has been demonstrated that water containing a relatively small number of harmful microbes can produce shellfish containing pathogen concentrations which will transmit disease. As reported by Wood (undated), the transmission of enteropathogenic diseases by polluted mollusks was first documented for typhoid fever in the latter part of the nineteenth century. Since then, Wood (undated) reports, contaminated shellfish have been associated with the transmission of a wide range of diseases including paratyphoid fever, cholera, viral hepatitis and many other gastro-enteric conditions.

The USEPA (1976) recommended that the "evaluation of the microbiological suitability of waters for recreational taking of shellfish should be based upon the fecal coliform bacterial levels. When possible, samples should be collected under those conditions of tide and reasonable rainfall when pollution is most likely to be a maximum in the area to be classified. The median fecal coliform value should not exceed an MPN of 14 per 100 ml and no more than 10 percent of the samples should exceed an MPN of 43". As stated in the USEPA (1976) report, the primary source of these guidelines was the internationally accepted microbiological criterion for shellfish water quality of 70 total coliforms per 100 ml, using a median MPN, with no more than 10 percent of the values exceeding 230 total coliforms per 100 ml. The fecal coliform guidelines recommended were simply derived from the fecal to total coliform ratios (approximately 20%) based on more than 3500 sets of data measured in the USA.

Table 2.10 Guidelines and Standards for Microbiological Water Quality for Shellfish Harvesting (per 100 ml) (Salas, 2002)

Country	Coliforms Total	Coliforms Fecal	Reference
USA (EPA)		14[a] 90% < 43	USEPA (1986), Dufour and Ballentine (1986)
USA (California)	70[b]		Cal. Sta. Water Res. Board (undated)
UNEP/WHO		80% < 10 100% < 100	WHO and UNEP (1978)
Mexico	70[b] 90% < 230		SEDUE (1983)
Peru	80% < 1000	80% < 200 200% < 1000	Ministerio de Salud (1983)
Puerto Rico	70[c] 80% < 230		Puerto Rico JCA (1983)
Venezuela	70[a] 90% < 230	14[a] 90% < 43	Venezuela (1978)
Japan	70		Environmental Agency (1981)
People's Republic of China	Coli index < 50	14	SEPA (1998)

a. Logarithmic average for a period of 30 days of at least 5 samples
b. Monthly average
c. At least 5 samples taken sequentially from the waters in a given instance

The above mentioned internationally accepted total coliform standard was originally established in 1925 (Committee on Evaluation of Safety of Fishery Products, 1991) based on typhoid epidemiological investigations conducted from 1914 to 1925 by the states and the Public Health Service of the USA. It was believed that typhoid fever would not normally be attributed to shellfish harvested from water in which "*not more than 50% of the 1 cc portions of water examined were positive for coliforms*" (FDA, 1989). This equates to 70 MPN total coliforms/100 ml, which is equivalent to the fecal material from one person diluted in 8 million cubic feet (226,700 m^3) of coliform-free water. Later, these standards were extrapolated to fecal coliforms on the basis that fecal coliforms were more accurate indicators of fecal contamination.

Table 2.9 shows that most Latin American countries have adopted this internationally accepted criterion with the exception of Peru that has a standard that 80% of samples be less than 200 fecal coliforms per 100 ml and 100% less than 1000 in shellfish harvesting waters.

Water quality aspects 47

WHO/UNDP current criterion on shellfish harvesting waters is that 80% of samples should be less than 10 fecal coliforms per 100 ml and 100% less than 100 fecal coliforms per 100 ml (Helmer *et al.*, 1991).

2.6.5 Standards for indigenous organisms

There are few standards for protection of indigenous organisms. Apparently, only Mexico and Peru have formulated such standards. They are shown in Table 2.11.

Table 2.11 Standards for Microbiological Water Quality for Protection of Indigenous Species (per 100 ml) (Salas, 2002)

Country	Coliforms Total	Fecal	Reference
Mexico	10,000[a] 80% < 10,000 100% < 20,000		SEDUE (1983)
Peru	80% < 20,000	80% < 4,000	Ministerio de Salud (1983)

[a] Monthly average

2.7 RECOMMENDATIONS

Tables 2.8 and 2.9 provide a range for the water quality planner principally for total and fecal coliforms as indicator organisms. Although more than 80 years have passed since their first application, a range of three orders of magnitude continues to exist around the world. The number of epidemiological studies that justify these total and fecal coliform standards is very limited, although the application of these standards can result in significant costs for control systems. As such, the simple adaptation of a particular set of standards is inappropriate without a thorough review of their origin and the local socio-economic circumstances.

The Cabelli (1983) studies provided for the first time a quantitative relation between health risk and indicator organisms (enterococci) level, although factors such as the general health and immunity of the local population imply that caution should be exercised in directly applying the relationships developed in other areas. Furthermore, in epidemiological studies conducted after those of Cabelli, correlation between sickness symptoms and bacterial indicator concentrations varied considerably.

As such, it is recommended that the countries, especially in developing countries where priorities often must be established for projects of first necessity in the context of limited economic resources, conduct local epidemiological studies directed at establishing the relationship between health risk and indicator organisms. There is significant controversy concerning the type of study that should be adopted. The majority of studies conducted to date are prospective cohort studies (Pruss, 1998) as were the Cabelli (1983) studies (protocol available in WHO, 1986). However, in the recent WHO efforts, only controlled randomized trial studies have been considered by the experts' group as more accurate (Pruss, 1996) and the only study of this type conducted in marine waters by Kay, Fleisher *et al.* (1994) in the United Kingdom, has been used as the basis for developing the WHO guidelines for recreational marine waters for enteric illness. The application of these guidelines to the tropical waters is of concern.

The cost of epidemiological studies is justifiable in the context of the large potential capital expenditures associated with control systems. Also, the adaptation of a particular risk level for human health should be based on the local socio-economic situation if it is to be viable.

Furthermore, at the global launching of the WHO Guidelines for Safe Recreational-water Environments during the XXVIII AIDIS Congress held in Cancun, Mexico on 30 October 2002, the following conclusions were made:

"...Concerns were expressed about the broad applicability of the WHO Guidelines to Latin America and the Caribbean. Issues discussed included: tropical waters, local endemic illnesses, susceptibility of children and the elderly, tourists, and length of exposure.

It was concluded that epidemiological studies should be conducted in the Region to evaluate the applicability of the WHO Guidelines to Latin American and the Caribbean temperate and tropical environments. It was also recommended that pilot studies be conducted applying the Annapolis Protocol for beach management. Generally, there was recognition of the need for the guidelines and an appreciation was expressed for the efforts of WHO/PAHO."

An International Experts Consultation to develop a protocol for epidemiological investigations in recreational bathing waters to determine the applicability of the WHO Guidelines for Safe Recreational Water Environments to the tropical waters and conditions of Latin America and the Caribbean (LAC) and elsewhere took place in Mexico City from the 28 to 30 November 2005. The Summary Report (Salas and Robinson, 2006) of the Experts Consultation presents the major issues, conclusions and recommendations of the world-renowned experts who participated and includes epidemiological bathing beach protocols worldwide with their associated questionnaires that were conducted in Great Britain, upon which the numerical values of the WHO Guidelines are based, in the USA, in Germany for freshwaters, in Brazil and in Trinidad & Tobago. The Epidemiological Protocol for LAC developed is available in Salas and Robinson (2008).

In the interim, it is recommended that the WHO Guidelines for Safe Recreational Water Environments be adopted in developing countries until further epidemiological investigations are conducted. Intestinal enterococci that best relates to disease burden should be adopted as the primary biological indicator for surveillance water quality monitoring of bathing beaches and the transition from other indicator organisms should begin as soon as possible based on the capabilities of the countries to do so.

3
Wastewater mixing and dispersion

3.1 INTRODUCTION

The manner in which wastewaters are discharged has a crucial effect on their environmental impacts. It also determines their public health impacts, as demonstrated by the relative health risks of different discharge and treatment and alternatives shown in Table 1.1. A discharge near a beach will contaminate it, will be visible to all, and will cause considerable health risks; a discharge from an open-ended pipe into a shallow bay with poor flushing will contaminate the bay. But these and other impacts can be almost completely eliminated by an effective outfall with an efficient diffuser that discharges into moderately deep water some distance offshore. This eliminates shoreline impacts (and hence most health risks), and the high dilution and efficient dispersion that occurs in coastal waters will reduce any impacts on receiving water quality and the ecosystem to minimal levels.

© 2010 IWA Publishing. *Marine Wastewater Outfalls and Treatment Systems.* By Philip JW Roberts, Henry J Salas, Fred M Reiff, Menahem Libhaber, Alejandro Labbe, and James C Thomson. ISBN: 9781843391890. Published by IWA Publishing, London, UK.

Designing an ocean outfall to meet these objectives requires an understanding of how wastewaters mix in coastal waters and what governs their fate and transport. This mixing has unique and complex characteristics because domestic sewage is less dense than seawater and is released from a diffuser into coastal waters that have their own complex hydrodynamics. The purpose of this chapter is to review and discuss the dominant mixing processes of ocean outfalls discharges.

Although mathematical models are now widely used for fate and transport predictions, the emphasis in this chapter is to explain basic processes and present practical methods to predict the various phases of mixing. The reason for this approach is that knowledge of basic mechanisms is essential to good outfall design, to informed use of more complex mathematical models, and as a check on their reliability. Indeed, the equations presented here will often suffice for fairly straightforward outfall designs and can, in some circumstances be more reliable than the predictions of more "sophisticated" mathematical models. Mathematical models commonly used for outfall mixing processes are discussed in Chapter 4.

The essential features of a typical ocean outfall discharge are shown in Figure 3.1. The wastewater is usually ejected horizontally as round turbulent jets from a multiport diffuser. The ports may be spaced uniformly along both sides of the diffuser or clustered in risers attached to the outfall pipe. See Chapter 9 for a discussion of diffuser and port designs.

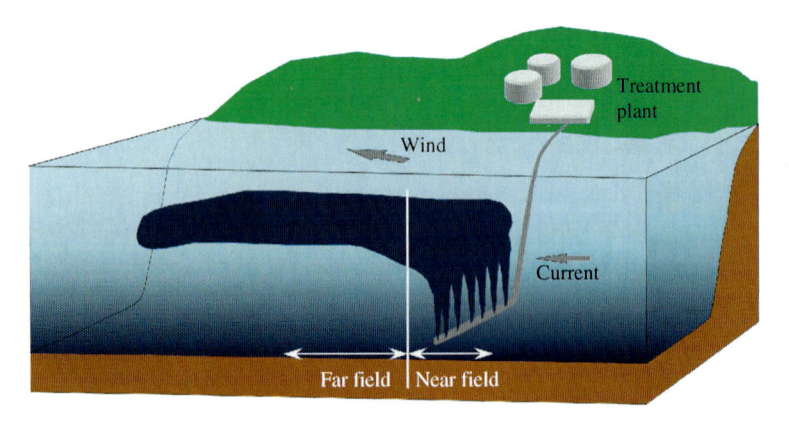

Figure 3.1 Typical behavior of wastewater discharged from an outfall into coastal waters

Buoyancy and oceanic density stratification play fundamental roles in determining the fate and transport of marine discharges. Because the density of domestic sewage is close to that of fresh water, it is very buoyant in seawater.

The jets therefore begin rising to the surface and may merge with their neighbors as they rise. The turbulence and entrainment induced by the jets causes rapid mixing and dilution. The region in which this occurs is called the "near field" or "initial mixing region" (these terms are defined in Table 2.2 and their hydrodynamic meanings are discussed in Section 3.3). If the water is deep enough, oceanic density stratification may trap the rising plumes below the water surface; they stop rising and begin to spread laterally. The wastefield then drifts with the ocean current and is diffused by oceanic turbulence in a region called the "far field." The rate of mixing, or increase of dilution, is much slower in the far field than in the near field. As the wastefield drifts, particles may deposit on the ocean floor and floatables may reach the ocean surface to be transported by wind and currents. Finally, large scale flushing and chemical and biological decay processes removes contaminants and prevents long-term accumulation of pollutants. In reality, there is not a sudden transition between the near and far fields as implied in Figure 3.1. The transition is gradual, but for predictive purposes it is convenient to model it as if it were instantaneous.

It is instructive to think of these processes in terms of their length and time scales. These scales increase as the effluent moves away from the diffuser, in a way that Brooks (1984) has referred to as a *"cascade of processes at increasing scales."* The most important of these processes, illustrated graphically in Figure 3.2, are:

- Near field mixing: Mixing caused by the buoyancy and momentum of the discharge; it occurs over distances of 10 to 1,000 m and times of 1–10 minutes;
- Far field mixing: Transport by ocean currents and diffusion by oceanic turbulence; it occurs over distances of 100 m to 10 km and times of 1 to 20 hours;
- Long-term flushing: Large scale flushing, upwelling or downwelling, sedimentation; it occurs over distances of 10 to 100 km and times of 1 to 100 days and longer.

Further complicating the situation is that these processes can overlap, and there may be transitions between the various phases.

The mixing mechanisms in each of the three major phases and methods to predict them are discussed in Section 3.3. Of these phases, near field mixing is the only one that is under the control of the designer. But it can also have a substantial effect on all of the subsequent transport phases, so it is dealt with in more detail than the others. For a review of the difficulties of modeling water quality and dispersion in coastal waters, see James (2002).

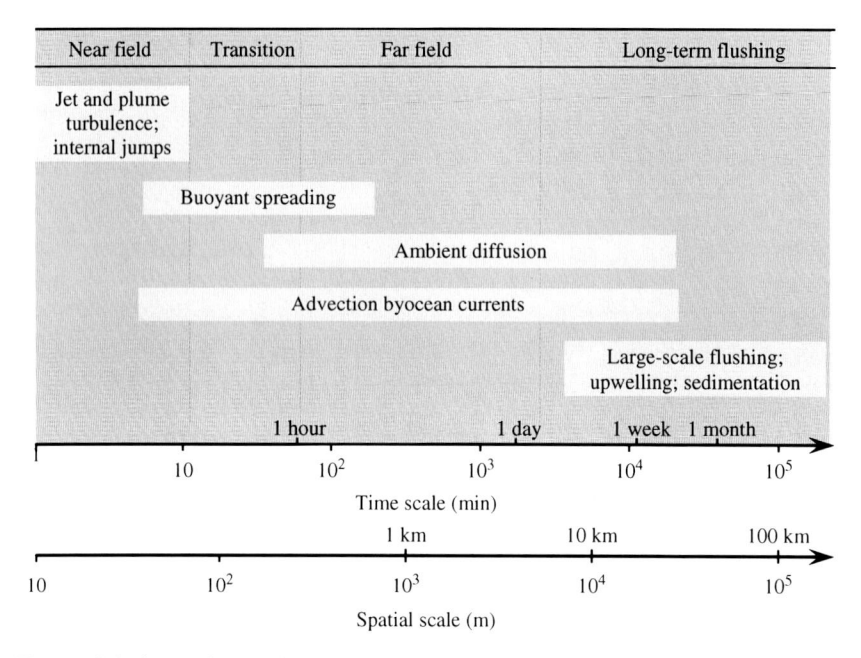

Figure 3.2 Approximate time and length scales of the main processes affecting ocean outfall discharges

Because of the wide range of time and length scales involved, it is not usually feasible to model them in one overall "omnibus" model. Instead, sub-models that are linked together are often used. The procedures to accomplish this are discussed here and in the following chapter.

3.2 FEATURES OF COASTAL WATERS

3.2.1 Introduction

Coastal waters have particular hydrodynamic characteristics that greatly influence the fate of pollutants discharged to them, as discussed in this section. For further details see review papers such as Csanady (1979, 1983b), Gross (1983), Winant (1980), and Allen (1980); the discussion below follows the reviews of Csanady.

3.2.2 Water motions

Coastal waters are driven primarily by winds and tides, although freshwater runoff from the land can also be an important forcing mechanism. Because of differing climate, bathymetry, and density stratification, responses to these forcing mechanisms vary widely. Up to about 10 km offshore the motions are strongly influenced by the coast. They may differ significantly from those farther offshore even though offshore motions, especially deepwater gyres, can drive circulation close to shore. All motions can be modified by the earth's rotation (Coriolis forces), and the sloping bottom. Although the complexity and unsteadiness of these motions make them difficult to predict, they result in near coastal waters being quite energetic and able to mix and disperse effluent quite efficiently. Evidence of this efficient mixing or "assimilative capacity" can be seen, for example, in the rapid disappearance of freshwater runoff away from shore.

The ultimate fate of wastewater released into coastal waters is determined by currents, turbulence, and shear. These result from motions that have widely differing time scales, including surface waves, tidal currents, wind-driven longshore currents, and large-scale mean circulation. These differ for "Atlantic" type continental shelves, which are shallow for considerable distances from the coast, and Pacific or Caribbean shelves, where the depth increases rapidly with distance offshore. Pollution problems are generally less critical for coastlines with steep bottom slopes.

Contrary to intuition, surface waves do not contribute substantially to mixing because they are closely irrotational. Waves may break, however, causing turbulence, mixing, and transport at the air-sea interface. And storms may cause wave-induced motions at the seabed that lead to resuspension of waste particulates. Wave-induced velocities decrease with water depth, but storm waves are capable of stirring up fine particles at depths up to 30 to 40 m.

Tides are often the major constituent of the kinetic energy of coastal currents. Typical tidal velocities over Atlantic shelves are around 20 cm/s. Particles experience elliptical displacements over a tidal cycle, the major axis of the ellipse being parallel to shore and typically several kilometers long. The minor axis of the tidal excursion vanishes right at the coast. At one or two km from the coast, cross-shore tidal excursions are restricted by the coast to distances less than about one kilometer. Obviously, outfalls should be longer than the onshore tidal excursion to prevent sewage from coming to shore.

These tidal oscillations make a major contribution to mixing because they distribute the effluent over a large area. As the coastal waters oscillate over an outfall, the water discharged is distributed over a considerable water mass that extends several kilometers alongshore and about one km wide. This process can be thought of as the waters being still and the outfall moved over an elliptical area of similar dimensions. This effect is discussed further and analyzed in Section 3.3.3.

3.2.3 Density stratification

Coastal waters that are deeper than about 20 m are usually density stratified, i.e. the seawater density varies with depth. This is due to salinity and/or temperature variations, and there are many equations and procedures from which seawater density can be calculated, for example, Bigg (1967), Millero *et al.* (1980), and Millero and Poisson (1981). In areas remote from freshwater inflows, such as estuaries, the stratification is mostly due to temperature, with warmer, less dense water near the surface, and colder, more dense water near the bottom. Near to significant freshwater sources, however, the stratification may result more from salinity variations. Salinity stratification can be much stronger than that due to temperature alone.

Although the absolute density variations are very small, of the order of a few percent, they have profound implications for wastewater behavior. Typical horizontal and vertical variations of density in coastal waters are shown in Figure 3.3. The planes (isolines) of constant density are approximately horizontal. Vertical variations often show a well-mixed surface layer a few meters thick (the surface mixed layer), a region with a rapid change of density (the pycnocline), and deeper water below that is weakly stratified. Effluent released in deep waters below a strong pycnocline will be trapped in the bottom layer.

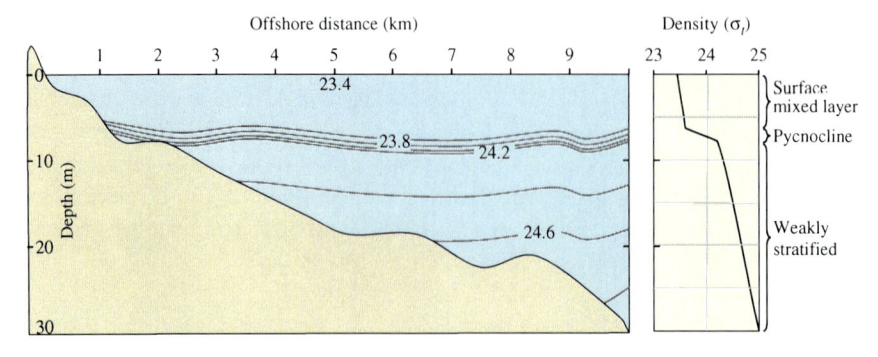

Figure 3.3 Typical horizontal and vertical density variations in coastal waters

A very important consequence of density stratification is that, because the planes of constant density are approximately horizontal, a trapped plume will remain submerged and not reach the shore. This is illustrated in Figure 3.4. The density stratification can also cause internal-mode tidal motions (internal waves), which can cause large vertical motions of the constant density surfaces and considerable cross-shore mass exchange. It has been suggested that internal waves may be a mechanism for bringing a submerged plume to the shoreline (Boehm *et al.*, 2002), but the authors know of no documented cases of this occurring. Another important consequence of density stratification is that it inhibits vertical mixing. The vertical thickness of waste fields can therefore remain almost constant for long distances from their source.

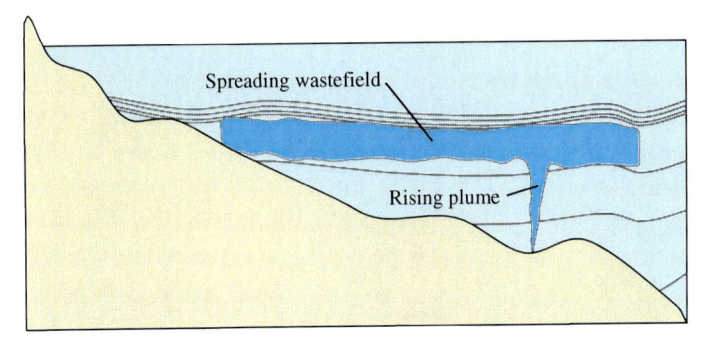

Figure 3.4 Trapping of plume by density stratification and prevention of shoreline contact

3.2.4 Winds

Wind-driven currents can dominate flows near coastlines, especially if tidal motions are small. They can have the same amplitude as strong tidal currents (30 cm/s or so) but occur much less regularly, depending on the passage of weather cycles. Although periodicities of a few days may occur, the winds can be quiescent for much longer.

Wind-driven currents under well-mixed conditions are approximately parallel to shore. They are not exactly parallel, however, so intense shear and cross-shore water movements can occur near the surface that may be compensated by an opposite flow at greater depth. Because wind-driven currents generally persist for a day or two, Coriolis forces due to the earth's rotation are also important.

Wind effects are strongest in waters that are deep enough to remain strongly stratified during the summer (deeper than about 30 m). In this case, the wind mostly affects the surface waters, which move faster than the lower levels and in a different direction than in a well-mixed water column. This can cause upwelling or downwelling and tilt the isolines of constant density. The most important aspect of this wind-induced upwelling or downwelling is a massive cross-shore exchange of water. The entire water mass can be effectively renewed in a nearshore band following intense episodes.

Intense mass-exchange episodes can also occur in nontidal waters that are not subject to the damping influence of tidal friction. Strong winds may propagate along shore as long (Kelvin) waves and cause mass-exchange episodes long after the wind event has passed. In coastal regions with strong tidal currents these waves do not seem to be frequent, and wind-induced upwelling and downwelling are relatively less intense.

The mean circulation over a coastal shelf is important in limiting the buildup of long-lived (conservative) waste constituents, such as heavy metals. Longshore mean velocities are generally of the order of three cm/sec which will effectively "flush" a coastal zone. Methods to estimate this flushing are given in Section 3.3.3.

The mean circulation of coastal waters is generally a boundary-layer component of the deep oceanic gyres, which impose a longshore sea level slope. This large-scale pattern is modified by wind-stress, which "sets up" portions of the shelf according to its topography, especially in semi-enclosed basins. The wind-related setup is a pressure field trapped within relatively shallow water, typically within the 50 m depth contour. Where outfalls are located on Atlantic-type shelves, the water is typically 10 to 30 m deep and the mean wind stress is generally stronger than the gravity force associated with the local longshore mean sea-level slope, so that mean flow is along the mean wind.

There are other components of coastal zone flow, such as those due to internal wave motions, edge waves and topographic waves, local flows produced by seamounts or canyons, discharges from coastal lagoons, large eddies in the wake of promontories or islands, river plumes and floating lenses of somewhat fresher or warmer water originating from a distant river. Although these add to the complexity of water movements, they are only occasionally as important as those discussed above.

Discharges may be into waters deeper than about 50 m where the seabed slope is steep. These occur in Pacific-type continental shelves or in the Caribbean. Motions due to surface waves are not important in such locations, but density stratification and vertical movements of the constant density surfaces are very important (Roberts, 1999b). Most of the kinetic energy in the bottom layer is provided by motions associated with transient pycnocline movements, such as internal tides, inertial oscillations, and, especially important, wind-induced coastal upwelling and downwelling cycles. While these motions will tend to transport pollutants away, and also promote mixing by turbulence and shear-diffusion, they may also bring bottom waters to the surface, usually very close to shore, in a strong upwelling event. The frequency of such events may be a design consideration.

Coastal environments can therefore differ widely from one outfall site to another. The typical values given above may not always apply, so reliable estimates of flow characteristics must be obtained by oceanographic measurements, preferably extending at least for a full year. Sometimes, and for fairly small discharges, there may be enough general information on oceanographic conditions to be certain that dispersion will take care of the disposal problem. For example, in an open coastal location, with strong tidal currents, which are usually known to the inhabitants of the sea coast, one can confidently predict effective dispersal over the range of tidal excursions. Long-term buildup of persistent toxic chemicals can be checked using a conservative mean (flushing) velocity. In most cases, however, oceanographic investigations will be needed, and extensive studies are becoming increasingly common, even for fairly small projects. The gathering of the oceanographic data needed for outfall design is discussed in Chapter 5.

3.3 MECHANISMS AND PREDICTION OF WASTEFIELD FATE AND TRANSPORT

3.3.1 Introduction

The three major phases of mixing shown in Figure 3.2: Near field, far field, and long-term flushing, are discussed in this section. The emphasis is on presenting simple practical equations that can be used to predict mixing in each phase and that aid in understanding the mixing processes and therefore result in effective outfall design. These simple equations will suffice for many outfalls, especially smaller, simpler, ones, but for larger or more complex cases, more extensive modeling may be needed. The mathematical models needed for this are discussed in Chapter 4.

3.3.2 Near field mixing

3.3.2.1 Introduction

Prediction of near field mixing requires understanding of the dynamics of jets, plumes, and buoyant jets. These are defined according to the notation of Fischer *et al.* (1979): a jet is a flow driven by the source momentum flux only, a plume is driven by the source buoyancy flux only, and a buoyant jet is driven by both momentum and buoyancy fluxes. Discharges from an outfall diffuser have both momentum and buoyancy and are therefore buoyant jets, but we will show that the buoyancy flux is usually dominant so that ocean outfall discharges can often be approximated as plumes. This leads to considerable analytical simplification and to useful equations for predicting plume behavior and dilution.

In this section, we derive and present these equations to predict near field behavior for the cases most relevant to ocean outfalls. The analysis follows that of Fischer *et al.* (1979) and others, in that we mainly use dimensional analysis with the source fluxes as independent variables combined with length scales and experimental data. Although this method of analysis may initially seem to be rather arcane, it is a powerful method that yields usable answers with little analysis.

We begin with the simplest discharge case, a single horizontal buoyant jet into a stationary, homogeneous environment, and gradually build to the most complex case of merging buoyant jets in a flowing, density-stratified environment.

We will extensively use images obtained by laser-induced fluorescence (LIF) laboratory experiments to illustrate the mixing processes. In this technique, a fluorescent tracer dye is added to the flow. A laser causes the dye to fluoresce and the emitted light is captured by a camera. By suitable calibration, quantitative tracer concentrations can be obtained, and the images can be color-coded to visualize and show the concentration (and therefore dilution) distributions in the flow. It is beyond the scope of the present book to describe this technique in detail, but descriptions are given in many publications, for example, Koochesfahani and Dimotakis (1985). In addition, we will show three-dimensional LIF (3DLIF) images obtained by the techniques described in Tian and Roberts (2003).

3.3.2.2 Horizontal buoyant jet in stationary, homogeneous environment

The simplest case is that of a single buoyant jet discharging horizontally into a stagnant, unstratified receiving water of depth H, as shown in Figure 3.5. This case also has considerable historical significance in that it was the first one studied for ocean outfalls by Rawn and Palmer (1929).

Because of its buoyancy, the jet follows a trajectory that curves upwards towards the water surface. As it rises, it entrains ambient fluid that mixes with and dilutes the effluent. After impacting the water surface, it makes a transition to a horizontal flow that spreads laterally where it may undergo an internal hydraulic jump and other mixing processes that result in additional dilution. We are most interested in the centerline dilution at the surface, S_m, the thickness of the spreading layer after the jump, h_n, and the dilution after the jump, S_n. As will be discussed later, the subscript n refers to the near field.

Wastewater mixing and dispersion 61

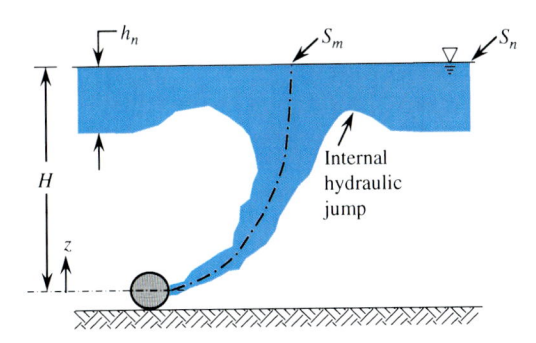

Figure 3.5 Horizontal buoyant jet into stationary homogeneous environment

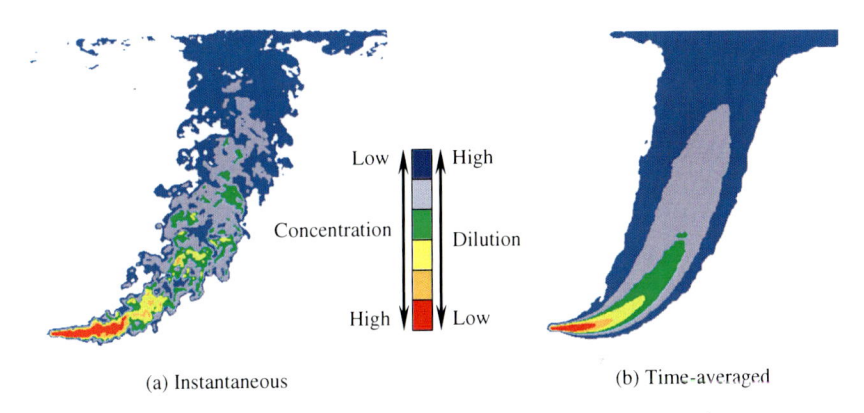

(a) Instantaneous

(b) Time-averaged

Figure 3.6 LIF images of a horizontal buoyant jet into stationary homogeneous environment

LIF images of a horizontal buoyant jet are shown in Figure 3.6. They are color-coded to show the concentration (and therefore dilution) distributions. Two images are shown: an instantaneous "snapshot" and a time-averaged image.

Tracer concentrations decrease (and therefore dilution increases) along the plume trajectory. The instantaneous image (Figure 3.6a) shows a patchy, intermittent concentration field, with regions of high and low concentrations separated by steep concentration gradients. This is contrast to the time-averaged image (Figure 3.6b) which shows concentrations varying smoothly in space. The instantaneous values can be considerably higher than averaged ones. The dilution occurs because the rising plume entrains ambient fluid into itself, which is then mixed by turbulent diffusion. This entrainment and mixing is quite efficient, and results in rapid dilution in the rising plume. Although the time-averaged image is the classical way in which turbulent jets and plumes are illustrated, it should be kept in mind that the averaged picture never exists in nature; it is an artifact of the averaging process. Nevertheless, water quality regulations are usually expressed in time-averaged values.

A buoyant jet is characterized by the exit velocity, u_j, the jet (nozzle) diameter d, the effluent density, ρ_o, and the receiving water density ρ_a. If the density differences are small (this is true for ocean outfalls, where the effluent density is typically 2.5% less than seawater) the densities can be combined into a single parameter:

$$g'_o = g \frac{\rho_a - \rho_o}{\rho_a} = g \frac{\Delta \rho_o}{\rho_a} \tag{3.1}$$

This is commonly known as the Boussinesq assumption, and g'_o is known as the modified acceleration due to gravity. A dimensionless parameter that can be formed from u_j, d, and g'_o is the densimetric Froude number, F_j:

$$F_j = \frac{u_j}{\sqrt{g'_o d}} \tag{3.2}$$

The significance of the Froude number is that it expresses the relative importance of buoyancy and inertial forces on the buoyant jet.

Although u_j, d, and g'_o are commonly used variables, more insight can be obtained by using the source kinematic fluxes of volume, Q_j, buoyancy, B, and momentum, M:

$$Q_j = \frac{\pi}{4} d^2 u_j \quad M = u_j Q_j = \frac{\pi}{4} d^2 u_j^2 \quad B = g'_o Q_j \tag{3.3}$$

Following Wright (1984) and others, we define the following length scales:

$$l_Q = \frac{Q_j}{M^{1/2}} = \left(\frac{\pi}{4}\right)^{1/2} d \quad l_M = \frac{M^{3/4}}{B^{1/2}} \tag{3.4}$$

The length scales are related to the Froude number and nozzle diameter by:

$$\frac{z}{d F_j} = \left(\frac{\pi}{4}\right)^{-1/4} \frac{z}{l_M} \quad \text{and} \quad F_j = \left(\frac{\pi}{4}\right)^{-1/4} \frac{l_M}{l_Q} \tag{3.5}$$

Their dynamical significance is that l_Q characterizes the distance from the origin over which the source volume flux influences the flow field; this length scale is essentially the nozzle diameter, d. l_Q is almost always very small for ocean outfalls discharging domestic sewage so Q_j can be usually be neglected as a dynamical parameter. More important is l_M, which characterizes the distance over which the source momentum flux is important relative to the buoyancy flux. For distances much less than l_M the flow is dominated by momentum, i.e. the flow behaves like a jet. For distances greater than l_M the dynamics are governed by buoyancy, i.e. the flow behaves like a plume. This leads to a very important conclusion: Given enough depth, a buoyant jet will always turn into a plume where the effect of the source momentum flux can be neglected. This will often be a good approximation for ocean outfalls, as discussed below.

Consider, for example, the dilution at any height z. An expression for this can be obtained by taking the local buoyant weight, $g' = g(\rho_a - \rho)/\rho_a = g\Delta\rho/\rho_a$ as the dependant variable:

$$g' = f(z, Q_j, M, B) \tag{3.6}$$

In Eq. (3.6) and in what follows, the flows are assumed to be always fully turbulent and therefore independent of viscosity and Reynolds number (Fischer et al., 1979). Following a dimensional analysis of Eq. (3.6) and some manipulation we can derive an equation for the minimum (centerline) dilution S_m:

$$S_m \frac{l_Q}{l_M} = f\left(\frac{z}{l_M}, \frac{z}{l_Q}\right) \tag{3.7}$$

If $z/l_Q \gg 1$ the effects of Q and l_Q are negligible and Eq. (3.7) becomes:

$$S_m \frac{l_Q}{l_M} = f\left(\frac{z}{l_M}\right) \quad \text{or} \quad \frac{S_m}{F_j} = f\left(\frac{z}{dF_j}\right) \tag{3.8}$$

These arguments were also given by Roberts (1977) who gathered experimental data on centerline dilutions and plotted them, with corrections for the increases in momentum flux due to the use of sharp edged orifice nozzles, as shown in Figure 3.7. It can be seen that, for $z/dF_j \gg 1$ (which is the same as $z/l_M \gg 1$), the effect of the source momentum flux is "forgotten" and the flow becomes a plume. Dilution is then predicted by (Fischer et al., 1979):

$$\frac{S_m}{F_j} = 0.107 \left(\frac{z}{dF_j}\right)^{5/3} \tag{3.9}$$

which is plotted in Figure 3.7 and describes the results for $z/dF_j > 10$. In other words, the source volume and momentum fluxes can be neglected for $z/dF_j > 10$ and the pure plume formula, Eq. (3.9), can be used. A more general semi-empirical equation that applies for $z/dF_j > 0.5$ has been suggested by Cederwall (1968):

$$\frac{S_m}{F_j} = 0.54\left(0.66 + 0.38\frac{z}{dF_j}\right)^{5/3} \qquad (3.10)$$

which is also plotted on Figure 3.7.

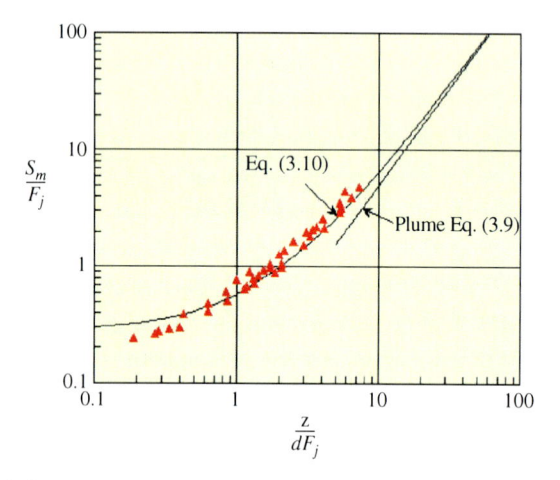

Figure 3.7 Centerline dilution of a horizontal buoyant jet into a stationary homogeneous environment

Ocean outfall discharges usually operate such that $z/dF_j > 10$ so a very important conclusion from Figure 3.7 is that they can usually be approximated as plumes. From now on, we will assume this to be the case; in other words, the source is characterized by its buoyancy flux only, and the effects of the source volume and momentum fluxes are dynamically insignificant. Note that this has an important design consequence: Because the density difference between the sewage and seawater is essentially fixed, dilution depends only on the flow per port, and not on the port diameter (and therefore exit velocity).

The thickness of the surface layer, h_n, can also be estimated by dimensional analysis. For a fully turbulent plume in water of depth H:

$$h_n = f(H, B) \quad \text{i.e.} \quad \frac{h_n}{H} = C_1 \qquad (3.11)$$

where C_1 is an experimental constant. Because B contains units of time, it cannot be eliminated from Eq. (3.11), so the layer thickness is independent of the buoyancy flux and is proportional to the water depth. Tian *et al.* (2004a) suggest the value of C_1 constant is 0.11, so Eq. (3.11) becomes:

$$\frac{h_n}{H} = 0.11 \tag{3.12}$$

The interaction of the plume with the free surface can also be important, as shown in Figure 3.8. The surface layer interactions were analyzed for three-dimensional flows (round jets) by Lee and Jirka (1981) and for two-dimensional flows (slot jets) by Jirka and Harleman (1979). They showed that, for a vertical discharge of low buoyancy, the flow may become unstable and the horizontal layer may be re-entrained back into the rising plume, resulting in lowered dilutions.

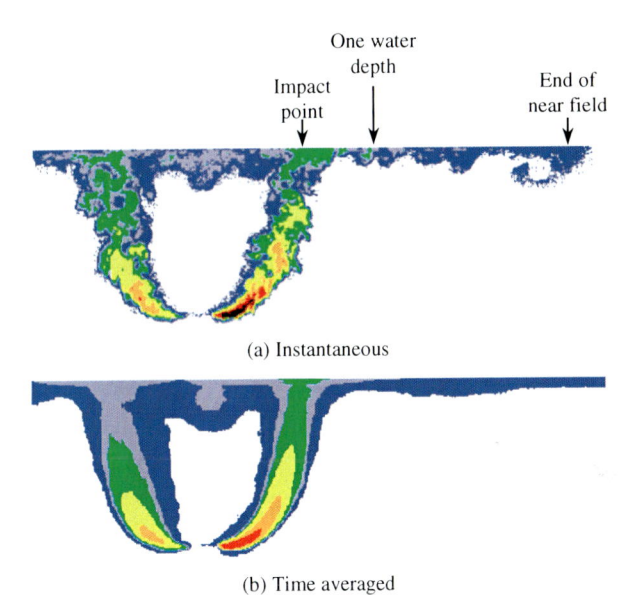

Figure 3.8 LIF images of horizontal opposing buoyant jets into a stationary homogeneous environment

More commonly for ocean outfall discharges, however, the horizontal layer is stable and the diluted effluent flows away as a surface density current. An internal hydraulic jump or spreading vortices may form (Figure 3.8) that entrain additional ambient fluid, further increasing the dilution in the spreading layer. According to Tian *et al.* (2004a), the near field dilution for a round plume source is a factor of about three higher than the surface centerline value, S_m, so Eq. (3.9) becomes, for the near-field dilution:

$$\frac{S_n}{F_j} = 0.32\left(\frac{z}{dF_j}\right)^{5/3} \tag{3.13}$$

3.3.2.3 Multiple buoyant jets in stationary, homogeneous environment

Next, we consider multiple buoyant jets discharged from a multiport diffuser, as shown in Figure 3.9. The spacing between the ports is s. (The diffuser is shown with the ports contained in Tee-shaped risers, but the same analysis applies for ports along the diffuser side walls.)

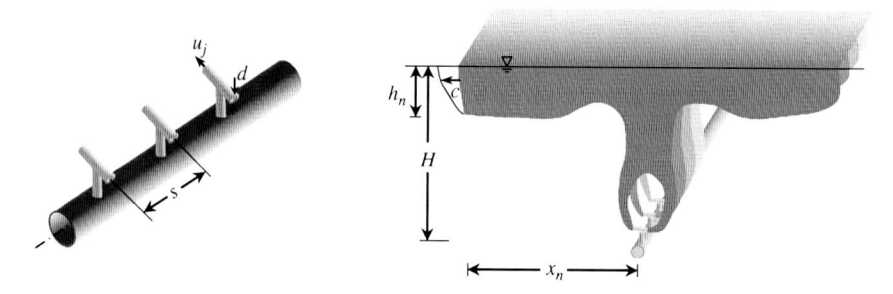

Figure 3.9 Plumes from a multiport diffuser into a stationary, homogeneous environment

The effect of port spacing depends on the ratio s/H. If the ports are widely spaced, i.e. $s/H \gg 1$, the individual plumes do not merge and they behave like the isolated, point plumes discussed above. As the ports are brought closer together, the plumes merge and dilution decreases. Eventually, when the plumes are very close together, they rapidly merge and behave as if discharged from a continuous slot; this is known as a line source.

The source fluxes from a multiport diffuser are better characterized by line-source parameters, i.e. fluxes per unit length of the diffuser, rather than fluxes per individual port. The point and line source fluxes are summarized in Table 3.1. Here, Q_T is the total flowrate discharged from the outfall and L is the diffuser length.

A 3DLIF image of plume merging is shown in Figure 3.10. According to the experiments of Tian et al. (2004a) the flow behaves like a line plume (fully merged) for $s/H < 0.3$, like a point plume (i.e. no merging) for $s/H > 1$, and with a transition in between. As for point plumes, the line plume characteristics can be derived by dimensional analysis. The resulting equations, expressed in terms of the flux variables, are summarized in Table 3.2 for both point and line plumes. For further details, see Tian *et al.* (2004a).

Table 3.1 Discharge fluxes from multiport diffusers

Variable	Point source	Line source
Volume flux	$Q_j = \frac{\pi}{4}d^2 u_j$	$q = \frac{Q_T}{L}$
Momentum flux	$M = u_j Q_j$	$m = u_j q$
Buoyancy flux	$B = g'_o Q_j$	$b = g'_o q$

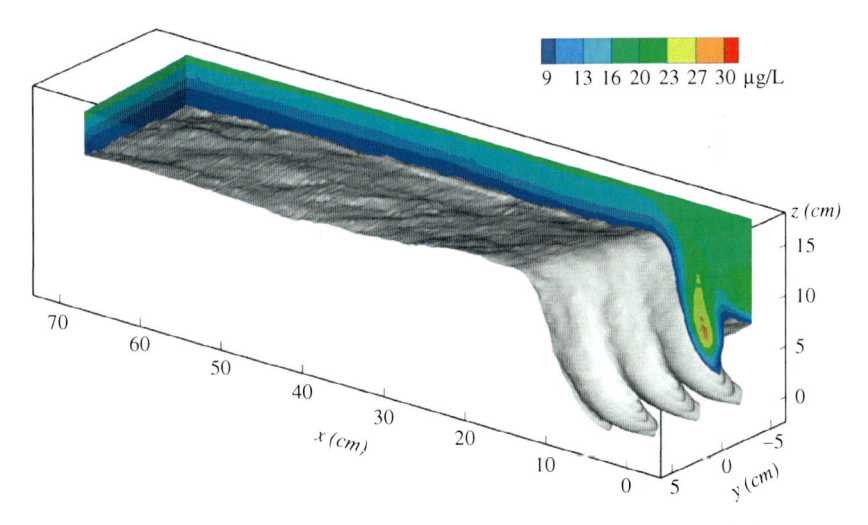

Figure 3.10 3DLIF image of merging plumes from a multiport diffuser (from Tian *et al.*, 2004a)

Table 3.2 Equations for near field properties of point and fully merged (line) plumes in a stationary, homogeneous environment (after Tian *et al.*, 2004a)

	Point plume ($s/H > 1$)	Line plume ($s/H < 0.3$)
Dilution	$\dfrac{S_n Q_j}{B^{1/3} H^{1/3}} = 0.26$	$\dfrac{S_n q}{b^{1/3} H} = 0.49$
Layer thickness	$\dfrac{h_n}{H} = 0.11$	$\dfrac{h_n}{H} = 0.36$
Length	$\dfrac{x_n}{H} = 2.8$	$\dfrac{x_n}{H} = 0.9$

These equations have important consequences for the effects of port spacing and diffuser length on dilution. For most multiport diffusers, dilution depends mainly on the buoyancy flux per unit length, b. Because the density difference between effluent and seawater is essentially constant, this means that dilution depends primarily on the volume flux per unit length, $q = Q_T/L$. Therefore, for a fixed flowrate (Q_T), the diffuser length (L) primarily determines the dilution, which increases as the diffuser length increases. If the diffuser length is fixed, dilution *increases* as more ports are added and the plume spacing is decreased, despite the increased plume merging. This applies until the ports are close enough that the line plume limit is reached, which occurs when $s/H \approx 0.3$. Adding more ports that are more closely spaced will not increase dilution any further.

This is contrary to the oft-stated recommendation that diffusers should be designed so that the plumes do not merge. But it should be noted that this only applies for fixed diffuser length; increasing the number of ports or the port spacing so that the diffuser length increases will result in higher dilution. Again there are limits, however. If the number of ports is maintained constant, increasing their spacing will increase dilution only until the spacing reaches the point plume limit ($s/H \approx 1$). Further increasing the spacing (and therefore making the diffuser longer) will not result in higher dilution.

3.3.2.4 Effect of flowing currents: single plume

The cases discussed above were for discharges into a stationary ocean. This is the worst case that results in lowest dilution. Usually there will be a current flowing over the diffuser as sketched in Figure 3.11. The current, of speed u, sweeps the plume downstream and increases its dilution.

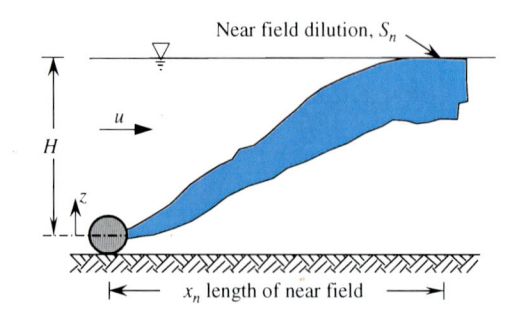

Figure 3.11 Single plume in an unstratified current

Time-averaged 3DLIF images of a vertical buoyant jet discharged into a crossflow is shown in Figure 3.12. The variation in centerline dilution can be seen in the longitudinal profile, Figure 3.12(a). The lateral profiles, Figure 3.12(b), show the characteristic "horse-shoe" shapes that result from paired trailing vortices generated by the interaction of the vertical momentum and buoyancy with the crossflow.

(a) Vertical plane through the centerline

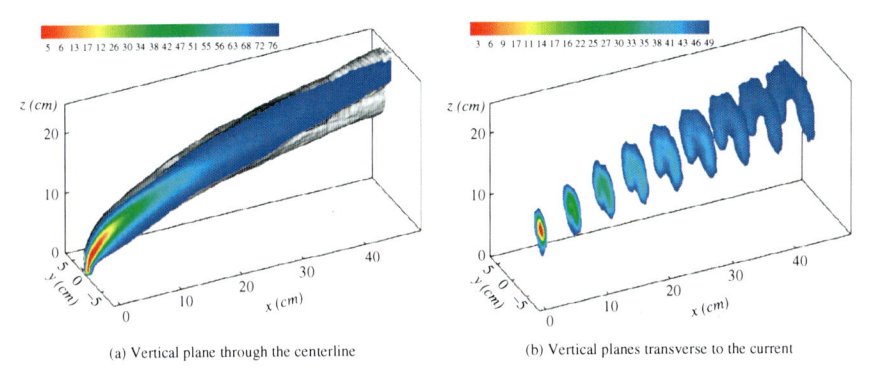

(b) Vertical planes transverse to the current

Figure 3.12 3DLIF images of a vertical buoyant jet in a current

Because the dominant parameter for ocean wastewater outfalls is the buoyancy flux, the most useful length scale that describes the effects of currents is:

$$l_s = \frac{B}{u^3} \tag{3.14}$$

For $z \gg l_s$, the flow is said to be an advected thermal (also called the buoyancy-dominated far-field, BDFF, by Wright (1977). The flow shown in Figure 3.12 has been analyzed by Chu (1979). For a plume, i.e. neglecting effects of the source volume and momentum fluxes, he obtained an expression for the centerline dilution S_m at any height z above the source:

$$\frac{S_m Q_j}{u l_s^2} = C_2 \left(\frac{z}{l_s^2}\right) \tag{3.15}$$

where C_2 is an experimental constant. Lee and Neville-Jones (1987) report measurements of surface dilution in laboratory and field experiments that they correlated with Eq. (3.15) by replacing z with the water depth, H. For the field experiments, they obtained a value of the constant equal to 0.32, so Eq. (3.15) becomes:

$$\frac{S_m Q_j}{u H^2} = 0.32 \tag{3.16}$$

The more recent experiments on multiport diffusers in flowing currents of Tian et al. (2004b) found, however, that Eq. (3.16) did not describe the measured dilutions even for very wide port spacings and a better fit to the data over the range $0.5 < s/H < 4.5$ for near field dilution was:

$$\frac{S_n q}{uH} = 0.55\left(\frac{s}{H}\right)^{-1/2} \quad (3.17)$$

The difference between Eqs. (3.16) and (3.17) may be due to merging of the plumes that are emitted from the upstream and downstream diffuser ports, as discussed further below.

3.3.2.5 Merging plumes in a flowing current

Merging plumes in a current are sketched in Figure 3.13. The dynamical effect of the current is expressed by a type of Froude number (Roberts, 1979):

$$F = \frac{u^3}{b} \quad (3.18)$$

where b is the buoyancy flux per unit diffuser length (Table 3.1) and the other parameters are as before. The discharge can be characterized by the port spacing and by either point or line fluxes of volume, momentum, and buoyancy (see Table 3.1).

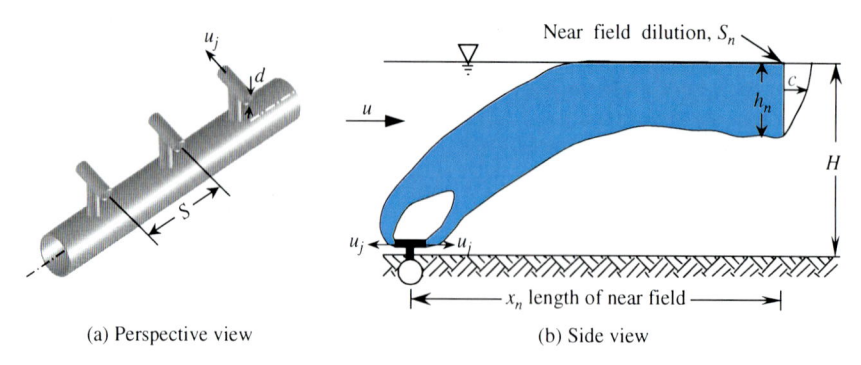

(a) Perspective view

(b) Side view

Figure 3.13 Definition diagram for plumes from a multiport diffuser in an unstratified current

Different flow regimes can exist, depending on the value of F. For $F \ll 1$ the flow is dominated by buoyancy, and for $F \gg 1$ the flow is dominated by the ambient current. Based on experiments with a slot source, Roberts (1979) found three possible flow regimes for a current perpendicular to the diffuser, as sketched in Figure 3.14.

(a) F < 0.2. Plume and upstream wedge

(b) 0.2 < F < 1. Forced entrainment and upstream wedge

(c) F > 1. Forced entrainment, no upstream wedge

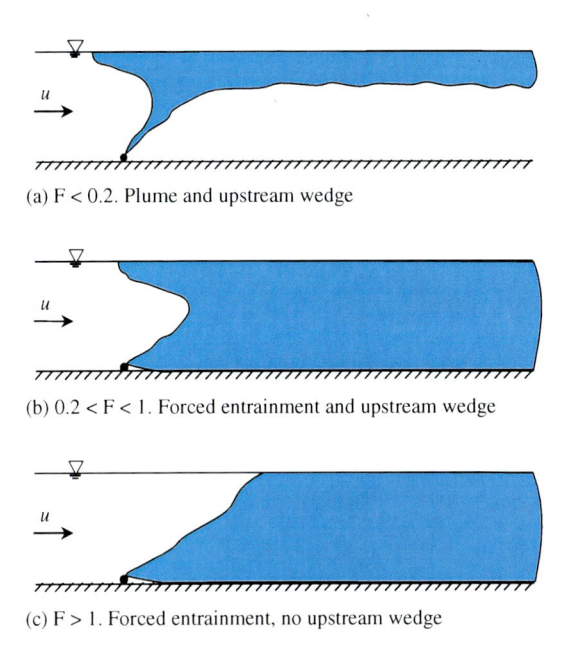

Figure 3.14 Flow regimes for a diffuser approximating a line plume in a current (after Roberts, 1979)

For weak currents ($F < 0.2$), the flow has a normal plume-like pattern and forms a surface layer that spreads upstream (as a wedge) and downstream. As the current speed increases to $F > 0.2$ the plume cannot entrain all of the oncoming flow and maintain a free plume pattern. It becomes attached to the lower boundary and mixes over the depth; this is sometimes called the forced entrainment regime. Finally, when $F > 1$, the upstream wedge is expelled. The same flow regimes were reported by Davidson *et al.* (1990) from experiments with multiport discharges. In contrast, experiments with multiport diffusers reported by Mendez-Diaz and Jirka (1996) did not indicate mixing over depth even for Froude numbers as high as $F = 8$. This observation is also supported by the experiments with a multiport diffuser in stratified flows of Roberts *et al.* (1989a), as seen in Figure 3.23. The implication is that mixing over depth with bottom attachment for a multiport diffuser with gaps between the ports does not occur until the Froude number exceeds a critical value that lies somewhere between 1 and 10.

The complexity of merging plume dynamics in a flowing current is evident from the instantaneous and time-averaged planar LIF images shown in Figure 3.15. The jets discharging upstream (counter-flowing jets) are quickly bent back by the current and experience rapid dilution. The jets discharging downstream (co-flowing jets) are somewhat sheltered from the current so their deflection and rate of dilution is less than that of the upstream jets.

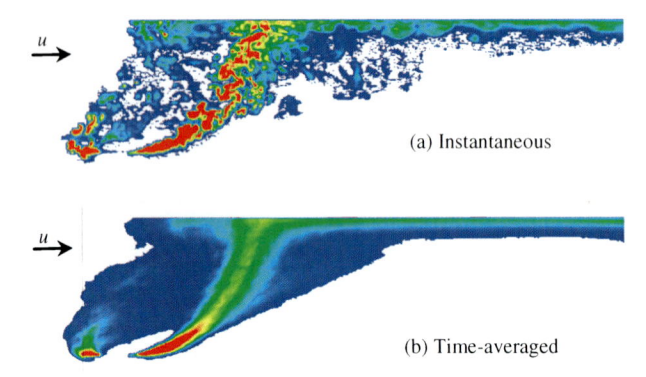

(a) Instantaneous

(b) Time-averaged

Figure 3.15 LIF images of plumes from a multiport diffuser in an unstratified current

Plume merging is shown in the 3DLIF images in Figure 3.16. Depending on the port spacing, the jets may first merge with their lateral neighbors and then with their upstream or downstream counterparts, or the upstream and downstream jets may first merge and then merge with their lateral neighbors. In all cases, however, lateral mixing and spreading will eventually result in a fully merged wastefield at the water surface, as shown in Figure 3.16. This full merging may or may not occur within the near field.

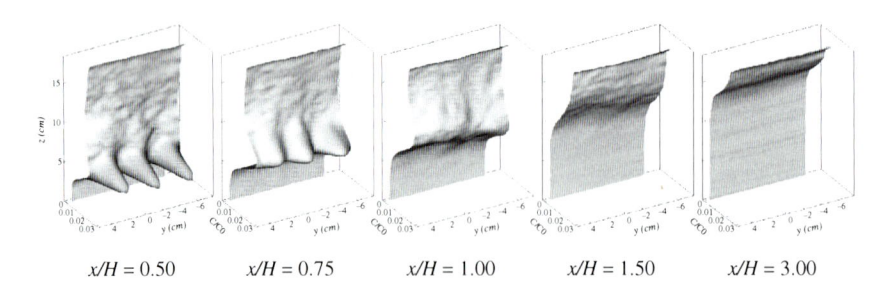

$x/H = 0.50$ $x/H = 0.75$ $x/H = 1.00$ $x/H = 1.50$ $x/H = 3.00$

Figure 3.16 Lateral tracer concentrations showing merging of multiport plumes with distance from the diffuser (Tian et al., 2004b)

The presence of a current flowing past a multiport diffuser introduces a new variable, the angle of the current relative to the diffuser axis, Θ. For a diffuser approximating a line plume, the near field dilution is expressed by (Roberts, 1979):

$$\frac{S_n q}{uH} = f(F, \Theta) \qquad (3.19)$$

Experimental results for currents flowing perpendicular, parallel, and at 45° to the diffuser are summarized in Figure 3.17, which is based on Roberts (1979), updated by the experiments of Tian *et al.* (2004b). For other results, including the length of near field and layer thicknesses, see Tian *et al.* (2004b).

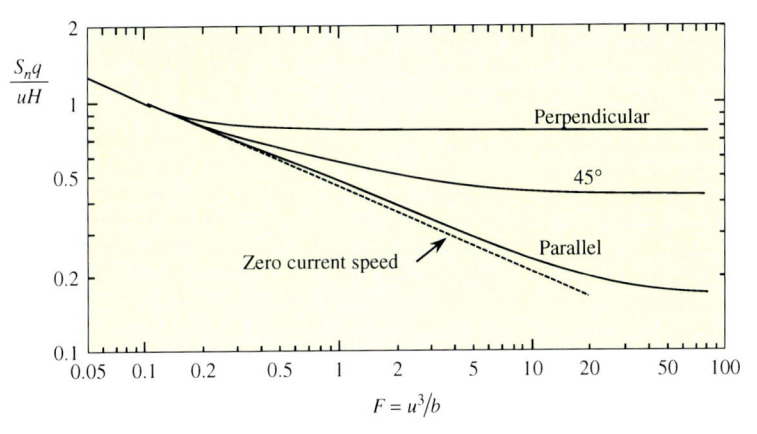

Figure 3.17 Near field dilution for a diffuser approximating a line plume in an unstratified current of various directions. From Roberts (1979) updated with Tian *et al.* (2004b)

As would be intuitively expected, a diffuser perpendicular to the current results in highest dilution, and one parallel to the current the lowest dilution. Even for a parallel current, however, the dilution is always higher than for zero current speed.

3.3.2.6 Single plume into stationary, stratified flow

Density stratification in the receiving water can have a profound effect on the rising plumes. As can be seen from the photograph of a laboratory plume issuing into a stationary, linearly stratified environment in Figure 3.18, it can trap the plume beneath the water surface. A definition sketch of this situation is given in Figure 3.19.

74   Marine Wastewater Outfalls and Treatment Systems

Figure 3.18 Photograph of horizontal buoyant jet in a stratified environment

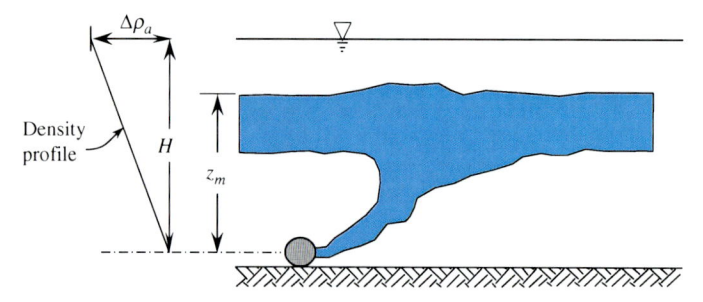

Figure 3.19 Definition sketch for horizontal buoyant jet in a stationary, stratified environment

If the ambient density stratification varies linearly with depth, it can be characterized by the buoyancy frequency, N:

$$N = \sqrt{-\frac{g}{\rho}\frac{d\rho_a}{dz}} = \sqrt{-\frac{g}{\rho}\frac{\Delta\rho_a}{H}} \qquad (3.20)$$

where $\rho_a(z)$ is the ambient density at depth z and $\Delta\rho_a$ is the density difference over the water depth, H. This introduces new point and line length scales that relate to the flow behavior (Fischer et al., 1979; Wright 1984, and others):

$$l_M = \frac{M^{3/4}}{B^{1/2}} \qquad l_B = \frac{B^{1/4}}{N^{3/4}} \qquad l_m = \frac{m}{b^{2/3}} \qquad l_b = \frac{b^{1/3}}{N} \qquad (3.21)$$

The effects of source momentum flux can be neglected for a point source when $l_M/l_B < 1$ (Wong and Wright, 1988), and for a line source when $l_m/l_b < 0.2$

(Roberts *et al.*, 1989a). According to the experiments of Daviero *et al.* (2006), dilution is given by:

$$\frac{S_n Q_j N^{5/4}}{B^{3/4}} = 0.90 \tag{3.22}$$

the near field length by:

$$\frac{x_n}{l_B} = 4.1 \tag{3.23}$$

and the rise heights and layer thickness by:

$$\frac{z_n}{l_B} = 2.7 \quad \frac{z_m}{l_B} = 3.5 \quad \frac{h_n}{l_B} = 1.6 \tag{3.24}$$

3.3.2.7 Merging plumes into stationary, stratified environment

A sketch of merging plumes from a multiport diffuser into a stationary, stratified environment is shown in Figure 3.20 and a photograph of laboratory plumes is shown in Figure 3.21. The rising plumes are again trapped by the density stratification, but the spreading layer occupies a much higher fraction of the plume rise height than does a single point plume (Figure 3.18).

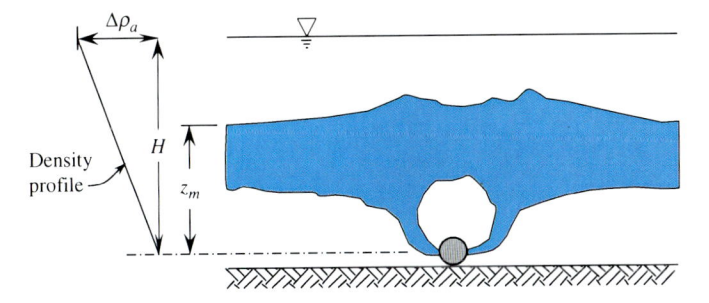

Figure 3.20 Definition diagram for merging plumes into stationary stratified environment

The effects of source momentum and port spacing are expressed by the length scale ratios l_m/l_b and s/l_b (Roberts *et al.*, 1989). The flow behaves like a line plume for $l_m/l_b < 0.2$ and $s/l_b < 0.3$. For that case, according to the recent experiments of Daviero and Daviero (2006), the near field dilution is given by:

$$\frac{S_n q N}{b^{2/3}} = 0.86 \tag{3.25}$$

and the length of the near field and the rise heights by:

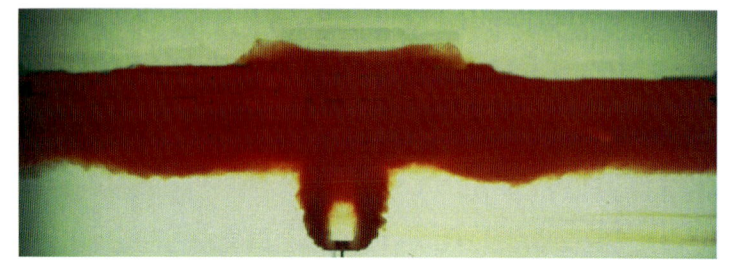

$$\frac{x_n}{l_b} = 2.3 \quad \frac{h_n}{l_b} = 1.5 \quad \frac{z_n}{l_b} = 1.7 \quad \frac{z_m}{l_b} = 2.5 \tag{3.26}$$

Figure 3.21 Photograph of merging plumes into a stratified environment

3.3.2.8 Merging plumes into a flowing, stratified current

A definition sketch of plumes from a multiport diffuser into a flowing, density stratified current is shown in Figure 3.22.

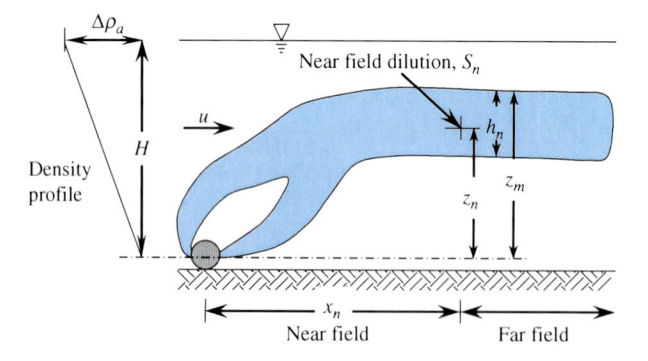

Figure 3.22 Merging plumes into a flowing, stratified current

Photographs of multiport diffuser discharges that approximate a line plume in density-stratified perpendicular currents of various speeds are shown in Figure 3.23. The Froude number (Eq. 3.18) is again the most important dynamical parameter that expresses the effect of the current, and, as for the unstratified case, different flow regimes can occur depending on the value of F. For zero current speed (Figure 3.21) the two plumes from each side of the diffuser partially merge before entering the horizontally spreading layer; they overshoot their equilibrium rise height before collapsing back to form a wastefield which occupies a substantial fraction of the total rise height.

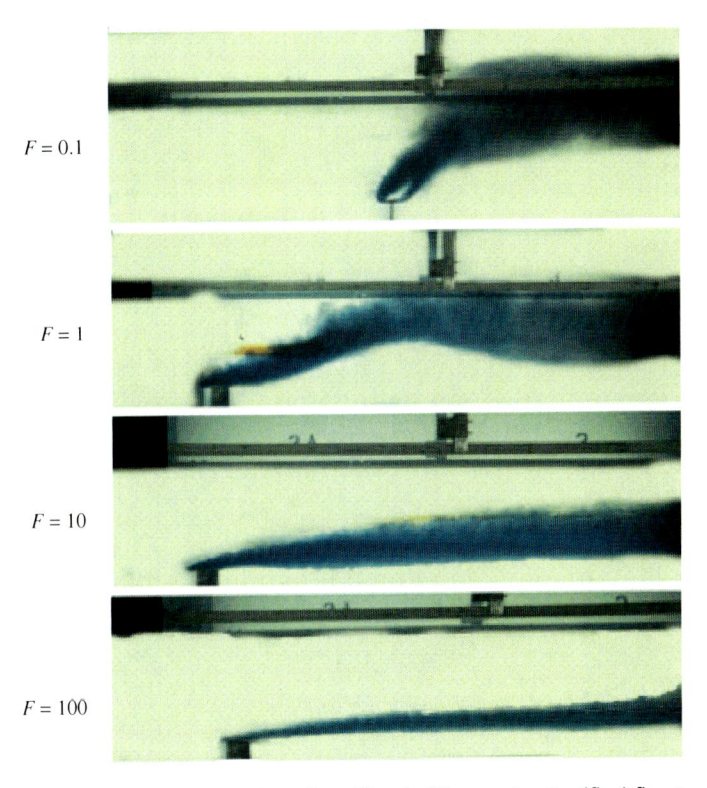

Figure 3.23 Photographs of multiport diffusers in stratified flows (after Roberts *et al.*, 1989)

The plume behavior depends on current speed (Figure 3.23). For slow currents ($F = 0.1$), the upstream layer is expelled and all of the flow is swept downstream. Increasing the current speed to $F = 1$ causes the plumes from the opposite sides of the diffuser to rapidly merge, and the wastefield takes on a wave-like form. Different mixing processes occur at different current speeds. At low current speeds (e.g. $F = 0.1$) the flow has the normal plume like pattern with the plume bent downstream. At higher current speeds (e.g. $F = 10$) the plume cannot entrain all of the oncoming flow while maintaining the free plume pattern and the wastefield bottom stays at the nozzle level. This is the forced entrainment regime (see also Figure 3.14 for the unstratified case), and it occurs when the Froude number exceeds a value that lies somewhere between 1 and 10 for the stratified case. The rise height and thickness decrease with increasing current speed in the forced entrainment regime.

For diffusers approximating a line plume, the near field dilution can be expressed as:

$$\frac{S_n q N}{b^{2/3}} = f(F, \Theta) \qquad (3.27)$$

where Θ is the angle of the current relative to the diffuser axis. Measurements of near field dilution reported in Roberts *et al.* (1989a), are reproduced in Figure 3.24.

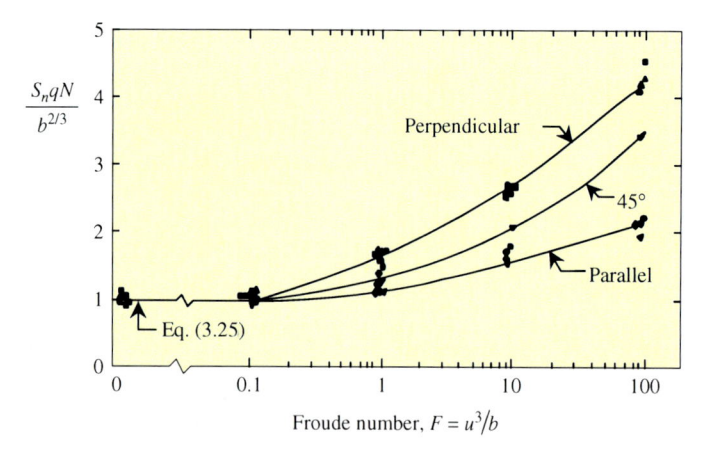

Figure 3.24 Near field dilution of multiport diffusers in stratified currents (after Roberts *et al.*, 1989a)

The near field dilution is unaffected by the current when $F < 0.1$ but thereafter increases with increasing current speed. Dilution is highest for a diffuser perpendicular to the current and lowest for one parallel. Even the parallel case, however, shows an increase in initial dilution with current speed and in no case does the current cause dilution to be lower than for zero current speed.

Dilution increases with distance from the diffuser until it reaches a limiting value, after which it remains constant. The location of the limiting value of dilution defines the end of the near field. The length of the near field for perpendicular currents for $F > 0.1$ can be reasonably estimated by (Roberts *et al.*, 1989b):

$$\frac{x_n}{l_b} = 8.5 F^{1/3} \qquad (3.28)$$

Wastewater mixing and dispersion

More recent experiments (Tian *et al.*, 2006) give a slightly smaller value of the constant equal to 8.0. Current directions other than perpendicular result in shorter near field lengths, so Eq. (3.28) is an upper limit. 3DLIF images presented in Tian *et al.* (2006) show that, for perpendicular currents, the effluent concentration profiles become very uniform laterally, even when the port spacing is very wide. This uniformity results from the merging and gravitational collapse of the individual plumes.

Measurements of rise height, z_m, are shown in Figure 3.25. The rise height decreases rapidly with increasing current speed for perpendicular currents in the forced entrainment regime (as can be seen in the photographs, Figure 3.23). The rise height also decreases for parallel currents, but not as rapidly as for perpendicular currents.

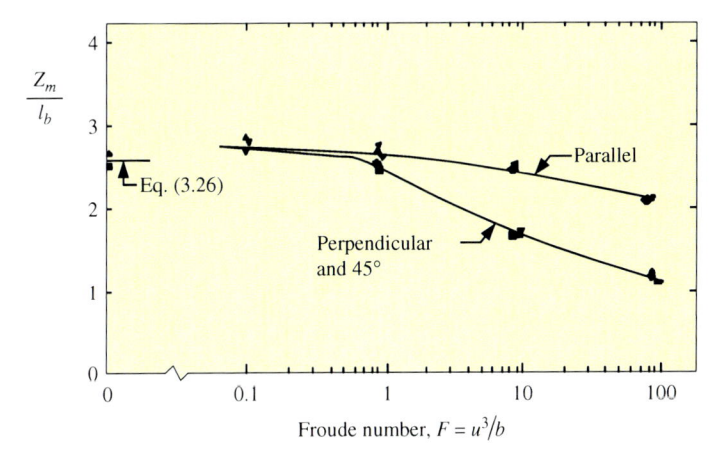

Figure 3.25 Rise height for multiport diffusers in stratified currents (after Roberts *et al.*, 1989a)

The width of the wastefield at the end of the near field is governed by lateral gravitational spreading. This is particularly important for diffusers parallel to the current and can result in a wide wastefield, as illustrated by the photographs in Figure 3.26. The spreading is linear, and the width w up to the end of the diffuser is given by (Roberts *et al.*, 1989b):

$$\frac{w}{x} = 0.70 F^{-1/3} \qquad (3.29)$$

Because of this spreading, the width of the plume at the end of the diffuser is typically of the order of the diffuser length, regardless of the current direction, for typical outfall conditions. This spreading also occurs when the current is perpendicular to the diffuser, but is not as rapid as when parallel.

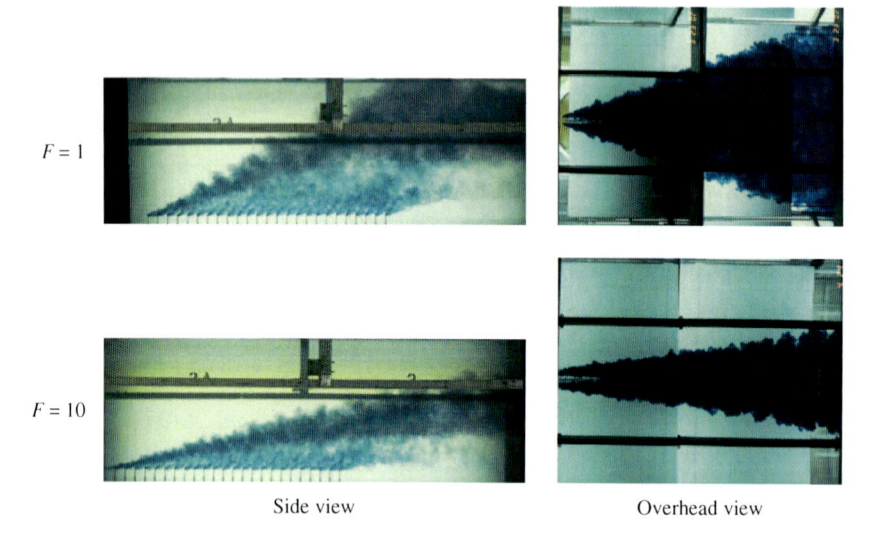

$F = 1$

$F = 10$

Side view   Overhead view

Figure 3.26 Photographs of multiport diffusers in parallel currents showing gravitational spreading (after Roberts *et al.*, 1989a)

3.3.2.9 Discussion and summary

A great number of equations were given above. They are summarized for stationary stratified and unstratified environments for point and line plumes in Table 3.3. These equations are useful for the limiting cases of line and point plumes under "worst-case" dilution conditions with no current. For other cases, such as intermediate port spacing, flowing currents, etc. see Roberts *et al.* (1989a, b, c), Tian *et al.* (2004a, b, 2006) and Daviero and Roberts (2006).

These equations, along and others for flowing currents, higher momentum fluxes, and wider port spacings, have been incorporated into a mathematical model NRFIELD, which is available from the US EPA. This model also incorporates the effects of nonlinear density stratification. It and other mathematical models for plume predictions are discussed in Chapter 4.

Table 3.3 Equations for predicting near field properties for point and fully merged (line) plumes in stationary environments. After Tian et al. (2004a) and Daviero and Roberts (2006)

Unstratified	Point plume ($s/H > 1$)	Line plume ($s/H < 0.3$)
Dilution	$\dfrac{S_n Q_j}{B^{1/3} H^{5/3}} = 0.26$	$\dfrac{S_n q}{b^{1/3} H} = 0.49$
Length of near field	$\dfrac{x_n}{H} = 2.8$	$\dfrac{x_n}{H} = 0.9$
Layer thickness	$\dfrac{h_n}{H} = 0.11$	$\dfrac{h_n}{H} = 0.36$
Rise height	H	H
Stratified	Point plume ($s/l_B > 3$)	Line plume ($s/l_B < 0.5$)
Dilution	$\dfrac{S_n Q_j N^{5/4}}{B^{3/4}} = 0.90$	$\dfrac{S_n q N}{b^{2/3}} = 0.86$
Length of near field	$\dfrac{x_n}{l_B} = 4.1$	$\dfrac{x_n}{l_b} = 2.3$
Layer thickness	$\dfrac{h_n}{l_B} = 1.6$	$\dfrac{h_n}{l_b} = 1.5$
Rise height	$\dfrac{z_m}{l_B} = 3.5$	$\dfrac{z_m}{l_b} = 2.5$

3.3.3 Far field mixing

3.3.3.1 Introduction

Following completion of initial mixing, the established wastefield drifts with the ocean currents and is diffused by oceanic turbulence in a phase of mixing often referred to as the "far field." The complexity of coastal currents makes it difficult to predict transport in this phase. In the past decade or so, however, the development of recording current meters, which has allowed collection of long time series of horizontal currents, has greatly increased our knowledge of coastal circulation processes. The use of these meters is now routine in major outfall studies, and the combination of data obtained from them with appropriate models has led to much improved predictions of wastefield behavior in the far field. Approaches to this problem, and their implications for design, are discussed in this section.

3.3.3.2 Oceanic turbulent mixing

Consider first the role of turbulent diffusion. The presence of density stratification inhibits vertical mixing, so that mixing of a submerged or surfacing field is primarily effected by lateral diffusion, as sketched in Figure 3.27. This process is usually estimated by application of Brooks (1960) solution to the turbulent diffusion equation with a variable diffusion coefficient.

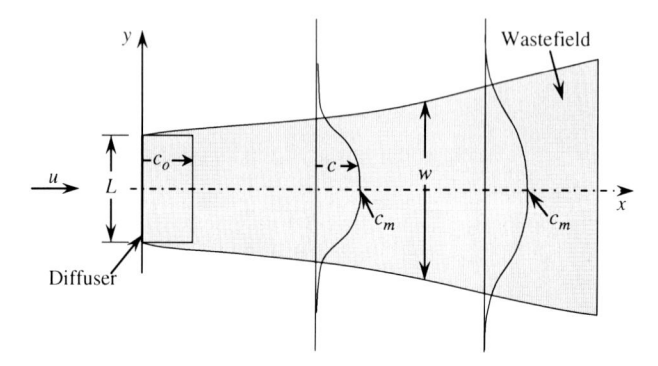

Figure 3.27 Diffusion from a continuous line source of finite length

The governing advective-diffusion equation is (Roberts and Webster, 2002):

$$u\frac{\partial c}{\partial x} = \frac{\partial}{\partial y}\left(\varepsilon\frac{\partial c}{\partial y}\right) - kc \qquad (3.30)$$

where c is the concentration of some contaminant contained in the wastewater, such as bacteria, whose initial concentration after near field dilution is c_o, and ε is a turbulent diffusion coefficient for transverse mixing. Equation (3.30) assumes steady-state conditions and neglects diffusion in the vertical and longitudinal directions. Bacterial decay is incorporated in the decay term, $-kc$, which corresponds to a first-order decay process with k the decay constant. Bacterial decay is more commonly expressed in terms of T_{90}, the time for 90% reduction in bacteria due to mortality, which is related to k by:

$$k = \frac{1}{T_{90}\log_{10} e} \qquad (3.31)$$

Solutions to Eq. (3.30) for various assumptions about the diffusion coefficient were obtained by Brooks (1960). A common assumption for coastal waters is that ε follows the "4/3 power law:"

$$\varepsilon = \alpha L^{4/3} \qquad (3.32)$$

where α is a constant whose value depends on the rate of energy dissipation, and L is the diffuser length. Values of α are given in Figure 3.5 in Fischer et al. (1979), and range (in cgs units) from 0.002 to 0.01 cm$^{2/3}$/s. For this case, Brooks obtained a solution to Eq. (3.30) for the centerline (maximum) concentration, c_m:

$$c_m(x) = c_o e^{-kx/u} erf \sqrt{\frac{3/2}{\left(1 + \frac{2}{3}\beta\frac{x}{L}\right)^3 - 1}} \qquad (3.33)$$

where $\beta = 12\varepsilon_o/uL$ and $erf()$ is the standard error function. For a conservative substance (i.e. zero decay rate), $k = 0$, and Eq. (3.33) can be expressed in terms of a "far-field dilution" $S_f = c_o/c_m$ that depends only on the travel time, t of the effluent to any location:

$$S_f = \left[erf\left(\frac{3/2}{\left(1 + 8\alpha L^{-2/3}t\right)^3 - 1}\right)^{1/2}\right]^{-1} \qquad (3.34)$$

This equation is plotted, along with solutions for far field dilution for other various common assumptions about the lateral diffusion coefficient, in Figure 3.28.

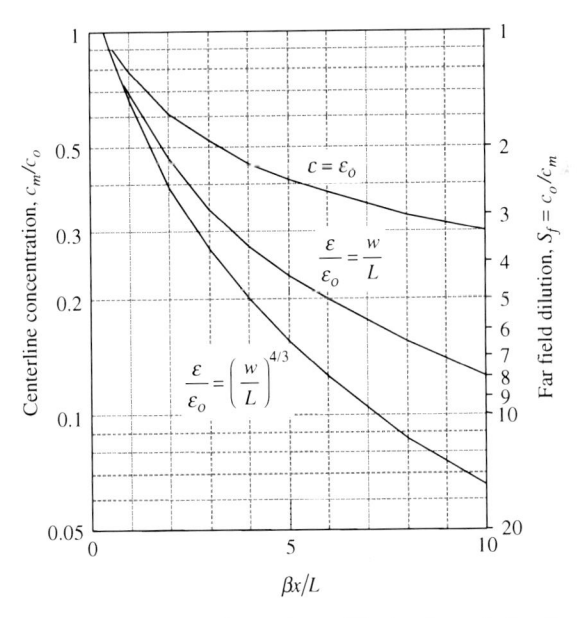

Figure 3.28 Centerline dilution for various diffusion laws (after Brooks, 1960)

The wastewater field width w at distance x is given by:

$$\frac{w}{L} = \left(1 + \frac{2}{3}\beta\frac{x}{L}\right)^{3/2} \tag{3.35}$$

To show the magnitudes of far field dilution for ocean outfalls, some typical values (using Figure 3.28 or Eq. 3.34), are given in Table 3.4 for the common 4/3 power law assumption. Dilutions are given for a short and long diffuser, assuming an upper value for $\alpha = 0.01$ cm$^{2/3}$/s (Fischer et al., 1979, Figure 3.5).

Table 3.4 Far field dilutions for diffusers of various lengths assuming 4/3 power law

Travel time t (hr)	Far field dilution S_f	
	Diffuser length, L (m)	
	$L = 35$	$L = 700$
1	2.4	1.0
3	7.4	1.4
10	35.5	3.2
20	95.9	6.9

An important conclusion from Table 3.4 is that dilution by oceanic turbulence can be quite effective for short diffusers, but is relatively minor for long diffusers. The physical interpretation of this result is that the time needed for the centerline concentration to be reduced is the time required for eddies at the plume edges to "bite" into it. For a wide field produced by a long diffuser, the eddies have farther to go so it takes them longer to get to the centerline. Much effort is often devoted during outfall designs to field studies aimed at measuring the rate of oceanic diffusion. Even allowing for the uncertainties in the calculations, this example suggests that these efforts are misguided for large outfalls. It is more important to know where the wastefield goes, rather than whether the far field dilution is 3 or 5, and ways to estimate this are discussed below.

3.3.3.3 Statistical far field model

The above analysis is for steady-state conditions. However, currents in coastal waters fluctuate widely under their various forcing mechanisms (Section 3.2). They often consist of a fluctuating component, u', whose magnitude is much larger than the mean drift, U. This can cause the wastefield to wander in the vicinity of the diffuser for an extended period before being flushed away. And the unsteadiness of the currents results in a continuously shifting plume whose location at any instant is best treated as a stochastic variable. As shown above, for a long source, the centerline concentrations may not be substantially reduced by turbulent diffusion for travel times of a few hours after release. Pollutant concentrations at any location then alternate between near-background and near-maximum levels. The first problem we face, then, is how to quantify the environmental impact of such a situation.

This problem has been considered in detail by Csanady (1983a). He considers that a measure of environmental impact consists of computations of background concentration, the maximum concentration, and the frequency of immersion of any point in the plume. Csanady suggests that the modeling approach be based on a division of the plume into contaminant puffs of distinct "ages." "Young" puffs are those which have traveled for a few hours after release and are advected by local currents; "old" puffs have traveled for days or more and contribute to what may be called the background concentration. The modeling of young puffs, which we identify with the far field, is discussed in this section.

The frequency of immersion of any point in the plume is termed the "visitation frequency" by Csanady. He presents methods to compute this quantity from the statistics of currents measured by a meter at a fixed location. A somewhat similar approach is given by Koh (1988), who refers to "advective transport probabilities."

Target point, $\vec{x}$

Plume centerline:
$$\vec{x}_c(t,t') = \int_t^{t+t'} u_L(\tau) \, d\tau$$

Wastefield at time $t + t'$ due to releases from t to $t + t'$.

Diffuser

Predict transport probability from current meter records:
$$\vec{x}_c(t,t') = \int_t^{t+t'} u_E(\tau) \, d\tau$$

Visitation frequency:
$$\gamma(\vec{x}, t_d) = \iint P(\vec{x},t) d\vec{x} dt$$

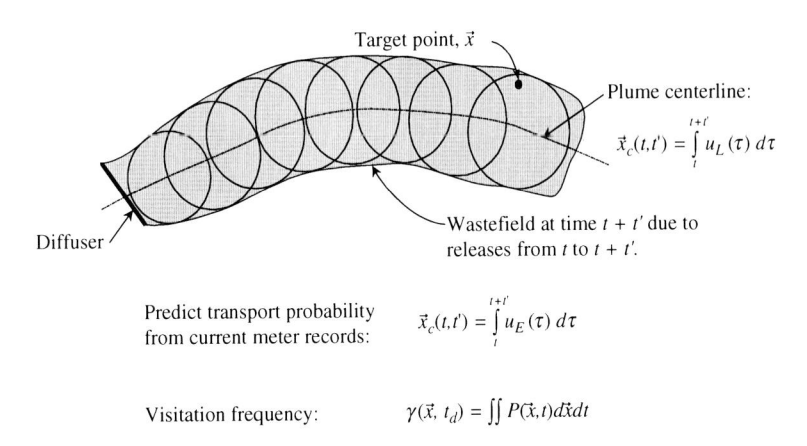

Figure 3.29 Statistical far field transport model, FRFIELD

A model based on this procedure, illustrated in Figure 3.29, was developed by Roberts (1999b). This is a statistical model in which the wastefield is discretized as a series of puffs released at the current meter sampling interval, typically 10 to

30 minutes. Each puff is allowed to grow by turbulent diffusion according to Eq. (3.35) and is followed up to a maximum time horizon. A grid is superimposed on the area, and if a puff overlays a grid square, this is counted as a "visit." The number of visits by a puff of age younger than the maximum time horizon is summed and divided by the total number of releases. An example of this procedure applied to the San Francisco outfall is given in Roberts and Williams (1992).

The visitation frequency γ at a target point $\vec{x}$ is defined as the fraction of time for which the plume center is within vector $\vec{w}/2$ of $\vec{x}$, where $|\vec{w}|$ is the expected plume width. This is the same as the probability of finding the plume within $\pm \vec{w}/2$ of $\vec{x}$. Consider a puff released at $\vec{x} = 0$ at time $t = 0$. The probability distribution function for the puff center at a later time t is $P(\vec{x}, t)$ such that the probability of the puff center being between $\vec{x}$ and $\vec{x} + d\vec{x}$ is $P(\vec{x}, t)d\vec{x}$. The probability that the puff is overlapping a point $\vec{x}$ is given by:

$$p(\vec{x}, t) = \int_A P(\vec{x}, t)d\vec{x} \qquad (3.36)$$

where the area of integration A is the area of the puff at time t. For a continuous source, the probability of impaction, or visitation frequency, of the plume for all prior releases is obtained by integrating over release time t'. For stationary conditions:

$$\gamma = \int_{-\infty}^{t} \int_A P(\vec{x}, t - t')d\vec{x}dt' = \int_0^{\infty} \int_A P(\vec{x}, t)d\vec{x}dt \qquad (3.37)$$

Csanady defines a dividing time t_d to distinguish between "young" and "old" puffs. In a tidal environment, it would be expected that t_d is of the order of the tidal period. Eq. (3.37) then becomes:

$$\gamma(\vec{x}, t_d) = \int_0^{t_d} \int_A P(\vec{x}, t)d\vec{x}dt \qquad (3.38)$$

which is the visitation frequency of all the "young" puffs of age less than or equal to t_d. Equation (3.38) is analogous to Eq. (26) of Csanady (1983a).

The displacement probability, $P(\vec{x}, t)$, of the puffs can be computed from their trajectories, if known. The location of the puff center at various times t, is:

$$\vec{x}_c(t) = \int_0^t u_L(t')dt' \qquad (3.39)$$

where u_L is the Lagrangian velocity of the puff center. Csanady (1983a) discusses how $P(\vec{x}, t)$, i.e. the statistics of $\vec{x}_c(t)$, can be computed from the statistics of u_L, and Koh (1988) discusses similar computations from simulated currents. The fundamental problem in air and water pollution, however, is that the Lagrangian velocities of the puffs are not generally known. Instead, what are commonly available are Eulerian velocity measurements at a fixed point such as from a current meter. The usual assumption, often applied in air pollution (Pasquill and Smith, 1983), is to infer Lagrangian displacement from an Eulerian record by the approximation:

$$\vec{x}_c(t) \approx \int_0^t u_E(t')dt' \qquad (3.40)$$

where u_E is the fixed point (Eulerian) record of velocity. The trajectory computed for various travel times t by Eq. (3.39) is a path line; the trajectory computed by Eq. (3.40) is known as a progressive vector diagram in oceanography. Clearly, the displacement predicted by Eq. (3.40) becomes increasingly unreliable as the distance from the source increases. Furthermore, Zimmerman (1986) has pointed out that regularly varying tidal flows over irregular topography can produce "Lagrangian chaos," that is, unpredictable and non-repeatable trajectories. Therefore, even if we had perfect Lagrangian information for an individual release, we could not use this to predict future trajectories even under identical forcing functions. For these reasons individual plume trajectories should not be inferred by these methods, but it is usually assumed that statistical inferences of the pattern and scale of the dispersion can be made. Support for this assumption is provided by List et al. (1990) who found good general agreement between diffusivities in coastal waters computed from drogues and from fixed current meters. Because of the complexity of the dispersion process and the relatively poor spatial resolution of coastal measurements, it is clearly not possible to pretend for great accuracy in any predictive method. But the statistical approach is very useful in assessing the probability of exceedance of some threshold concentration at particular locations such as near the coastline.

Outfall studies usually produce current meter records that consist of discrete measurements at a fixed point with a fixed sampling interval Δt. To use these data, we discretize the plume as a series of puffs released at a rate equal to the sampling frequency $(\Delta t)^{-1}$. The location of a puff released at time $t_0 = n\Delta t$ after travel time $T = m\Delta t$ is then assumed to be given by the discrete form of Eq. (3.40):

$$\vec{x}\langle t_o | T \rangle = \sum_{i=n}^{i=n+m} \vec{u}_i \Delta t \qquad (3.41)$$

where $\vec{u}_i(t)$ is the local measured velocity at time $t = i\Delta t$. This computation is repeated for all releases during the whole data record, and each puff is followed up to the maximum travel time or "time horizon," t_d. This procedure would typically include thousands of releases, each release being tracked at each time step as it travels. The area around the diffuser is overlain with a grid, and if a puff is within $\pm \vec{w}/2$ of a grid node, this is counted as a "visit." The number of visits by a puff of age younger than t_d is summed and divided by the total number of releases to obtain the visitation frequency at that location. Koh (1988) presents similar computations in which the probability of travel of the plume centerline into a grid square is computed.

As the plume travels it is diffused and grows due to oceanic turbulence. We use a gradient diffusion model for this process rather than a particle tracking model due to its computational simplicity and the difficulties of keeping track of the huge number of particles that would be necessary for an extended simulation over many months. It is assumed that the diffusion coefficient is proportional to the 4/3 power of the plume size (Eq. 3.32). List et al. (1990) observed good general agreement with this relationship for Southern California coastal waters.

The decay of peak concentration is assumed to be given by Brooks (1960) solution to the diffusion equation, Eq. (3.33). This applies to a continuous line source whose concentration is reduced by lateral diffusion only. For an isolated puff growing in three-dimensions the diffusion rate will be greater and the plume dilution will increase away from the puff center. We neglect these effects and use the conservative assumptions that the dilution in the puff is given by the continuous solution and is constant across the puff.

The puff size w is allowed to grow by diffusion as it travels according to Eq. (3.35), which can be written as:

$$w = L\sqrt{1 + 12\left(\frac{s_y}{L}\right)^2} \tag{3.42}$$

where s_y is the standard deviation of the lateral concentration distribution:

$$s_y = \frac{L}{\sqrt{12}}\sqrt{\left(1 + \frac{8\varepsilon_o t}{L^2}\right)^3 - 1} \tag{3.43}$$

The actual dilution S at any location when the plume is present is the product of the near field dilution S_n and the far-field dilution S_f:

$$S = S_n \times S_f \tag{3.44}$$

where S_n is the dilution at the end of the near field (computed, for example, by the methods of Section 3.3.2). The corresponding contaminant concentration is then given by:

$$c = \frac{c_{oo}}{S} \qquad (3.45)$$

where c_{oo} is the contaminant concentration in the effluent leaving the treatment plant. The concentration estimated by Eq. (3.45) is the maximum expected at any location. This will occur very infrequently, however, and time-average concentrations will be much lower.

An example of visitation frequencies computed using this approach are shown for Mamala Bay, Hawaii, in Figure 3.30 for travel times up to 12 hours. The plots give a good visualization of wastefield impact. The contours elongate in the East/West directions in accordance with the direction of the main current components. These currents are strongly tidal, with the result that the wastefield is swept back and forth in the vicinity the diffuser. This causes the visitation frequency to diminish rapidly with distance from the diffuser with very low probability of shoreward impaction. The wastefield is spread over an area whose dimensions are equal to the maximum tidal excursion within a time scale of about six hours. The visitation frequency at ~5 km from the diffuser is about 1%, or conversely the plume is *not* present for about 99% of the time. Note that a visitation frequency of one percent is about seven hours per month. Expansion of the contours for travel times longer than about six hours is primarily caused by the mean drift.

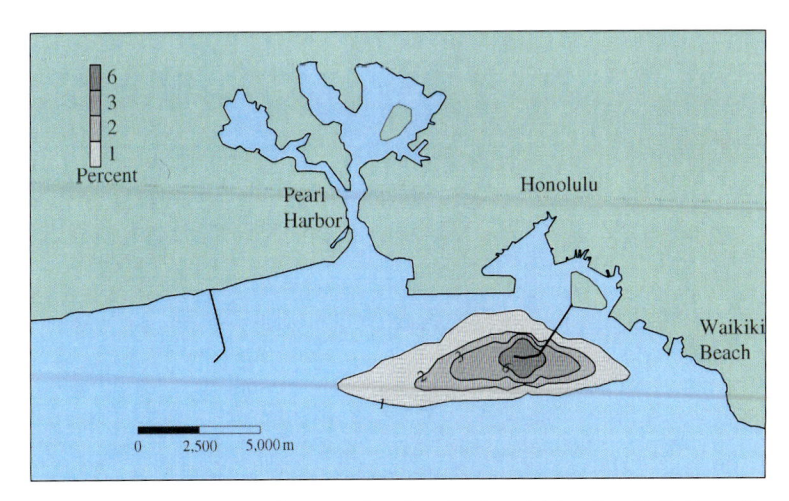

Figure 3.30 Visitation frequencies for the Sand Island outfall, Hawaii (after Roberts, 1999b)

Modeling bacteria is usually more important than visitation frequencies. This can be done by allowing bacteria to decay according to a first order process (see also Eq. 3.31) so that the bacterial concentration after travel time t is given by:

$$\frac{c}{c_o} = 10^{-\frac{t}{T_{90}}} \tag{3.46}$$

where T_{90} can vary diurnally. The bacterial concentrations are predicted by Eq. (3.33). Modeling of this type is discussed in Chapter 4, and the predicted bacterial concentrations for Mamala Bay corresponding to Figure 3.30 are shown in Figure 4.17.

An obvious goal of outfall design would be to move the diffuser far enough offshore that the outer contours do not intersect the shoreline for any travel time. The probability of effluent reaching the shore will then be vanishingly small, and shoreline bacterial standards will likely be met without chlorination. It is highly desirable to meet shoreline bacterial standards without chlorination, or, if this is not possible, to estimate the frequency and amount of chlorination required.

Shoreline bacterial concentrations estimated in this way are more realistic and lower than those obtained by traditional methods that assume that currents flow steadily and directly onshore. This method underestimates travel times to shore and overestimates the probability of shoreline impingement. This is because currents rarely remain steady for long nor do they maintain their onshore direction as they get close to shore. The traditional approach can lead to overly conservative outfall designs.

Even the approach advocated here is probably quite conservative. This is due to several factors. First is plume submergence, which considerably reduces the probability of shore impaction as explained in Section 3.2.3 (see Figure 3.4). Second, nearshore currents are usually slower and more parallel to the coast than currents farther offshore. Thus, if currents measured offshore are used to predict nearshore currents (as is usually the case), the travel times to shore (if the shore is reached at all) will be considerably underestimated. The longer actual travel times afford greater opportunity for bacterial decay and diffusion.

Plots of visitation frequencies are very useful when comparing alternative discharge locations. An example is given in figure 10 of Roberts (1986), which shows four candidate discharge sites. For only two of these was zero onshore transport predicted. The visitation frequencies are also very useful for assessing and showing impacts on areas of particular significance, such as shellfish areas.

3.3.4 Long-term flushing

Finally, we consider the long-term buildup of contaminants in the vicinity of the discharge, or coastal "flushing" which occurs on long time scales (Figure 3.2). In the previous section, we divided the plume into "young" and "old" puffs. The young puffs, whose travel times are of order a day or less, contribute most of the local bacterial impacts and can be analyzed by means of the statistical model presented above. The "old" puffs are subject to considerable decay and diffusion and generate a concentration that can be considered to be a "background" mean concentration field in the vicinity of the diffuser. The level of this concentration is governed primarily by flushing due to the mean drift, horizontal diffusion, and, for non-conservative substances, chemical and biological decay. One approach to predicting the physical dilution caused by these processes is to estimate it from a solution to the two-dimensional diffusion equation (Csanady, 1983a; Koh, 1988). We here consider a simpler method, however, which is particularly useful for comparing the relative orders of magnitude of the various processes. This is a mass-balance box model (Csanady, 1983b) as shown in Figure 3.31.

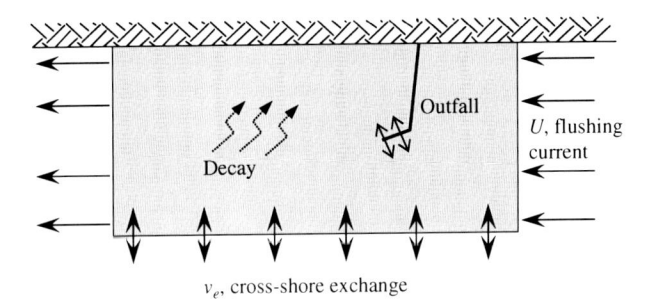

Figure 3.31 Box model for estimating long-term buildup of contaminants (after Csanady, 1983b)

Tidal currents distribute the effluent over an area, or "box" whose dimensions are approximately equal to the tidal amplitude. These dimensions are approximately $X = u_t T/2$ and $Y = v_t T/2$, in the alongshore and cross-shore directions, respectively, where u_t and v_t are the amplitudes of the tidal currents, and T is the tidal period. Csanady (1983b) calls this area the "extended source region." It would be comparable to the outer edge of the visitation frequency contours, for example Figure 3.30.

Long-term average current speeds are usually much slower than instantaneous values. They lead to an average dilution equal to UhY/Q_T, where Q_T is the total effluent flowrate, h the average depth of the plume over the extended area, and U the long-term average "flushing velocity."

This can be extended to include the other processes by applying a mass balance to the box. This yields a "long-term average dilution" S_p:

$$S_p = \frac{UhY}{Q_T} + \frac{v_e hX}{Q_T} + \frac{khXY}{Q_T} \tag{3.47}$$

The first term on the right hand side is the dilution due to flushing by the mean current. The second is dilution due to cross-shore mixing. This is parameterized by v_e, a mass transfer "diffusion velocity," which can be assumed equal to the standard deviation of the cross-shore tidal fluctuations (probably an underestimate). The third term is "dilution" due to chemical or biological decay, where k is a first-order decay rate. It can be seen that the total effective dilution is the sum of these individual dilutions.

Consider a typical outfall problem. Suppose we have a discharge $Q_T = 5$ m³/s into a tidal current whose alongshore amplitude is $u_t = 0.25$ m/s, and cross-shore amplitude is $v_t = 0.08$ m/s, and cross-shore rms velocity is $v_e = 0.04$ m/s. Suppose the average current speed (the flushing velocity) is $U = 0.06$ m/s. For a semi-diurnal tide, the period T is about 12 hours. Suppose further that the average depth (thickness) of the wastefield is 15 m, and the bacterial decay rate (averaged over 24 hours) is $T_{90} = 10$ hours, corresponding to $k \approx 6 \times 10^{-5}$ s^{-1}.

Then the extended source area (size of the box in Figure 3.31) is:

$$X = u_T T/2 = 0.25 \times 12 \times 3600/2 \approx 5400 \text{ m} \approx 5.4 \text{ km and}$$

$$Y = v_T T/2 = 0.08 \times 12 \times 3600/2 \approx 1700 \text{ m} \approx 1.7 \text{ km}$$

and the dilutions are (Eq. 3.47):

$$\text{Due to the mean current}: \quad \frac{UhY}{Q_T} = \frac{0.06 \times 15 \times 1700}{5} \approx 300$$

$$\text{Due to cross-shore exchange}: \quad \frac{v_e hX}{Q_T} = \frac{0.04 \times 15 \times 5400}{5} \approx 650$$

$$\text{Due to decay}: \quad \frac{khXY}{Q_T} = \frac{6 \times 10^{-5} \times 15 \times 5400 \times 1700}{5} \approx 1650$$

The total effective dilution, the sum of these dilutions, is about 2,600.

These are obviously only approximate order of magnitude calculations, but they are very useful for estimating long-term impacts. They can be applied to other substances such as toxic materials to estimate their potential accumulation.

3.4 CONCLUSIONS

This chapter has focused on the main processes that determine the fate and transport of wastewater discharged from ocean outfalls, means to predict them, and their implications for outfall design. It will usually be possible to design an outfall that achieves initial dilutions of 100:1 and greater. The conclusions from the far field modeling examples are typical of many coastal discharges. An unsteady and spatially variable current field is a very dispersive environment. It results in rapid decrease in visitation frequency, mean contaminant concentration levels, and other measures of bacterial impact with distance from the diffuser. This will ensure that any measurable environmental impact is confined to a small area around the discharge, even with minimal treatment. This has been confirmed in many outfall field studies. Mean current circulation patterns and other mechanisms will usually prevent significant accumulation or "background levels" of contaminants around the discharge. The calculations presented in this chapter confirm the importance of diffusion and dispersion in coastal waters, and why they are usually more important than treatment in minimizing the environmental impact of a marine disposal system.

4
Modeling transport processes and water quality

4.1 INTRODUCTION

Mathematical (i.e. computer) models are now widely used to predict the fate and transport of ocean discharges. Because of the very wide range of length and time scales of the various mixing processes (see Figure 3.2), it is not possible to simulate them with one overall "omnibus" model. Linked sub-models of the various phases are therefore usually used. In this chapter, we discuss the main modeling approaches that are commonly used and give examples of them. Modeling has a very extensive literature, and because of the complexity of the transport mechanisms, different methods have been adopted to model them. Therefore we can only here give an overview of the main issues and modeling approaches. We also give examples of modeling to illustrate the methods and especially their implications for outfall design.

© 2010 IWA Publishing. *Marine Wastewater Outfalls and Treatment Systems.* By Philip JW Roberts, Henry J Salas, Fred M Reiff, Menahem Libhaber, Alejandro Labbe, and James C Thomson. ISBN: 9781843391890. Published by IWA Publishing, London, UK.

4.2 NEAR FIELD AND MIXING ZONE MODELS

Models of near field processes (see Section 3.3.2) are sometimes called plume models. They fall into three major categories as shown in Table 4.1. The three model types are discussed in this section.

Table 4.1 Near field model types

Types	Comments	Examples
Length-scale models	Based on dimensional analysis and experimental data. Interpolation between flow cases	NRFIELD (RSB) CORMIX
Entrainment models:	Useful for nonlinear stratification, non-uniform velocity	
Eulerian	Fixed control volume	Fan and Brooks (1969)
Lagrangian	Moving control volume	UM3 – Visual Plumes
Turbulence models	CFD (Computational Fluid Dynamics) Based on turbulence closure assumptions. Not yet widely used	Hwang and Chiang (1995) Tang et al. (2008)

4.2.1 Length-scale models

Length scale models are based on analyses similar to those presented in Section 3.3.2. The relevant length scales and their ratios are first computed to determine the type of discharge involved and to classify the flow case. Plume properties are then computed using appropriate semi-empirical equations or by calling an appropriate entrainment model (Section 4.2.2). Examples are NRFIELD and Cormix (Section 4.3).

4.2.2 Entrainment models

The entrainment hypothesis was first suggested by Morton et al. (1956) and has since been applied to a variety of engineering and natural flows, as reviewed in Turner (1986). It is particularly relevant here as it has found great utility for predicting the jet and plume-type flows typical of ocean discharges. Below we summarize the essential features and limitations of these models; for details, the original references should be consulted, and for recent extensive reviews of entrainment models, see Jirka (2004, 2006).

The concept of entrainment, as applied to a simple round rising plume in a stationary environment, is shown in Figure 4.1. The rising plume entrains external fluid that then mixes with and dilutes the plume fluid. The entrainment hypothesis (Fischer et al., 1979) states that fluid is entrained at the plume radius b with a velocity u_e that is proportional to the mean centerline velocity, u_m:

$$u_e = \alpha u_m \tag{4.1}$$

where α is the entrainment coefficient (whose value is different for jets and plumes). The rate of change of volume flux Q in the plume with distance s is then given by:

$$\frac{dQ}{ds} = 2\pi \alpha b u_m \tag{4.2}$$

Equations (4.1) and (4.2) are the essence of the entrainment hypothesis, and form the basis for most entrainment models.

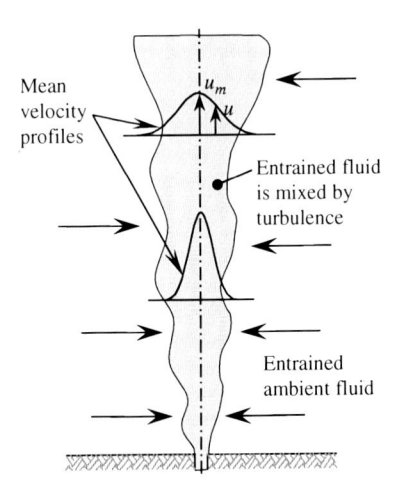

Figure 4.1 Entrainment and dilution in a simple plume

Entrainment models fall into two categories, Eulerian and Lagrangian, that differ basically in the definition of the control volume. In an Eulerian model the conservation equations are solved on a fixed control volume, and in a Lagrangian model on a moving control volume. Lagrangian and Eulerian integral flux models give similar results given similar assumptions.

An early application to outfalls was the Eulerian buoyant jet model of Fan and Brooks (1969). They analyzed unstratified and stratified environments as shown in the definition diagram, Figure 4.2. Fan and Brooks assumed that the profiles of velocity, density deficit, and tracer concentration were self-similar and Gaussian.

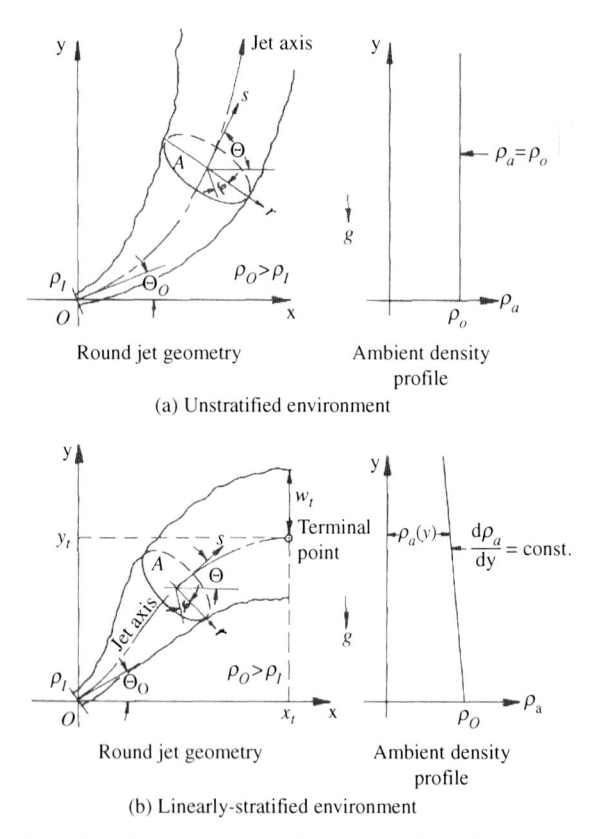

Figure 4.2 Definition diagram for entrainment model of Fan and Brooks (1969)

For example, the velocity profile is given by:

$$u = u_m(s)e^{-\frac{r^2}{b^2}} \tag{4.3}$$

where u is the local velocity at radial distance r from the jet centerline, and $u_m(s)$ is the maximum (centerline) velocity at distance s along the jet trajectory. The flow is therefore assumed to be radially symmetric. The equation for continuity, Eq. (4.2), then becomes:

$$\frac{d}{ds}\int_0^\infty u 2\pi r dr = 2\pi \alpha b u_m \tag{4.4}$$

Modeling transport processes and water quality

On substituting Eq. (4.3) and integrating this becomes:

$$\frac{d}{ds}(u_m b^2) = 2\alpha u_m b \qquad (4.5)$$

Similar equations are derived for conservation of horizontal and vertical components of momentum, buoyancy, and tracer. Horizontal momentum is conserved; vertical momentum is not conserved if there are density differences that cause buoyancy forces. The resulting equations, along with the equation for the geometry of the centerline, are then solved numerically along the jet trajectory. Because the conservation equations are integrated over the jet cross-sections, entrainment models are also known as integral models.

Lagrangian models are also widely used. The following example is based on the Lagrangian model UM3 that is incorporated into the U.S. EPA model suite Visual Plumes (Section 4.3.2) and follows Baumgartner *et al.* (1994). For a thorough discussion of Lagrangian models, see Lee and Chu (2003).

The control volume (called the plume element in UM3) is shown in Figure 4.3. The effluent fluid does not flow across the boundaries, although ambient fluid is entrained. This element is tracked through space by integrating through time, the independent variable, to continuously recalculate its position and other variables.

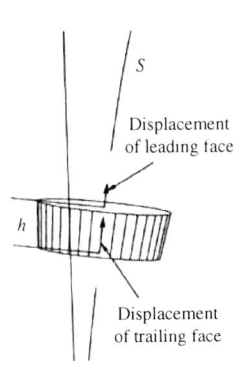

Figure 4.3 The round plume element in a Lagrangian model

The plume is assumed to be in steady state so that successive elements follow the same trajectory, i.e. the plume envelope remains invariant while elements move through it, changing their geometry and position with time. The envelope initially coincides with the edge of the port and the subsequent plume volume is calculated from the entrained volume and the assumed element geometry.

UM3 uses two entrainment models: the Taylor entrainment hypothesis (Morton et al., 1956) and the projected area entrainment (PAE: Frick, 1984) hypothesis. The PAE hypothesis is a statement of forced entrainment – the rate at which mass is incorporated into the plume due to a current. The element geometry determines the surface area, to which the Taylor entrainment term is proportional, and the projected area, to which forced entrainment is proportional. Overlap and curvature of the plume element is also incorporated (Figure 4.4). The element length changes with time due to the different velocities of the leading and trailing faces.

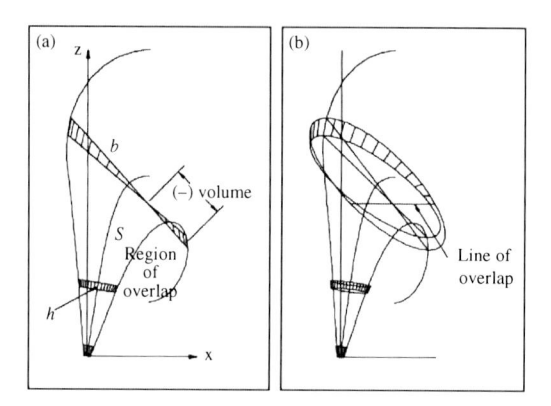

Figure 4.4 The round plume element at three stages of development

The mass conservation (continuity) equation for the plume element is

$$\frac{dm}{dt} = -\rho_a \vec{A}_p \cdot \vec{U}_a + \rho_a A_T v_e \tag{4.6}$$

where m is the element mass, ρ_a the ambient density, $\vec{A}_p$ is the upstream area vector of the plume element, $\vec{U}_a$ is the ambient current vector at the depth of the plume center line, v_e is the Taylor entrainment velocity, and A_T is the scalar Taylor plume element surface area. The first term on the right of Eq. (4.6) represents projected area entrainment and the second term the Taylor entrainment. The Taylor entrainment velocity is:

$$v_e = \alpha \left| \vec{U}_s - \vec{U}_a \right| \tag{4.7}$$

where α is an entrainment coefficient. Thus the entrainment velocity is proportional to the velocity difference (shear) between the plume element and the ambient fluid.

These equations show the general ideas behind UM3. It uses vector routines to define the different contributions to entrainment and can predict fully three-dimensional trajectories.

Entrainment (integral) models have been applied to a wide variety of jet and plume flows. They now include effects of merging of round plumes, ambient flows, and ambient stratification. They are simple and efficient to use and have great utility as an engineering tool.

The limitations of entrainment models should be kept in mind, however. The entrainment coefficient is often assumed to be constant, but even for the simplest case of a buoyant jet (Figure 4.2) it varies along the trajectory from a jet-like value at the origin to a plume-like value at large distances. To account for this, variable entrainment coefficients that depend on local conditions have been suggested (Fischer *et al.*, 1979). With crossflow, the mechanism of entrainment is less well-understood, and more complex entrainment mechanisms have been postulated. For example, the model of Schatzmann (1979) has five coefficients to account for various factors such as curvature. Other difficulties include uncertainties in the dynamics of merging and interactions of multiple plumes, especially those discharged from both sides of a multiport diffuser (see, for example the photograph in Figure 3.21), transition to horizontal flow at a free surface or terminal rise height, and turbulence collapse. It should be noted that these features are implicitly included in length-scale models based on experiments, such as NRFIELD. Given the uncertainties in entrainment models, we repeat the admonition given in Fischer *et al.* (1979) that the user should always perform checks on their model predictions by comparing them with predictions of the asymptotic jet and plume solutions such as those presented in Chapter 3.

4.2.3 CFD models

Computational fluid dynamics (CFD) modeling is being increasingly applied to a wide variety of turbulent flows in nature and engineering. There are several major CFD techniques; for a review, see Sotiropoulos (2005).

One method is direct numerical simulation (DNS). The unsteady, three-dimensional Navier-Stokes equations are solved over scales that are small enough to resolve the entire spectrum of turbulence, from the largest eddies, whose size is comparable to the flow domain, to the smallest Kolmogorov scales where turbulent energy is dissipated as heat. In principle, DNS could model turbulent flows with virtually no modeling uncertainties. Because the computational resources it requires increase dramatically with Reynolds number, however, it has mainly been applied to relatively simple, low Reynolds number flows. DNS is therefore not (at least within the foreseeable future) a practical modeling tool for simulating flows at engineering-relevant Reynolds numbers.

A more realistic approach is Large Eddy Simulation (LES). The spatially filtered unsteady Navier-Stokes equations are solved to resolve motions larger than the grid size, and smaller-scale motions are modeled with a sub-grid model. In flows where large-scale structures dominate the accuracy is close to DNS as the influence of the small scales on the large scales is small (hence the sub-grid scale model performance is not so important). However, for high Reynolds number flows of practical engineering interest very high grid resolutions and supercomputers are still required.

The most common CFD models are Reynolds-decomposition models. Flow quantities are decomposed into time-averaged and fluctuating values and the Navier-Stokes equations are then time averaged, producing what are known as the Reynolds-averaged Navier-Stokes (RANS) equations. Assumptions are made about the new terms that arise from this averaging. Examples are the "constant eddy-viscosity" model and turbulent mixing-length models. Probably the most common is the k-ε model that assumes an empirical relationship between turbulent kinetic energy, k, and the rate of energy dissipation, ε. Various adaptations of this model have been applied to a wide range of engineering flows.

There have not been many applications of CFD to jet and plume-type flows. Hwang and Chiang (1995) and Hwang et al. (1995) simulated the initial mixing of a vertical buoyant jet in a density-stratified crossflow. They employed a RANS model with a buoyancy modified k-ε model. Blumberg et al. (1996) and Zhang and Adams (1999) used far-field CFD circulation models to calculate near field dilutions of wastewater outfalls. Law et al. (2002) used a revised buoyancy-extended k-ε turbulence closure to investigate the dilution of a merging wastewater plume from a submerged diffuser with 8-port rosette-shaped risers in an oblique current. Davis et al. (2004) used the commercial codes ANSYS and FLUENT to simulate several case studies of effluent discharges into flowing water, including a line diffuser, a deep ocean discharge, and a shallow river discharge. They concluded that CFD models are becoming a viable alternative for diffuser discharges with complex configurations.

The paucity of CFD applications to near field mixing is because of the major challenges that they face. These arise from the geometrical complexity of realistic multiport diffusers, the large difference between port sizes and the characteristic length scales of the receiving waters, buoyancy effects, plume merging, flowing current effects, and surface and bottom interactions. To overcome these difficulties, Tang et al. (2008) applied a three-dimensional RANS model using a domain decomposition method with embedded grids to model diffusers with realistic geometries. They employed an algebraic mixing length model with a Richardson-number correction for buoyancy effects. They applied their model to simulate negatively buoyant wall jet flows and found good overall agreement with experiments and to the turbulent mixing of thermal discharges from single-port and multiport diffusers into a prismatic channel and a natural river. The CFD model showed the complex three-dimensional features of the flows, including jet merging and the interplay between the jets and the ambient flow.

Although promising, the complexity of CFD models and the effort required to set them up and run suggests that entrainment and length-scale models will continue to be used for many years. This is particularly true for the cases of interest to outfall design discussed in this book that may involve running thousands of simulations with different oceanographic and discharge conditions (see Section 4.7.2).

4.3 SOME NEAR FIELD AND MIXING ZONE MODELS

4.3.1 Introduction

Some commonly used models for near field and mixing zone analyses are briefly discussed below.

4.3.2 Visual Plumes

Visual Plumes (VP) is a near field and mixing analysis suite developed by the U.S. EPA that is freely available.[1] Operating instructions, tutorials, and some theoretical discussion may be found in the Visual Plumes manual (Frick *et al.*, 2003) and in Frick (2004). In the United States, VP is often used to assess outfall performance in mixing zones to obtain or write NPDES permits, but it is also used for other purposes, such as:
- Near- and far-field plume modeling;
- Outfall design;
- Beach bacteria estimates;
- Plume contouring such as concentration, temperature, salinity;
- Brine and desalination discharges.

VP features include (for an extended discussion see Frick *et al.*, 2003):
- User specified units and automatic units conversion;
- Multi context columns with inconsistent units;
- Sensitivity analysis with changing inputs;
- Time-series inputs of effluent and ambient variables;

---

[1] http://www.epa.gov/ceampubl/swater/vplume/index.htm

- Conservative tidal background-pollutant build-up;
- A multi-stressor pathogen decay "Mancini model" to predict coliform mortality based on temperature, salinity, solar insolation, and water column light absorption.

VP can model submerged and surface buoyant discharges from single port and multiport diffusers (with merging plumes) into flowing, stratified waters. VP is a model platform with a graphical user interface that supports several independent models, including:
- UM3, a Lagrangian model for single and merging plumes;
- DKHW, an Eulerian model for single and merging plumes;
- NRFIELD (RSB), a semi-empirical model for multiport diffusers;
- PDSW, an Eulerian model for surface discharges.

The basic steps in using VP are shown in Figure 4.5. After entering the input data, any of the main models can be executed. In addition, other specialized sub-models and extensions can be invoked with any model, including the Brooks algorithm for far field diffusion, and the Mancini model for bacterial decay.

Input to VP is accomplished through tabs for the diffuser, ambient, and program settings. The diffuser and effluent variables are inputted to the diffuser tab, an example of which is shown in Figure 4.6. Variables include:
- Number of ports, diameter and depth;
- Port orientation to the vertical and horizontal;
- Total flowrate;
- Start, end, and time interval for time-series;
- Mixing zone distance;
- Contour plotting variables;
- Effluent salinity and temperature (or density);
- Effluent pollutant concentration.

Ambient variables are inputted on the ambient tab, an example of which is shown in Figure 4.7. Variables can be specified at multiple depths and can include:
- Current speed and direction;
- Salinity and temperature (or density);
- Background pollutant concentration;
- Decay rate;
- Far-field current speed and direction;
- Diffusion coefficient.

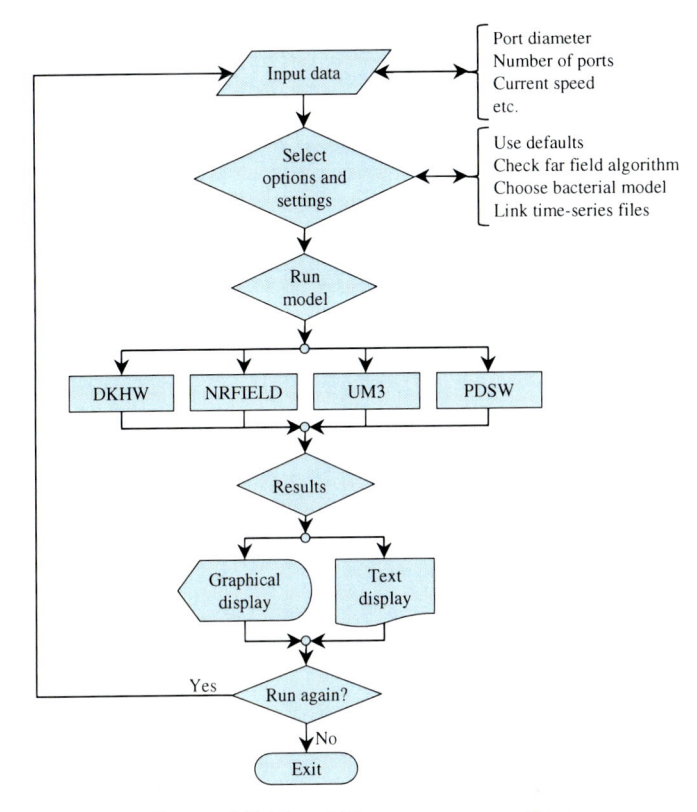

Modeling transport processes and water quality

Figure 4.5 Visual Plumes run procedure

Runs can be made in sequence, or as all possible combinations of cases.

Model outputs can be displayed in text format or graphically. A typical graphical output is shown in Figure 4.8 for a horizontal buoyant jet in a linear stratification (similar to that shown in Figure 3.18). Outputs for two models, UM3 and DKHW, are shown to illustrate how results for different cases and/or models can be compared. The graphs show the plume boundaries, the centerline trajectory in plan and elevation, average and centerline dilutions, and vertical density profile. The dilution (or tracer concentration) results can also be shown as contour plots, which are useful for comparing regions of elevated concentration to mixing zone dimensions.

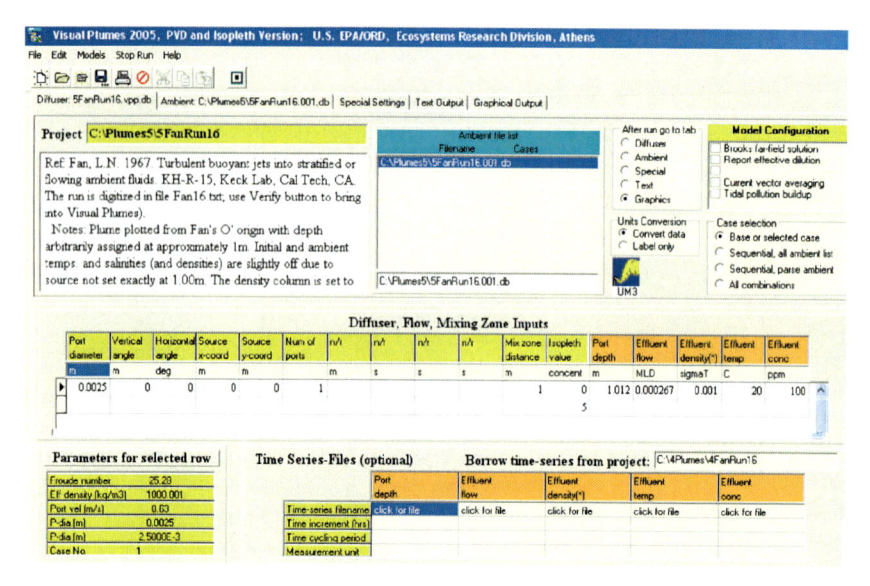

Figure 4.6 VP diffuser tab showing typical inputs

Figure 4.7 VP ambient tab showing typical inputs

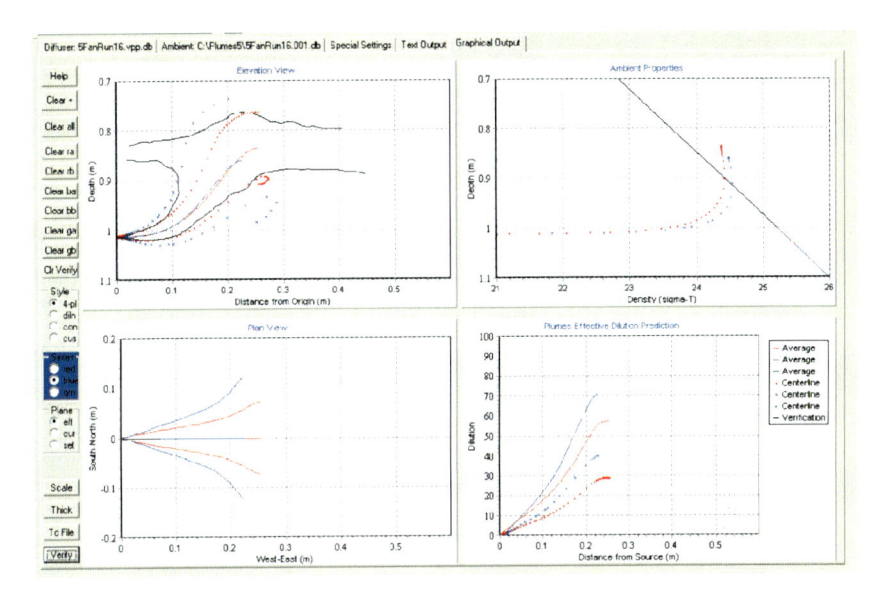

Modeling transport processes and water quality 107

Figure 4.8 Typical Visual Plumes graphics output

4.3.3 NRFIELD

NRFIELD is based on the experimental studies on multiport diffusers in density-stratified currents of arbitrary direction described in Roberts *et al.* (1989a,b,c) later updated with Tian *et al.* (2004a,b; 2006) and Daviero and Roberts (2006). These experiments cover a wide range of parameters typical of actual ocean outfalls. NRFIELD is available as part of Visual Plumes or as a stand-alone version. NRFIELD is specifically oriented to marine wastewater discharges and has been applied to many outfalls around the world.

The primary predictions of NRFIELD are the wastefield characteristics at the end of near field as defined in Figure 3.22. These include near field dilution, S_n, rise height to the top of the wastefield, z_m, thickness h_n, and length of the near field, x_n. The model includes the effects of: arbitrary current speed and direction (including parallel currents), density stratification, port spacing, source momentum flux, discharges from both sides of the diffuser (and the resulting merging of the plumes from both sides), re-entrainment and additional mixing in the spreading layer, direct plume impingement in parallel currents, and lateral gravitational spreading.

As with all models, the limitations of NRFIELD should be kept in mind. First, it is based on experiments with linear density stratifications. NRFIELD solves for the rise height in a non-linear stratification by an iterative procedure in which the density profile is linearized over the rise height. Comparisons with field and other laboratory measurements show this to be usually a good assumption. Second, the experiments are based on a uniform current, i.e. the current has constant speed and direction over the rise height. Coastal currents can sometimes exhibit strong shear in speed and direction. To account for this, depth-averaged currents over the plume rise height can be used. Third, the experiments are based on straight multiport diffusers. It should be kept in mind that the entrainment models are also ultimately based on coefficients derived from small scale laboratory experiments. Comparisons of NRFIELD predictions with field studies on outfalls in Boston, Rio de Janeiro, and San Francisco, showed the major wastefield characteristics to be well predicted.

NRFIELD executes very quickly and so can be used to run thousands of cases with long time series of current speed and direction, density stratifications, wastewater flowrate. This enables a much more reliable estimate of the statistical properties of wastefield such as rise height and dilution. This approach was taken in the modeling studies in support of the successful application for a waiver from secondary treatment for the San Diego outfall. For an example, see Section 4.7.2.

4.3.4 Cormix

CORMIX[2] is a proprietary expert system originally developed at Cornell University. It classifies the momentum and buoyancy of the discharge in relation to boundary interactions to predict mixing behavior. Boundary interactions can be surface or bottom contact or terminal layer formation in a stratified ambient. The hydrodynamic simulation system contains regional flow models that are based on integral, length scale, and passive diffusion approaches to simulate the hydrodynamics of near-field and far-field mixing. The algorithms provide rapid simulations for mixing zone problems with scales of meters to kilometers and time scales of seconds to hours.

Cormix has a graphics interface with tabs for input. The system analyses the data to determine the type of discharge involved and computes the appropriate length scales whose ratios classify the flow case. The results of experiments and field data are then used to generate algebraic expressions that make the predictions. Depending on the flow classification, CORMIX chooses the predictive equations to use. In some cases, it uses an integral model similar

---

[2] www.mixzon.com

to DKHW for the initial phase of plume development, and empirical algebraic expressions for the rest. CORMIX gives predictions for a wide array of discharge conditions and diffuser configurations. It simplifies most multiport diffusers to an equivalent slot discharge that has the same buoyancy and momentum fluxes as the actual diffuser. Output is in tabular and graphical format. More recent versions of CORMIX have expanded the graphical capability of the model and improved the hydraulic calculations. Because of the classification of flows into discrete cases, its predictions can have discontinuities and large changes in response to small changes in input parameters.

CORMIX has extensive input and graphics capabilities that are too extensive to discus here. For details, consult the website and manuals.

4.3.5 VISJET

VISJET was developed at the University of Hong Kong (Lee and Chu, 2003).[3] It is an interactive virtual reality-based initial mixing prediction model that combines the Lagrangian model JETLAG (Lam *et al.*, 2006) with visualization technology. VISJET is Windows-based and can be used for a wide range of effluent discharge and ambient environmental conditions. It has a number of graphic features that can be useful for studying environmental impact, plume merging in a stratified crossflow, and outfall design. It can be used to simulate and visualize merging buoyant jets from a rosette-type diffuser (for example, Figure 4.24) in a crossflow.

VISJET has been coupled with a far field model using a Distributed Entrainment Sink Approach (DESA) to predict effluent mixing and transport in the intermediate field. For further discussion, see Section 4.6.

The capabilities of VISJET are too extensive to discuss here, and the model's website should be consulted for more details.

4.4 FAR FIELD MODELS

4.4.1 Hydrodynamic models

In the past few decades hydrodynamic models of coastal circulation have matured and are being increasingly used to predict the fate and transport of coastal discharges. Most models have been two-dimensional (depth-averaged) and this may be adequate for fairly shallow (unstratified) waters. But in deeper

---

[3] http://www.aoe-water.hku.hk/visjet/

waters, especially if there are wind-shear effects, baroclinic processes, and density stratification, three-dimensional models are needed. In contrast to near field models, far field hydrodynamic models require extensive data input. These include currents, bathymetry, winds, density stratification, tides, and their spatial and temporal variability. Most models assume the vertical pressure distribution to be hydrostatic, and this is adequate for most coastal processes of interest. The models are either finite element, finite difference, or finite volume, of which finite difference is the most common. The models should be combined with field studies to ensure reliable results.

Ocean circulation models can be combined with mass transport models to predict contaminant transport. Examples are bacteriological pollution in nearshore areas due to storm water runoff (Carnelos, 2003) and marine outfalls during different flow conditions such as flood and ebb tides (Liu *et al.*, 2007). Hydrodynamic models have also been used to predict near field plume behavior (Blumberg *et al.*, 1996; Zhang, 1995).

Some commonly used ocean circulation models are listed in Table 4.2. These models are applicable to oceans, coastal waters, lakes, rivers, and estuaries. Some are commercial and some are open source (free). Free surface, terrain-following (sigma or *s*-coordinates) ocean models emerged about 20 years ago from the need to model turbulent processes in surface and bottom boundary layers and to simulate flows in estuaries and coastal regions. These efforts led to the development of models such as POM, ECOM, and ROMS. These models use curvilinear orthogonal horizontal coordinates, a horizontal numerical staggered "Arkawa-C" grid, and a vertical staggered grid with either a sigma or a more general *s*-coordinate system. Although the basics of the models are similar, there are considerable differences in numerics and parameterizations.

A widely used 3D hydrodynamic model is the Princeton Ocean Model (POM) (Blumberg and Mellor, 1983, 1987). POM is a nonlinear, fully three-dimensional, primitive equation, finite difference model that solves the heat, mass, and momentum conservation equations of fluid dynamics. The model includes the Mellor and Yamada (1982) level 2.5 turbulence closure parameterization.

Modeling transport processes and water quality 111

Table 4.2 Examples of hydrodynamic ocean circulation models

Name	Assumptions	TCM	Extensions/Application	Comments
Delft3D (FVM)	SW, HY	k-ε, k-L, algebraic, constant	GG, WQ, SD, W, PT, MD, EC	Delft Hydraulics. Commercial; pre- and post-processing packages are available
POM (FDM)	BO, HY	Mellor-Yamada	None	Blumberg and Mellor (1987). Open source, no pre or post-processing packages
ECOM	Based on POM	Mellor-Yamada	WQ, SD, W	HydroQual. Blumberg and Mellor (1987). Open source, updated version of POM, post processing packages are available
ROMS (FDM)	BO, HY	Mellor-Yamada, k-ε, k-w	TR, WQ, PT, ST, W	Rutgers University (1994). Open source, well documented. Updated POM, pre and post processing packages available
Mike (FVM)	BO, HY	Constant eddy viscosity, Smagorinsky subgrid, k-ε, mixed Smagorinsky/k-ε	TR, EC, ST, MT, WQ, PT	DHI Group. Commercial, pre- and post-processing graphical user interface packages are available
Telemac 3D (FEM)	BO, HY and non-HY	Mixing length	SD, WQ, W, GF, HA	Sogreah Consult. Commercial, pre- and post-processing packages are available
ELCOM (FDM)	BO, HY	Eddy-viscosity or mixed layer	CAEDYM (WQ)	CWR, University of Western Australia, open source, well documented

BO = Boussinesq Approximation
EC = Ecology and Water Quality Model
EFT = Ekman Flow Theory
FEM = Finite Element Method
FDM = Finite Difference Method
FVM = Finite Volume Method
GG = Grid Generator
HA = Harbor Agitation
HY = Hydrostatic Assumption
WQ = Water Quality
LWT = Long Wave Theory

MT = Mud Transport
PT = Particle Tracking
RANS = Reynolds-Averaged Navier-Stokes
RLT = Rigid Lid Theory
SD = Sediment Transport Module
SW = Shallow Water Approximation
TCM = Turbulence Closure Model
TR = Transport Module
UF = Groundwater Flow
W = Wave

Most models assume incompressibility and are hydrostatic and Boussinesq, so that density variations are neglected except where they are multiplied by gravity in the buoyancy force terms. The basic equations of a hydrodynamic model (such as POM) are based on continuity, momentum, and thermodynamics including temperature and salinity, and an equation of state as follows:

$$\nabla \cdot \vec{V} + \frac{\partial W}{\partial z} = 0 \tag{4.8}$$

$$\frac{\partial U}{\partial t} + \vec{V} \cdot \nabla U + W \frac{\partial U}{\partial z} - fV = -\frac{1}{\rho_0} \frac{\partial P}{\partial x} + \frac{\partial}{\partial z}\left[\varepsilon_M \frac{\partial U}{\partial z}\right] + F_X \tag{4.9}$$

$$\frac{\partial V}{\partial t} + \vec{V} \cdot \nabla V + W \frac{\partial V}{\partial z} - fU = -\frac{1}{\rho_0} \frac{\partial P}{\partial y} + \frac{\partial}{\partial z}\left[\varepsilon_M \frac{\partial V}{\partial z}\right] + F_Y \tag{4.10}$$

$$\frac{\partial \theta}{\partial t} + \vec{V} \cdot \nabla \theta + W \frac{\partial \theta}{\partial z} = \frac{\partial}{\partial z}\left[\varepsilon_H \frac{\partial \theta}{\partial z}\right] + F_\theta \tag{4.11}$$

$$\frac{\partial S}{\partial t} + \vec{V} \cdot \nabla S + W \frac{\partial S}{\partial z} = \frac{\partial}{\partial z}\left[\varepsilon_H \frac{\partial S}{\partial z}\right] + F_S \tag{4.12}$$

$$\rho = \rho(\theta, S) \tag{4.13}$$

where ε_M is the vertical eddy diffusivity for momentum and ε_H is the vertical eddy diffusivity for heat and salt. The horizontal viscosity and diffusion terms are:

$$F_x = \frac{\partial}{\partial x}\left[2A_M \frac{\partial U}{\partial x}\right] + \frac{\partial}{\partial y}\left[A_M \left(\frac{\partial U}{\partial y} + \frac{\partial V}{\partial x}\right)\right] \tag{4.14}$$

$$F_y = \frac{\partial}{\partial y}\left[2A_M \frac{\partial V}{\partial y}\right] + \frac{\partial}{\partial x}\left[A_M \left(\frac{\partial U}{\partial y} + \frac{\partial V}{\partial x}\right)\right] \tag{4.15}$$

$$F_{\theta,S} = \frac{\partial}{\partial x}\left[2A_M \frac{\partial(\theta,S)}{\partial x}\right] + \frac{\partial}{\partial y}\left[A_H \frac{\partial(\theta,S)}{\partial x}\right] \tag{4.16}$$

$$A_M = C\Delta x \Delta y \frac{1}{2}\left|\nabla V + (\nabla V)^T\right| \tag{4.17}$$

A_M and A_H are horizontal eddy diffusivities that damp small-scale computational noise. Horizontal momentum diffusion is commonly assumed to be equal to horizontal thermal diffusion where the primary mixing process is eddy diffusion. The Smagorinsky diffusivity, A_M, is small for computations with high resolution and small velocity gradient.

A turbulence closure sub-model is defined in terms of kinetic energy (q^2). The turbulence field is described by prognostic equations for the turbulence kinetic energy and turbulence length scale as:

$$\frac{\partial q^2}{\partial x} + \vec{V} \cdot \nabla q^2 + W\frac{\partial q^2}{\partial z} = \frac{\partial}{\partial z}\left[K_q \frac{\partial q^2}{\partial z}\right] + 2K_M\left[\left(\frac{\partial U}{\partial z}\right)^2 + \left(\frac{\partial V}{\partial z}\right)^2\right]$$
$$+ \frac{2g}{\rho_o}K_H\frac{\partial \rho}{\partial x} - \frac{2q^3}{B_1 l} + F_q \qquad (4.18)$$

and

$$\frac{\partial(q^2 l)}{\partial t} + \vec{V} \cdot \nabla(q^2 l) + W\frac{\partial(q^2 l)}{\partial z} = \frac{\partial}{\partial z}\left[K_q \frac{\partial(q^2 l)}{\partial z}\right]$$
$$+ 1E_1 K_M\left[\left(\frac{\partial U}{\partial z}\right)^2 + \left(\frac{\partial V}{\partial z}\right)^2\right] + \frac{1E_1 g}{\rho_o}K_H\frac{\partial \rho}{\partial z} - \frac{q^3}{B_1}\tilde{W} + F_1 \qquad (4.19)$$

$$K_M \equiv lqS_M; \quad K_H \equiv lqS_H; \quad K_q \equiv lqS_q \qquad (4.20)$$

where S_M, S_H and S_q are vertical eddy diffusivities.

The governing equations are solved numerically using the finite difference method on a staggered grid. The model uses time-dependent wind stress and heat flux forcing at the surface, zero heat flux at the bottom, free-slip lateral boundary conditions, and quadratic bottom friction.

Three-dimensional models are probably needed for waters deeper than about 30 m or so that are stratified. This is because the currents can be strongly sheared, not only flowing at different speeds over depth but in different directions also; two-dimensional models would not capture this variability. But for reliable results, three-dimensional models require extensive data on currents and density at the boundaries and intensive efforts to set up and verify. For these reasons they are not commonly used for smaller outfall projects, but may be part of larger ones.

4.4.2 Nesting technique

Specification of the boundary conditions for hydrodynamic models of open coastal waters is major problem. Detailed information is needed on the variations of currents and water level, stratification, and other parameters and their temporal variations around these boundaries. Due to computational restrictions, it is usually not practical to model an area large enough that the area of interest is independent of these boundary conditions. Therefore, a common approach is to model a large area with a coarse grid and to embed a finer-scale model within it. The grid size of the smaller model is small enough to resolve scales of interest to outfall dispersion. The fine-grid model derives its boundary conditions from the larger model and is said to be nested within it.

An example of nesting is the models used to predict the transport of pollutants from the Grand River, Michigan, to adjacent beaches in Lake Michigan (Nekouee et al., 2007). Circulation and thermal structure in the whole lake was modeled in three dimensions on a coarse grid (2 km resolution) using a version of POM (POMGL) that has been developed and adapted for the Great Lakes (Schwab and Bedford, 1994; Beletski and Schwab, 2001). The model employs a terrain-following vertical coordinate system (sigma-coordinate) with 20 vertical levels (sigma levels, which represent a proportion of a vertical column) with finer spacing near the surface and the bottom. A nested model with a domain of 6×24 km and a 100 m grid size was implemented around the river mouth (Figure 4.9). The model uses the 3D currents and temperatures extracted from the coarse grid model along its open boundaries. The models are calibrated by comparing their predictions with measured currents. The nested model provides more accurate hydrodynamics in higher resolution as a basis for a mass transport model.

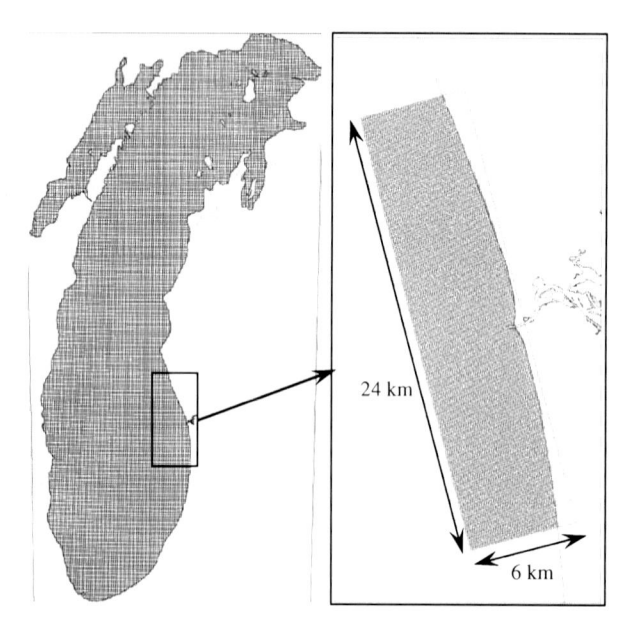

Figure 4.9 Whole-lake simulation with a 2 km grid (left) and the nested simulation with a 100 m grid (right) (Nekouee et al., 2007)

4.5 WATER QUALITY MODELS

4.5.1 Introduction

The two main mechanisms for transporting dissolved constituents in coastal waters are transport due to the flow velocity, usually called advection, and turbulent diffusion, due to random velocity fluctuations. Advection is a very important transport mechanism, so reliable knowledge of currents is essential. This may be obtained by a hydrodynamic model, as described in the previous section, current measurements, or, best of all, both. Other chemical and biological transformations may also occur.

A water quality model predicts the far field transport of constituents contained in the wastewater and their chemical or biological transformations. The objective is to predict pollutant concentrations and their temporal variations, in other words water quality. Two main model types are used, gradient diffusion and particle tracking, either of which may be Eulerian or Lagrangian. The models are reviewed briefly below.

4.5.2 Eulerian models

Eulerian concentration models numerically solve the mass transport equation on a fixed grid. The beginning point is usually the time-averaged mass conservation equation:

$$\frac{\partial c_k}{\partial t} + u\frac{\partial c_k}{\partial x} + v\frac{\partial c_k}{\partial y} + w\frac{\partial c_k}{\partial z} = \frac{\partial}{\partial x}\left(\varepsilon_x \frac{\partial c_k}{\partial x}\right) + \frac{\partial}{\partial y}\left(\varepsilon_y \frac{\partial c_k}{\partial y}\right) + \frac{\partial}{\partial z}\left(\varepsilon_z \frac{\partial c_k}{\partial z}\right)$$
$$\pm S_k + W_k \qquad (4.21)$$

where c_k is the concentration of the kth constituent, S_k is production of formation of species due to chemical or other processes, and W_k is a source term. ε_x, ε_y, ε_z are turbulent diffusion coefficients in the z, y, and z directions, respectively. The diffusion coefficients may be calculated from a hydrodynamic model. In order to solve Eq. (4.21), information is first needed on the current field (assuming the water quality constituents don't influence the current field). This is usually obtained from a hydrodynamic model such as discussed in Section 4.4.1. The hydrodynamic model is run first, and then the resulting current field is inputted to the water quality model. Equation (4.21) is then solved numerically for the pollutant constituents of interest with a separate module. An example is Delft3D-WAQ.

Models of biogeochemical processes to predict, for example, nutrients, eutrophication, and dissolved oxygen (DO) via the source, sink, and reaction terms in Eq. (4.21), can be very complex. An example is CAEDYM[4] (Computational Aquatic Ecosystem DYnamic Model) which is a freely available model developed at the University of Western Australia that simulates the major processes influencing water quality. It is a set of routines for primary and secondary production, nutrient and metal cycling, oxygen dynamics, and sediment movement. It may be run independently or coupled with hydrodynamic models such as ELCOM. The biological model includes up to seven phytoplankton groups, five zooplankton groups, six fish groups, four macroalgae groups, three invertebrate groups, and models for seagrass and jellyfish. Phytoplankton growth is dependent on both environmental factors and internal nutrient storage. Each phytoplankton group has its own set of coefficients to define its characteristics associated with the maximum growth rate, limiting factors, and internal nutrient storage. DO processes include exchange at the air/water interface and the sediment/water interface, oxygen production and consumption, water column biochemical oxygen demand, nitrification, and respiration of higher organisms. The models for metals such as iron, manganese, and aluminum are similar and simulate processes that include oxidation, reduction, release from sediments and settling and resuspension.

4.5.3 Lagrangian models

Eulerian gradient concentration models based on Eq. (4.21) are widely used in coastal areas and estuaries. They give erroneous results near the source, however, due to high numerical diffusion, and cannot resolve concentrations on scales smaller than the grid size. Lagrangian models bypass these problems and are particularly useful for ocean outfalls. They either follow puffs of material that are allowed to advect and diffuse according to some solution of the diffusion equation (for example, FRFIELD, Section 3.3.3.3), or track multiple particle releases.

Particle tracking models are also known as random walk particle tracking (RWPT) models. The discharge is represented by a number of particles that are advected (transported) by the local current with a simple random walk formulation to represent turbulent diffusion. The flow field (currents) can be either supplied by a 3D hydrodynamic model or measured by a recording device such as an ADCP. To determine concentration, a three dimensional grid is overlain on the simulated area and the concentration in each grid cell is computed from the number of particles in the cell.

---

[4] //www.cwr.uwa.edu.au/services/models.php?mdid = 3

The particle-based approach is robust and has a number of advantages over Eulerian models. It is not subject to artificial diffusion near sharp concentration gradients. It retains exact mass conservation. The particles can be assigned properties, such as mass and age, which makes the method particularly well suited to bacterial predictions. The sources are easily represented and can be readily coupled to the near field model (see the next section for further discussion of coupling). But it also has shortcomings. It is not well suited to prediction of water quality parameters such as nutrients that involve chemical reactions between constituents. The numbers of particles may be restricted by memory and computation time. RWPT models have been extensively used in ocean outfall design and predicting wastewater transport in coastal areas; examples are the models of Chin and Roberts (1985) and Zhang (1995).

A recent example is the particle tracking model used for predicting nearshore bacteria in Lake Michigan that was discussed above. The model is illustrated in Figure 4.10.

Figure 4.10 Conceptual diagram for the proposed hybrid model

The domain is represented as an array of square boxes (Bennett and Clites, 1987; Beletsky *et al.*, 2006). The particles are introduced at the end of the near field and are advected, diffused, and decayed over the nested domain. The z-coordinates are transformed to the sigma coordinate system before the tracers are transported, and after particle propagation are transformed back. The procedure is illustrated in Figure 4.11.

At each time step the particle displacement is computed from an advective, deterministic component and an independent, random Markovian component. The deterministic component is taken from the POMGL simulations. The random velocities that represent diffusion are defined as a function of the turbulent diffusion coefficients as:

$$\begin{Bmatrix} u_i \\ v_i \\ w_i \end{Bmatrix} = \begin{Bmatrix} u_{adv}(x,y,z,t) \\ v_{adv}(x,y,z,t) \\ w_{adv}(x,y,z,t) \end{Bmatrix} + \begin{Bmatrix} u'_i \\ v'_i \\ w'_i \end{Bmatrix} \qquad (4.22)$$

where

$$\begin{Bmatrix} u'_i \\ v'_i \\ w'_i \end{Bmatrix} = \frac{1}{\sqrt{\Delta t}} \begin{Bmatrix} \xi_i \sqrt{2\varepsilon_h} \\ \xi'_i \sqrt{2\varepsilon_h} \\ \xi''_i \sqrt{2\varepsilon_v} \end{Bmatrix} \qquad (4.23)$$

```
Start
  │
  ▼
┌─────────────────────────────────────────────────┐
│            Hydrodynamic model                   │
│   ┌──────────────────────────────────┐          │
│   │    Simulate the whole lake       │          │
│   └──────────────────────────────────┘          │
│   ┌──────────────────────────────────────────┐  │
│   │ Input the boundary conditions on the     │  │
│   │ nested domain                            │  │
│   └──────────────────────────────────────────┘  │
│   ┌──────────────────────────────────────────┐  │
│   │ Output the ambient temperature and       │  │
│   │ current results                          │  │
│   └──────────────────────────────────────────┘  │
└─────────────────────────────────────────────────┘
  │
  ▼
┌─────────────────────────────────────────────────┐
│              Near field model                   │
│   ┌──────────────────────────────────────────┐  │
│   │ Input the ambient and discharge variables│  │
│   └──────────────────────────────────────────┘  │
│   ┌──────────────────────────────────────────┐  │
│   │   Compute dilution and trajectory        │  │
│   └──────────────────────────────────────────┘  │
└─────────────────────────────────────────────────┘
  │
  ▼
┌─────────────────────────────────────────────────┐
│            Particle tracking model              │
│   ┌──────────────────────────────────────────┐  │
│   │  Introduce particle at the end of the NF │  │
│   └──────────────────────────────────────────┘  │
│   ┌──────────────────────────────────────────┐  │
│   │  Input velocities from hydrodynamic model│  │
│   └──────────────────────────────────────────┘  │
│   ┌──────────────────────────────────────────┐  │
│   │       Advect, diffuse and decay          │  │
│   └──────────────────────────────────────────┘  │
│   ┌──────────────────────────────────────────┐  │
│   │    Convert particles to concentration    │  │
│   └──────────────────────────────────────────┘  │
└─────────────────────────────────────────────────┘
  │
  ▼
 End
```

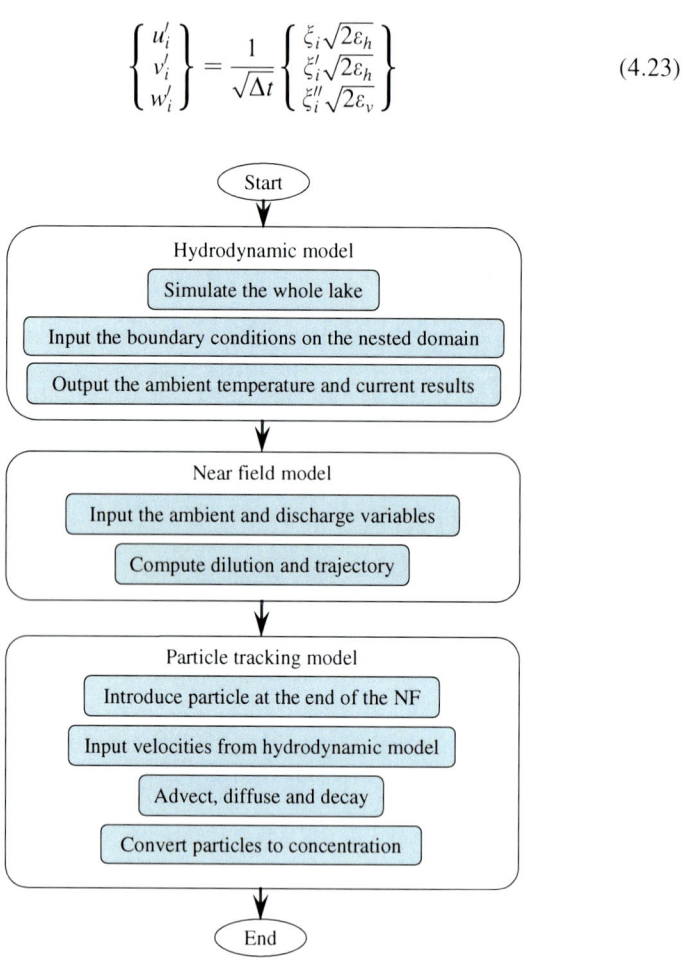

Figure 4.11 Flow chart of the proposed hybrid method

The particle coordinates at the end of each time step are:

$$x(t+1) = x(t) + u(t)\Delta t + \xi_i\sqrt{2\varepsilon_h \Delta t} \qquad (4.24)$$

$$y(t+1) = y(t) + v(t)\Delta t + \xi_i'\sqrt{2\varepsilon_h \Delta t} \qquad (4.25)$$

$$z(t+1) = z(t) + w(t)\Delta t + \xi_i''\sqrt{2\varepsilon_v \Delta t} \qquad (4.26)$$

where ξ, ξ', ξ'', and γ are independent random numbers in the range $[-1, 1]$ (Chin and Roberts, 1985), and ε_h and ε_v are the horizontal and vertical diffusion coefficients, respectively. In some studies two other terms (e.g. in the x direction $(d\varepsilon_h/dx)\Delta t$ and $(\varepsilon_h/h)(dh/dx)\Delta t$) have been added to the stochastic displacements. These terms represent artificial velocities due to the rate of change of the horizontal dispersion coefficient and to uneven bathymetry (Dimou, 1992; Suh, 2006).

4.6 MODEL COUPLING

The need for separate near and far field models brings about a somewhat complex and difficult problem – how to introduce the output from the near field model into the far field model? This procedure is known as coupling, in which the flow quantities, such as volume, momentum, and pollutant mass are transferred between the models, possibly in both directions.

The problem is illustrated in Figure 4.12. The near field dynamics are characterized by entrainment and small-scale turbulence. The plume entrains fluid that induces a current around the diffuser. For a typical long multiport outfall, the velocity of this current is approximately $0.3b^{1/3}$ (Roberts, 1979) where b (Table 3.1) is the buoyancy flux per unit length of the diffuser. This will usually be a few cm/s and its magnitude decreases with distance from the diffuser so it will generally be negligible compared to ambient currents. Also, the net momentum flux of a diffuser is almost always zero (the momentum of the jets from the opposing sides of the diffuser cancels out). Therefore, typical wastewater outfalls do not significantly affect coastal circulation patterns (this may not be true for large cooling water discharges from power plants). The coupling is therefore usually considered to be one way, i.e. local currents affect the discharge, but not vice versa. The main question then is how, and where, to introduce the effluent flow and its pollutant mass into the far field model. A number of studies have considered this problem.

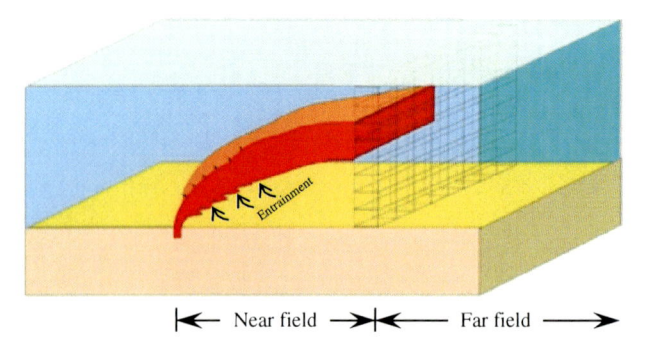

Figure 4.12 Model coupling (after Bleninger, 2006)

Chin and Roberts (1985) coupled a near field model with a far field particle tracking model (see Section 4.5.3). Their approach is illustrated in Figure 4.13. The source is defined as a vertical rectangular plane whose width is assumed equal to the diffuser length, L (or the width at the end of the near field computed by the methods of Section 3.3.2.7) and height equal to the wastefield thickness at the end of the near field. Mass flux is normal to the plane and distributed over it. The rectangular source configuration can be further discretized into rectangular elements as shown. Chin and Roberts used ULINE (a predecessor to NRFIELD) to predict the near field plume properties. A similar approach was used by Zhang (1995) who adapted a version of NRFIELD and coupled it to the far field hydrodynamic model ECOM. Zhang discusses different means of introducing the effluent into the far field grid.

The coupling problem has also been avoided by using a far field model to compute near field properties directly. Blumberg et al. (1996) applied the hydrodynamic model ECOMsi to predict the near field characteristics of the Boston outfall. The source was put into a bottom grid cell. They found good agreement with predictions of plume rise height and dilution compared to ULINE. Zhang and Adams (1999) also used ECOMsi to predict near field plume characteristics, and compared the results to NRFIELD (RSB) predictions. They found that rise height and dilution were reasonably predicted. Because far field models do not properly reproduce the physics of the near field processes, they can only give approximate estimates of near field plume properties.

Connolly et al. (1999) used a hybrid modeling approach to predict bacterial impacts from outfalls in Mamala Bay, Hawaii. They constructed a three-dimensional hydrodynamic model using ECOM to simulate advective and dispersive processes in the bay. NRFIELD was run with the measured density stratification and currents (see Section 4.2.7) to predict the time-variable structure of the effluent plume in the water column. The predicted near field characteristics were directly inputted into grid cells at the predicted plume rise height following the methodology of Zhang and Adams (1999).

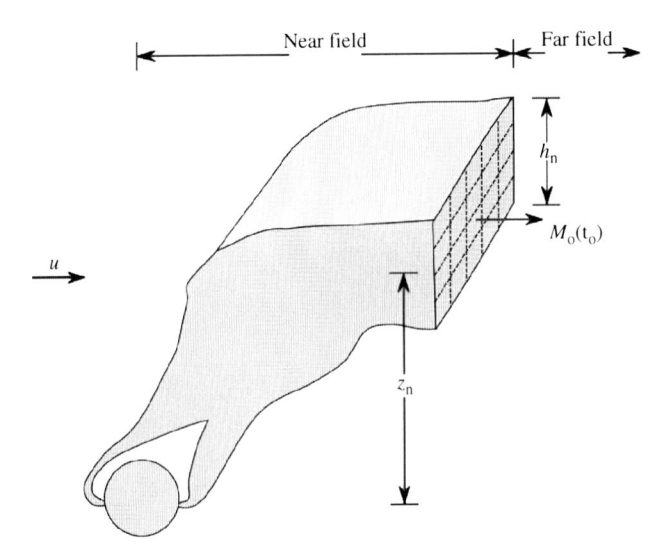

Figure 4.13 Coupled particle tracking model of Chin and Roberts (1985)

The far field statistical puff model described in Section 3.3.3.3 can be easily coupled to a near field model. This is particularly useful for predicting bacterial impacts. The near field model is first run to predict plume properties using time series of density stratification and currents. For each time step, the bacteria concentration is computed as the concentration in the raw wastewater divided by the near field dilution. This becomes the source concentration co for Brooks' far field model. "Puffs" of effluent with this concentration are then advected by the ocean currents and followed for a fixed period. As the puffs travel, the bacteria diffuse and decay according to Eq. (3.33). Thousands of releases can be readily simulated to build up a statistical picture of the spatial variation of bacterial levels around the discharge. For an example of this procedure, see Section 4.2.7.

Dynamic, i.e. two-way, linkage between the near and intermediate fields was addressed by Choi and Lee (2007). They applied a distributed entrainment sink approach (DESA) to model the intermediate field by coupling a 3D far field model with a Lagrangian near field model (JETLAG). The action of the plume on the surrounding flow is modeled by a distribution of sinks along the jet trajectory, and a diluted source flow is injected at the predicted terminal rise height. This establishes a two-way dynamic link at grid cell level between the near and far field models. The predictions agreed well with laboratory experiments, including the interaction of a confined rising plume with ambient stratification, and the mixing of a line plume in a cross-flow.

Bleninger (2006) describes an approach in which output from the near field model CORMIX is linked to a far field hydrodynamic model, Delft3D. Bleninger assumes passive, i.e. one-way, coupling. The source is introduced into the far field grid cells as a volume flux that is equal to the source volume flux multiplied by the near field dilution with a contaminant concentration equal to the source concentration divided by the near field dilution. Although this preserves the contaminant mass flux, it does not satisfy volume continuity as the entrained flow is not removed from any cells. As discussed above this is usually a good assumption for marine wastewater outfalls.

A similar approach was used by Roberts and Villegas (2006) for the Cartagena outfall (see Section 4.7.4) except that NRFIELD was used to predict the near field dilutions directly from the measured field data. Far field diffusion and bacterial decay was computed from the water quality module Delft3D-WAQ.

Ding *et al.* (2008) discuss the application of a 3D hydrodynamic model (ELCOM) with an embedded near field model to the design of the proposed tunneled outfall for Los Angeles. The near-field model needs the local ambient conditions, particularly the velocity and density profiles. These are extracted from the simulation at the beginning of a time step from a vertical column of grid cells above the diffuser. The near field model is then applied and the average dilution along the plume trajectory in each cell is computed. Water is withdrawn from each of these cells based on this dilution and mixed with the effluent to form the effluent plume that is then passed to the cell above. Finally, the diluted effluent is inserted into the cell where the near field model indicates trapping or surfacing. Flow rate, temperature, salinity, and tracer concentrations within the inserted inflow are determined by mass conservation.

Swanson and Isaji (2008) simulated total and fecal coliforms and enterococcus impacts from the same outfall using a Lagrangian particle tracking far field model. NRFIELD was first run using actual measured data that included stratification from thermistor strings and currents from ADCPs. Particles were then injected at the predicted plume rise heights. One million particles were released for each simulation and each particle was assigned a mass based on the effluent flow rate and predicted dilution. The currents in the far field were obtained from the extensive ADCP data that was available, with interpolation between the mooring locations. Bacterial decay was assumed to depend on solar radiation and depth. The horizontal diffusion coefficient was assumed to be constant and equal to about 8 m^2/s, and the vertical diffusion coefficient was assumed to be dependent on local density stratification, equal to 40 cm^2/s during low stratification, and 0.3 cm^2/s during high stratification. Note that vertical diffusion is much slower than horizontal because of density stratification.

Suitable coupling between the near and far field models is essential for reliable prediction of outfall impacts. If near field dilution is not accounted for, predicted far field dilutions will be much too low, leading to considerable overestimates of environmental impacts. If the correct plume rise height is not used, far field transports can be significantly in error. As this review has shown, numerous approaches to the coupling problem have been adopted. Simplest is coupling to a far field Lagrangian model, where satisfying continuity of entrained flow is not an issue. Even if continuity is not strictly satisfied, however, the effect on local currents is usually small. Because near field plume behavior is very dependent on local density stratification, the models that take a "hybrid" approach and use actual measured density stratifications, rather than modeled values, are particularly useful.

4.7 NUMERICAL MODELING CASE STUDIES

4.7.1 Introduction

Case studies of numerical modeling of ocean outfalls are discussed at various points in this book. Here we summarize some to illustrate modeling techniques and the issues involved and particularly their implications for outfall design and environmental impact.

4.7.2 Mamala Bay

Mamala Bay, Hawaii, was the subject of intensive studies related to the potential impacts and treatment upgrades of two wastewater outfalls (Colwell *et al.*, 1996). As previously discussed, Connolly *et al.* (1999) developed a three-dimensional hydrodynamic model of the flow in the bay and around the island of Oahu. Extensive field measurements were undertaken, including measurements of currents at several locations by ADPCs and electromagnetic current meters, and density stratification by recording thermistor strings. The following is a summary of the modeling discussed in Roberts (1999a,b).

The bathymetry, location of the outfalls and instrument mooring sites are shown in Figure 4.14. An ADCP and a moored thermistor string that recorded temperatures at several depths through the water column were placed near the Sand Island outfall at mooring D2. The data are discussed in Hamilton *et al.* (1995) and Petrenko *et al.* (2001). The current speed and direction and density stratification over the water column varied widely over a range of time scales from minutes to months and longer. This would cause the near field plume behavior to vary widely, and rapidly.

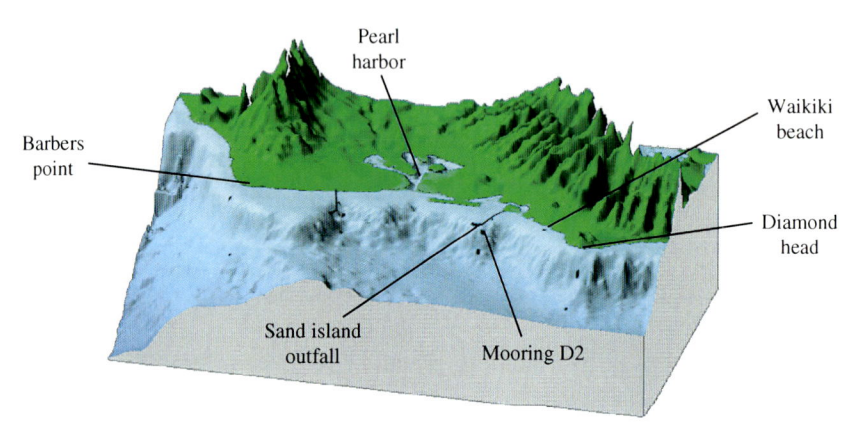

Figure 4.14 Bathymetry, outfalls, and mooring locations in Mamala Bay

To model this effect, the data were used as input to NRFIELD using the procedure shown in Figure 4.15. Time series of current speed and direction (measured by the ADCP), density profile (measured by the thermistor string), and wastewater flowrate were input at 30 minute intervals. Over 20,000 simulations for a period of almost one year were run.

Some typical measurements of currents and temperature difference over the water column and the resulting plume behavior are shown in Figure 4.16. The currents are tidal with significant unsteady low frequency content. The temperature difference over the water column varies widely, from essentially zero to 5°C over a few hours. This is caused by the divergence of the diurnal tidal wave around the island and then convergence in Mamala Bay which causes a large vertical movement of the thermocline. This results in the density stratification varying rapidly from strongly stratified to homogeneous. Finally, the wastewater flow rate also varies diurnally. These simulation results are the same as those used by Connolly *et al.* (1999) with a hydrodynamic model that were discussed in Section 4.6.

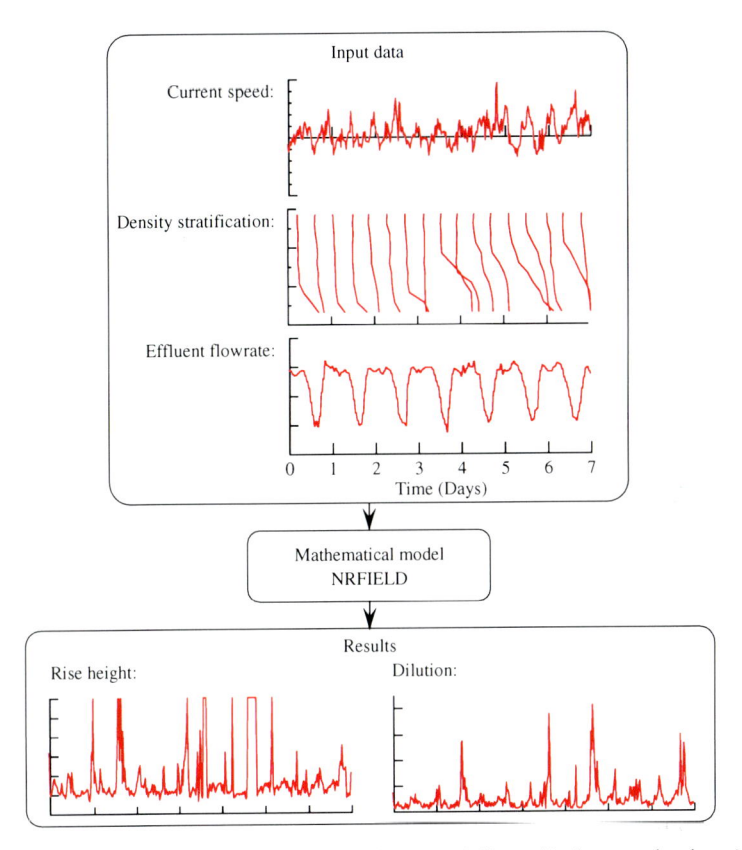

Modeling transport processes and water quality 125

Figure 4.15 Schematic depiction of near field modeling with time-series input data

The varying conditions significantly affect the near field plume hydrodynamics. Dilutions range from about 100:1 to several thousand and rise heights from deeply submerged to surfacing. The high dilutions occur when peak current speed, weak stratification, and low flow rates coincide. Conversely, low dilutions occur at slack water, high stratification, and high flow rates. The highest predicted dilution exceeded 4,000, and the median was 320. The median rise height was 25 m (the water depth is about 70 m, so the plume is submerged about 45 m below the water surface). The plume is submerged most of the time, surfaces briefly when the water column becomes homogenous, and then submerges again. The plume was predicted to be submerged 89% of the time, or conversely to surface about 11% of the time. For reasons discussed in

126   Marine Wastewater Outfalls and Treatment Systems

Roberts (1999a) this is believed to be a conservative estimate of plume surfacing. These results show the extreme variability of dilution and the impossibility of assigning a single number to it.

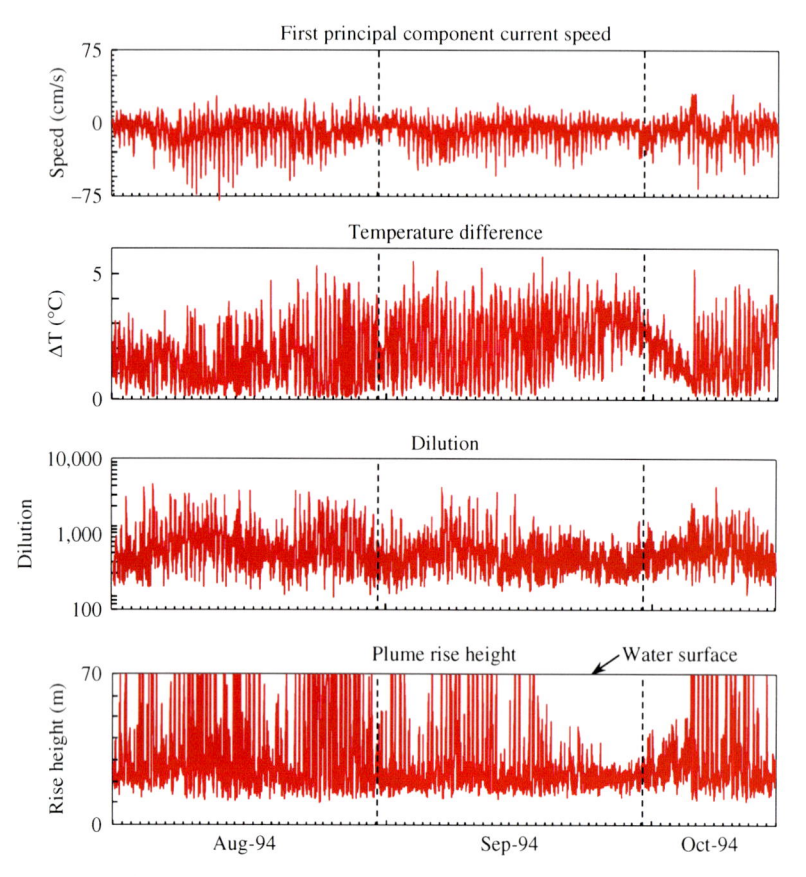

Figure 4.16 Measured currents, temperature difference over water column, dilution, and plume rise height for the Sand Island, Hawaii, outfall

Because surfacing plumes are most likely to impact the shoreline, their far field transport and shoreline impact was then modeled with the statistical model discussed in Section 3.3.3.3. When the plume was predicted to surface, it was advected as a puff using the measured currents in the near surface ADCP bin. The resulting visitation frequencies were shown in Figure 3.30. To predict bacteria, far field diffusion was computed by Eq. (3.33) and bacteria were assumed to decay according to a first order process (Eq. 3.46) with T_{90} varying diurnally from 6.9 hrs during daytime and infinity (i.e. no decay) at night. The resulting frequency with which total coliforms exceed 10,000 per 100 ml are shown in Figure 4.17. The bacteria levels decrease very rapidly with distance from the diffuser and the probability of exceeding shoreline bacterial standards is very small. Similar plots for an exceedance level of 1,000 per 100 ml are shown in Roberts (1999b) where it is concluded that the bathing water standard of less than 20% exceedance (the standard at that time) would be met right over the diffuser. For another example of this procedure, and other assumptions concerning the diurnal variability of T_{90}, see the discussion of the Cartagena outfall in Chapter 14.

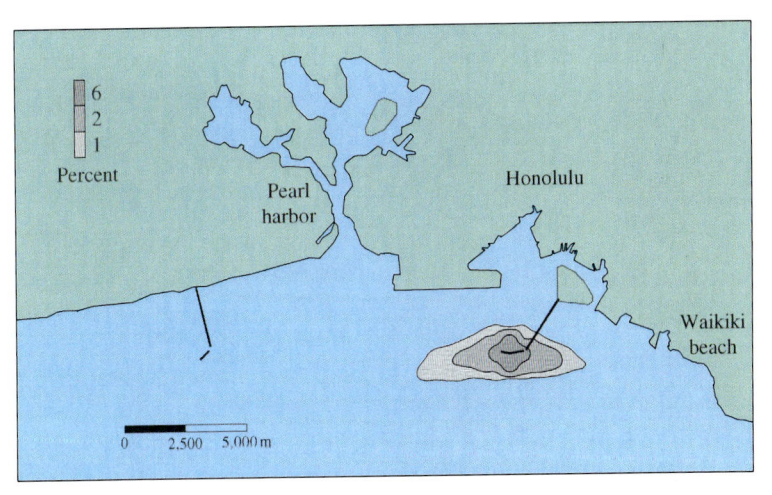

Figure 4.17 Frequency of total coliforms exceeding 10,000 per 100 ml for the Sand Island outfall, Hawaii (after Roberts, 1999b)

Finally, the long-term flushing model (Section 3.3.4) was applied. It was predicted that flushing, horizontal diffusion, and decay would result in high dilutions with no significant contaminant accumulation.

4.7.3 Rio de Janiero

An example of application of a hydrodynamic model is Carvalho (2003) who used a two-dimensional depth-averaged model to simulate circulation of the bay and coastal waters near Rio de Janiero as shown in Figure 4.18. He predicted the behavior of the Ipanema beach outfall with a near field model and far field dispersion with a Lagrangian particle tracking model. Bacterial decay was predicted as a function of several variables, including plume submergence (Carvalho et al., 2006). Examples of simulated coliforms at various periods in the tidal cycle are shown in Figure 4.19.

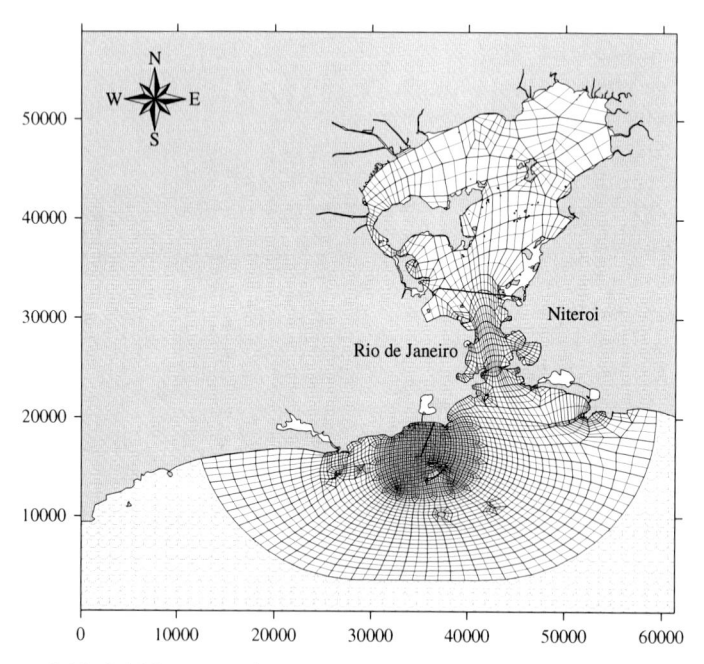

Figure 4.18 Grid for numerical hydrodynamic model of the Ipanema outfall (Carvalho, 2003)

The dispersion of the sewage field was evaluated for oceanographic characteristics typical of the different seasons. It was predicted that the Brazilian primary water contact standards would be met in all cases at the beaches. The plume can extend for long distances, more than 25 km, along the coast, however, and can be trapped by stratification and restrained by the Cagarras Island.

4.7.4 Cartagena

The proposed Cartagena, Colombia, outfall is discussed as a case study in Chapter 14. Its bacterial impacts were simulated using two approaches: Lagrangian far field modeling using measured currents at one location, and three-dimensional hydrodynamic of the adjacent coastal waters. The Lagrangian and near field time-series modeling are discussed in Section 14.2.4. The hydrodynamic modeling is discussed below.

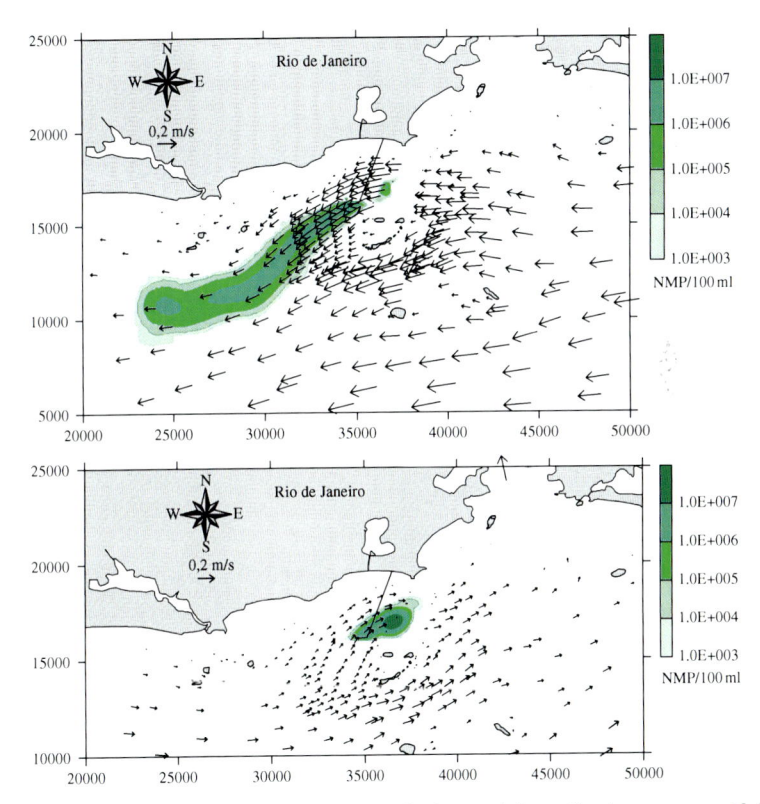

Figure 4.19 Simulations of coliforms discharged from the Ipanema outfall (Carvalho, 2003)

Two three-dimensional hydrodynamic models were used: a large-scale model and a nested model. The model domains are shown in Figure 4.20. The large scale model, ELCOM, was used to predict regional-scale currents in the Caribbean. Its domain was large enough that the open-ocean boundary conditions would not affect the results in the vicinity of Cartagena. The small scale, nested, model was Delft3D whose purpose was to predict the fate and transport of the wastefield in the vicinity of the proposed outfall. The horizontal and vertical sigma-coordinate grids for Delft3D are shown in Figure 4.21.

130   Marine Wastewater Outfalls and Treatment Systems

The boundary conditions for the nested model consisted of the measured currents spatially modified according to the ELCOM predictions. Locally measured winds and water levels were used in addition to the measured currents to drive the model. Modeling was conducted for the two main seasonal conditions: dry and wet that included the highest occurrence of onshore directions and would therefore be "worst-case" conditions.

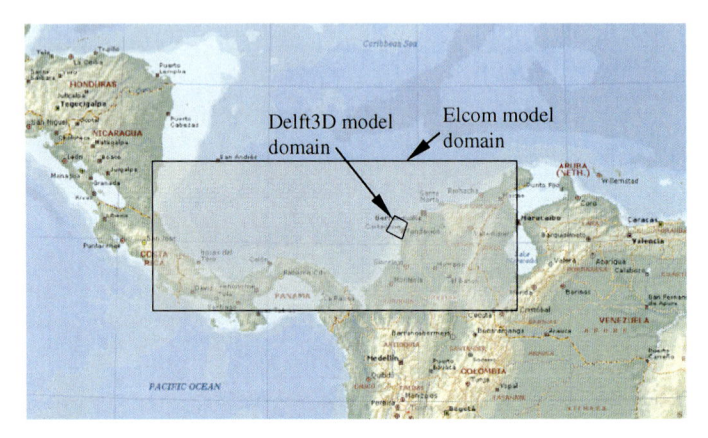

Figure 4.20 Hydrodynamic model domains for Cartagena, Colombia

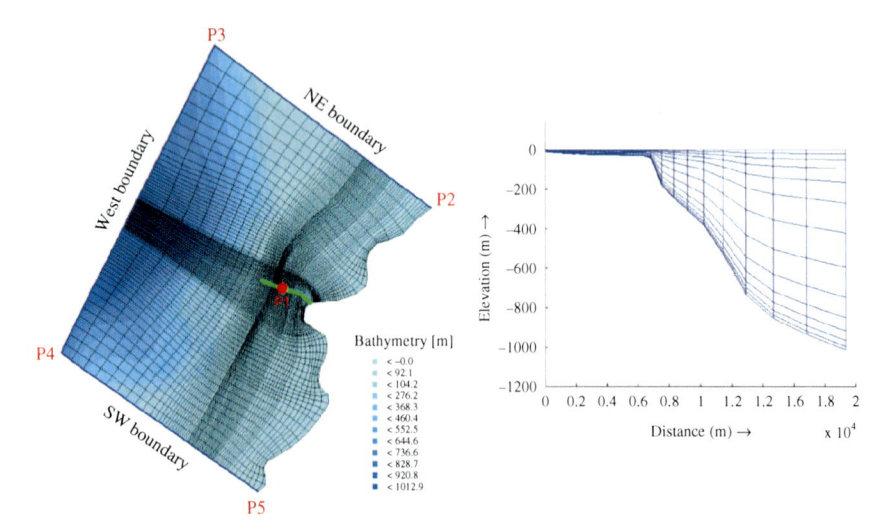

Figure 4.21 Nested model domain for Cartagena (left) and vertical grid cell configuration (right)

The hydrodynamic predictions were then employed in a water quality model (Delft3D-WAQ) to predict bacterial fate and transport. The source conditions for the water quality model were obtained from NRFIELD.

Some typical time series of predicted total coliform levels at six locations around the outfall over a 28 day period are shown in Figure 4.22. It can be seen that the levels are generally low and are very intermittent. They are close to zero for long periods that are interspersed with relatively high values when the plume is over the target location. Bacterial levels at the shore are always very low or zero.

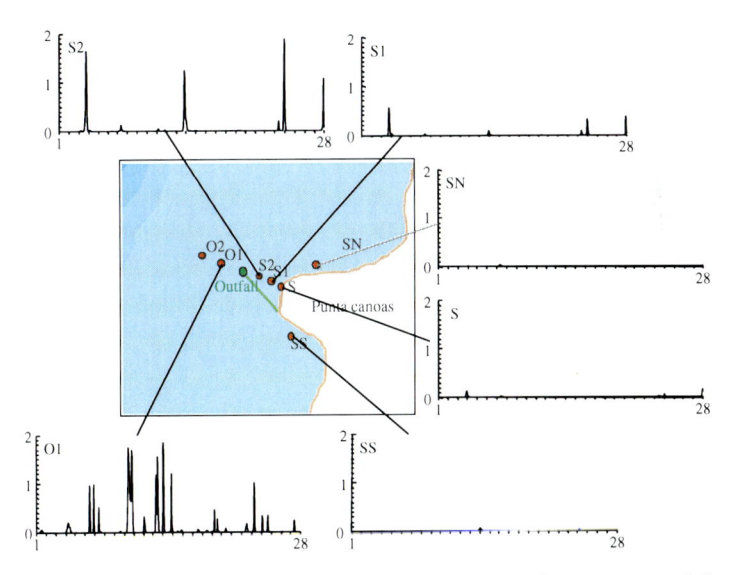

Figure 4.22 Predicted total coliforms around the Cartagena outfall. Units are per 100 ml $\times 10^4$

The exceedance frequencies and geometric mean values of total coliforms at the shoreline were predicted to be much lower than the WHO and California water quality standards given in Tables 2.8 and 2.9. These conclusions are the same as those derived from the modeling studies using a Lagrangian approach based on the measured currents at one location (see Figure 14.4).

It should be noted that the Lagrangian and Eulerian modeling approaches used here are quite different.

In the Lagrangian approach, the actual currents measured by the ADCP were used to generate plume trajectories (progressive vector diagrams). Thousands of simulations were performed to generate a statistical picture of the bacterial impact around the source. The advantage of this method is that the measured currents are used. It is not necessary to model them and all of the observed variability of the currents is automatically included. Relative diffusion (assuming the 4/3 power law of diffusion) is easily incorporated along with time-varying bacterial decay. Wind effects are incorporated inasmuch as the measured currents include them. However, the current measurements do not usually extend to the water surface, so surface wind transport may be underestimated. This effect was modeled by adding a wind-induced surface drift (Roberts, 2005). The other disadvantage of this approach is the assumption of spatial homogeneity of the currents. In reality they are not, and they also have a random component, so individual plume trajectories cannot be predicted although average displacements can be estimated. Therefore, the decrease of bacterial impacts with distance near the diffuser (for example, Figures 4.17 and 14.4) should be reasonably estimated, and shoreline impacts are probably overestimated. The conservative estimates, use of actual current data, and relative ease of making the calculations, make the Lagrangian approach a useful engineering tool. For further discussion of the theory and application of Lagrangian statistical modeling, see Roberts (1999b).

The Eulerian approach numerically solves the hydrodynamic equations on a fixed grid to predict currents and stratification. The advantage of this approach is that the full hydrodynamic variability in space and time, including near-surface currents, can, in principle, be predicted. The difficulty is that extensive boundary condition information, such as current speed and direction and density over depth, their variability with time and along the boundaries, and the temporal and spatial variability of the winds over the surface, are required to drive the model and produce reliable results. Coastal waters are usually very much under sampled, especially spatially, so such detail is almost never available, and even with good boundary data, currents are difficult to model. This is a significant problem for outfalls, as the transport on scales that most affect local bacterial impacts is advection-dominated. Correct prediction of the current field, in other words, where the water goes, is therefore essential for reliable predictions of bacterial impacts.

Notwithstanding the above comments, the two modeling approaches led to the same conclusion: Shoreline bacterial standards will be met by a large margin with only preliminary wastewater treatment.

4.7.5 Boston

Extensive numerical modeling of the Boston outfall has been performed. See Blumberg et al. (1996) and Signell et al. (2000).

4.8 PHYSICAL MODELS

Physical models, that is small-scale hydraulic models, are sometimes used in complex cases that cannot be reliably modeled by mathematical models. They are mainly restricted to near field processes; physical models of far field processes or currents are rarely used for open coasts. They have been used for semi-enclosed estuaries, for example Helliwell and Webber (1973), but have now been mainly replaced by numerical models. Physical models to simulate internal outfall hydraulics, particularly of saltwater purging in tunneled outfalls, are often used, for example Adams *et al.* (1994) and Lee *et al.* (2001). Examples of near field physical models for outfalls are Isaacson *et al.* (1983), Roberts and Snyder (1993a,b), and Couriel (1993). Near field models are typically done in towing tanks with the diffuser towed through stationary water. This is a convenient way to simulate a flowing stratified environment, which is difficult to produce in a laboratory.

An example is the physical model (Roberts and Snyder, 1993a) of near field plume behavior of the Boston outfall. The wastewater may contain stormwater runoff up to a peak design flow of 1270 mgd (56 m^3/s). To accommodate such a huge flow, the outfall, shown schematically in Figure 4.23, is a deep rock tunnel with an internal diameter of approximately 24 feet (7.3 m) and a length of approximately 9 miles (14 km).

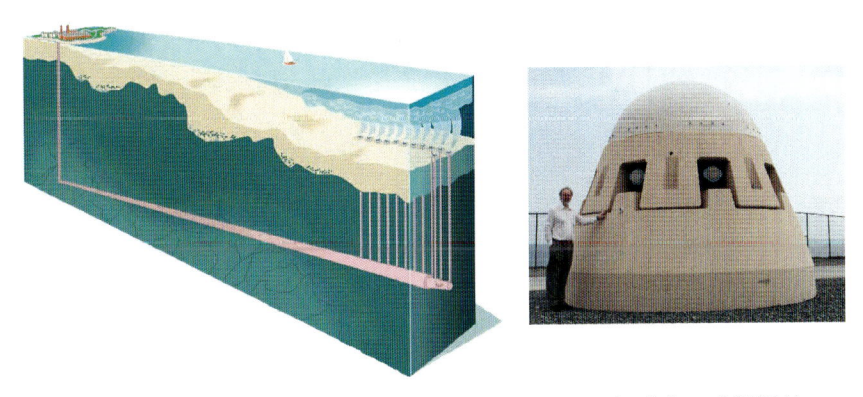

Figure 4.23 Boston outfall schematic and photograph of riser (MWRA)

The tunnel terminates in a diffuser 6600 ft (2012 m) long that consists of risers that extend to the sea floor where they are capped with multiport outlets. The risers are about 250 ft (76 m) long from the tunnel to the seabed and are constructed by drilling downwards with an oil drilling rig. This is expensive

134   Marine Wastewater Outfalls and Treatment Systems

(on the order of 1.5 million for each riser), and a considerable cost savings would be achieved if the number of risers could be reduced (without impairing the dilution capability of the diffuser). Little was known about the behavior of such discharges, especially in currents flowing parallel to the diffuser axis (which could occur often) with the possibility of "shading" of downstream risers. A study of a section of the diffuser was therefore performed in a density-stratified towing tank. The objectives were to determine the minimum number of risers consistent with the dilution requirements and to establish the characteristics of the wastefield formed for the final design under typical oceanic conditions.

Photographs from the model study showing the plumes in a stationary stratified environment are shown in Figure 4.24. It was found that, due to merging of the rising plumes, entrainment was restricted if too many ports were used. This is contrary to conventional diffuser wisdom, and resulted in the design having eight ports per riser, rather than 10 or 12. As a result of this model study, the number of risers was reduced from 80 to 55, with considerable cost savings. The model results were subsequently confirmed in field studies on the outfall (Roberts *et al.*, 2002).

Figure 4.24 Photographs of plumes from the physical model study of the Boston outfall (Roberts and Snyder, 1993a)

4.9 DISCUSSION

The fate and transport of wastewater discharged from ocean outfalls is complex and difficult to model. Consequently, conservative assumptions and approaches are commonly used, with the result that outfalls readily meet environmental standards and criteria. But modeling is becoming more realistic, and less conservative. For example, the results shown in Figure 4.17, that use measured time-varying currents and stratification, show much lower shoreline bacterial levels than would be predicted by the traditional method of assuming a constant onshore current and constant near field dilution. Similarly, time-series modeling of near field mixing, for example Figure 4.16, shows highly variable dilution whose median value is much greater than the lowest worst-case estimate. Worst-case conditions are extremely improbable, and their use could lead to overly conservative outfall designs and treatment levels.

Because of the widely varying scales of the major transport processes, sub-models of near and far field processes are needed. Many model types exist, but whatever models are used they must be properly coupled. Near field models are either length-scale, entrainment, or CFD. Presently (and for the foreseeable future) most models are length-scale or entrainment, or a combination of the two. Water quality models are either Eulerian or Lagrangian, and account for diffusion either by concentration gradient diffusivity or by tracking multiple particle releases. The currents and circulation patterns needed by these models are either measured directly by moored instruments, such as ADCPs, or predicted by two- or three-dimensional hydrodynamic models. Two dimensional hydrodynamic models are useful for fairly shallow waters, but three-dimensional models are needed for deeper, stratified waters. However, three dimensional models require extensive field data to produce reliable results, and are complex and expensive to set up. Most outfall projects won't justify the considerable expense of three-dimensional models and the difficulty of implementing them. Hybrid approaches that combine oceanographic measurements with near and far field models, and Lagrangian models that use measured currents are especially useful for engineering purposes. Good data are essential to any successful modeling effort. The means to acquire such data are discussed in the next chapter.

Even though numerical models may not accurately predict water quality in an absolute sense, they can be very useful in predicting *relative* effects, for example the difference in water quality resulting from different treatment levels.

5
Field surveys and data requirements

5.1 INTRODUCTION

Many types of information and davata are needed for planning, designing, constructing, and operating marine outfalls. Some are used with mathematical models to predict environmental impacts and to ensure that environmental criteria are met, and some to ensure its reliable operation over many years. To accomplish these objectives, data are needed on currents, density stratification, winds, waves, and bathymetry, among others.

Information is sometimes available from previous studies or measurements made for other purposes. These data are often insufficient or of the wrong type, however, and it is almost always necessary to conduct specially designed studies. Careful planning is essential to ensure that the correct type of data is obtained, and diligent quality assurance and control (QA/QC) must be exercised

© 2010 IWA Publishing. *Marine Wastewater Outfalls and Treatment Systems.* By Philip JW Roberts, Henry J Salas, Fred M Reiff, Menahem Libhaber, Alejandro Labbe, and James C Thomson. ISBN: 9781843391890. Published by IWA Publishing, London, UK.

to ensure the veracity of the data obtained. The campaign must be tailored to the specific needs of the outfall project, such as the mathematical modeling discussed in Chapters 3 and 4, and the structural design issues discussed in Chapters 8, 9, and 10. These campaigns can be expensive, costing from tens of thousands of dollars for a small outfall to more than a million dollars for a major project. But experience tells us that it is money well spent. Properly conducted studies at the early stages of outfall planning can significantly smooth the project's progress and ultimate success; conversely, poorly executed field studies can delay or even derail otherwise good projects.

Oceanographic instrumentation has advanced rapidly in recent years, and the cost has decreased to the point where adequate surveys can be readily made even in developing countries. But field campaigns can be very extensive and complex, and are quite site and project dependent. Therefore, we can only provide here some general guidance on the types of data needed and the means for obtaining it.

5.2 PHYSICAL OCEANOGRAPHY AND METEOROLOGY

5.2.1 Introduction

Physical oceanographic data are essential. They are used for the outfall structural design, prediction of the fate and transport of the wastewater, and in mathematical models to ensure that environmental criteria, such as near field dilution and shoreline bacterial levels, are met. Particularly important is information on bathymetry, currents, waves and tides, and the temperature, salinity, and density structure of the water column.

5.2.2 Currents

The use of current data falls onto several categories:

- Analysis and understanding of the hydrodynamics of the coastal region;
- Calibration of coastal hydrodynamic circulation models (Chapter 4);
- Prediction of initial dilution (Section 3.3.2);
- Prediction of far field transport (Sections 3.3.3 and 3.3.4);
- Prediction of horizontal or vertical pipeline movement, erosion beneath a seabed pipeline, removal of material around a buried pipeline, sedimentation around and burial of the diffuser (Chapters 8 through 11);
- Planning for the outfall installation.

The coastal ocean is a complex hydrodynamic regime whose special characteristics were discussed in Section 3.2. The main driving forces of coastal currents are tide, winds, and major ocean currents such as the Gulf Stream, the North Equatorial and South Equatorial Currents of the Atlantic, the Humboldt Current, and the Kuroshio Current of the Pacific. At any particular location, some or all of these drivers may dominate and some may be negligible. Bathymetry will always significantly affect local circulation patterns and storm surge during hurricanes can generate very strong local currents.

Tides often dominate. They can cause very strong currents, such as in southeastern Alaska, where tides can exceed 15 m and generate currents as high as 17 knots (8.7 m/s) in narrow entrance fjords. Strong currents can also occur where tides pass around islands, through archipelagos with narrow passages, and into or out of entrances to atolls. In open coastal waters tidal currents are typically much slower, however.

Currents can also be caused by waves. Wave-generated currents that run away from shore in an approximately perpendicular direction are known as rip currents and those that that run parallel to the shore as littoral drift. The velocities of these currents are influenced by the wave energy (which determines the onshore runup) and the geomorphology of the coastline. The currents can be strong enough to transport sufficient seabed material to change the water depth and bottom profiles by several meters, posing a significant threat to outfall stability.

Currents often show considerable variation with horizontal distance and over depth. Current speeds generally decrease somewhat with increasing depth, and their direction can also change, so measurements through the water column are desirable, especially in deeper water. Variations in the horizontal depend on the local topography. For a long straight shoreline, most variations occur in the offshore direction rather than alongshore. For a complex topography, however, such as with large depth changes, islands, or a nearby estuary mouth, they can vary in any direction. The required number of current meters will depend on the degree of spatial variation along with the needs to verify mathematical models and to provide their boundary conditions. The authors have worked on projects involving one or two current meters up to projects involving 18 meters.

Ocean currents also vary considerably in time over a wide range of scales from minutes to years. To capture seasonal variations, they should ideally be measured as long time series for one year with measurement intervals as short as 15 minutes. If this is not possible, they should be measured at least hourly for shorter periods that cover diurnal tidal variations. If continuous measurements for one year are not possible, they should be made during the months most representative of the various seasonal conditions.

140   Marine Wastewater Outfalls and Treatment Systems

The means of obtaining long time series of ocean currents has changed radically over the past twenty years or so. Previously, it was common to obtain point measurements with current meters such as electromagnetic instruments. To obtain readings over the water column, several instruments were required, strung at various depths on a cable. These have now been almost entirely supplanted by the Acoustic Doppler Current Profiler (ADCP).

ADCPs and some typical installations are shown in Figure 5.1. They can measure current speed and direction throughout the water column (except near the surface and bottom) at specified intervals and record the data internally. They can be deployed on the seabed looking upwards, or from the surface on a buoy or a moving boat looking downward. Some units operate horizontally to measure the lateral variation of currents at one depth, such as across a river or narrow estuary. The data are retrieved either by recovering the instrument by a diver and replacing it, by an acoustic modem, or by cable to a surface buoy. By transmitting the data to a surface buoy, it is also possible to telemeter the data to shore in real time. ADCPs are self contained and powered by a battery for up to (and sometimes exceeding) one year.

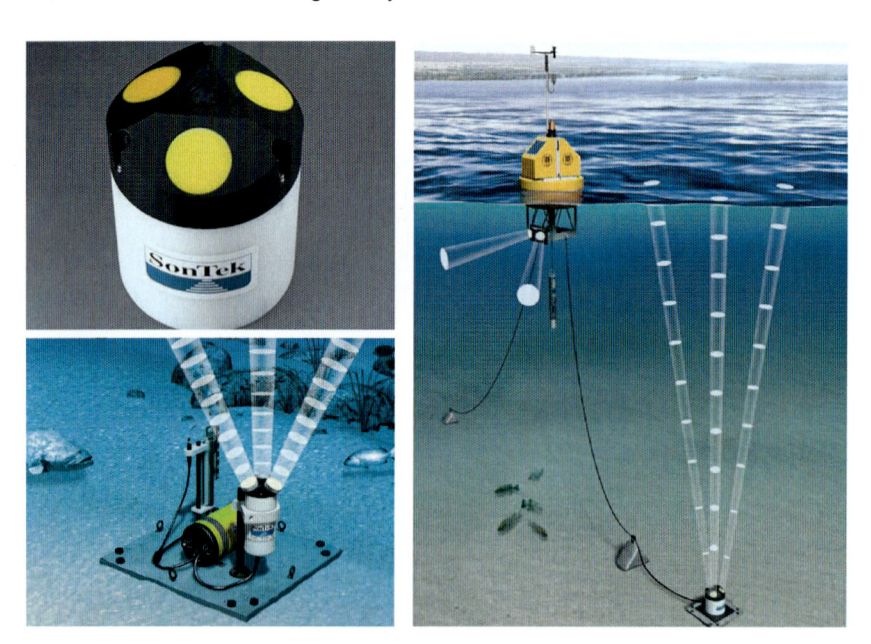

Figure 5.1 Typical ADCP and installations (Images courtesy of SonTek/YSI, a division of YSI, Inc.)

Field surveys and data requirements 141

ADCPs operate by sending sound waves (pings) of known frequency through the water column. By measuring the Doppler frequency shift and time of travel of echoes from scatterrers such as particles or bubbles in the water column, they can measure velocity through the water column in up to 128 layers (bins). Different sound frequencies are used for different profiling water depths. Higher frequency units, up to 3 MHz, are used for shallow water a few meters deep with bin sizes as small as 15 cm, and lower frequency, down to 0.25 MHz, for depths up to 200 m, with bin sizes around 2 m. Sensors can be added to measure other parameters, such as temperature, conductivity, depth, turbidity, and waves. The cost of a unit depends on how it is configured, what sensors are added, battery options, etc, but is of order US 10,000 and up. This is considerably cheaper and easier to deploy than a string of point meters such as electromagnetic instruments. ADCPs can be configured to surface buoys that measure wave and meteorological parameters and the combined data telemetered to shore in real time as shown in the right-hand image in Figure 5.1.

The richness of the data that can be obtained by an ADCP is illustrated in Figure 5.2. This shows current vectors at various depths and times obtained off the coast of Hawaii. Such data frequently reveals the complexity of coastal currents, which can flow in different directions at different depths. The spatial variation of currents over a wide area can be obtained from a moving boat equipped with a downward-looking ADCP; an example is shown in Figure 5.3. The use of ADCPs has expanded our understanding of coastal circulations considerably and their data can be used directly in models to predict the near and far field transport of an outfall plume and bacterial contaminations as discussed in Chapters 3 and 4.

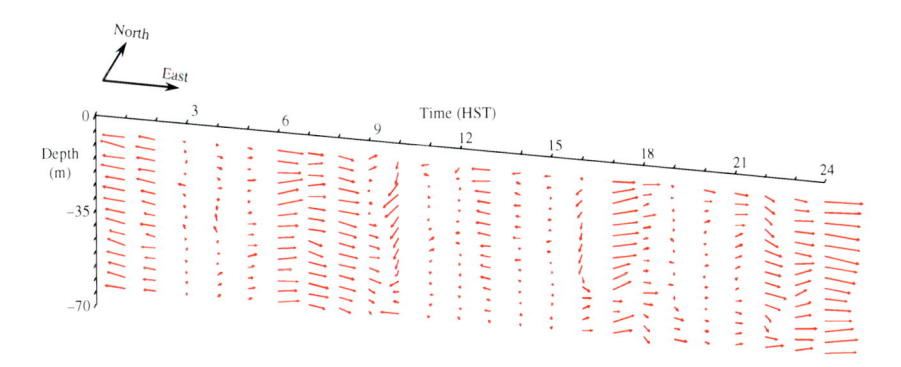

Figure 5.2 ADCP data obtained off the coast of Oahu, Hawaii

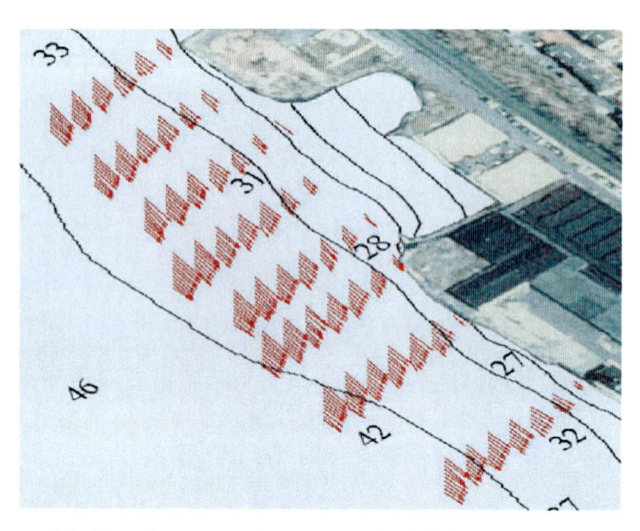

142   Marine Wastewater Outfalls and Treatment Systems

Figure 5.3 Data from vessel-mounted ADCP (Ocean Surveys, Inc.)

Considerable insight into current behavior can be obtained by visualizing the current data in different ways. For example, Figure 5.4 shows five polar scatter diagrams of currents at one depth over about one month in the Pacific Ocean near San Francisco, California. These plots are a useful way to see a quick "snapshot" of the current speeds and directions and their major characteristics. The diagram near the diffuser location shows the principal current axes (labeled PC1 and PC2) and the first principal axis at the other locations. The first principal axis maximizes the kinetic energy of the currents when projected onto them; the second principal axis is orthogonal to the first. It is often useful to decompose the currents onto these axes. The component of the currents along the first principal axis is the first principal component. The diffuser is oriented perpendicular to this axis to obtain the maximum beneficial effect of the currents on initial dilution (see Chapter 3). The first principal axes of the currents at each mooring are shown at each site. Several characteristics of the currents can be concluded from Figure 5.4. The first principal components point towards the mouth of the Golden Gate (because they are dominated by the tidal current flowing through the Golden Gate). The magnitude of the currents decrease with distance from the Golden Gate, and onshore current speeds are slow.

Although polar scatter diagrams indicate the currents' spatial variation, they tell us nothing about their temporal variation. A common method of showing current time series is the feather plot (Figure 5.5). This shows the current vectors at one location and depth by placing the origin of each vector sequentially at its measurement time on the horizontal axis. The tidal variation of the currents is now readily apparent. More sophisticated digital analyses are used for time series analyses of ocean currents, such as computation of energy spectra, rotary spectra, filtering to extract low or high frequency content, spatial correlations, polar histograms, etc. Such analyses are beyond the scope of this book, but some examples in the context of ocean outfalls are given in Roberts (1999a,b).

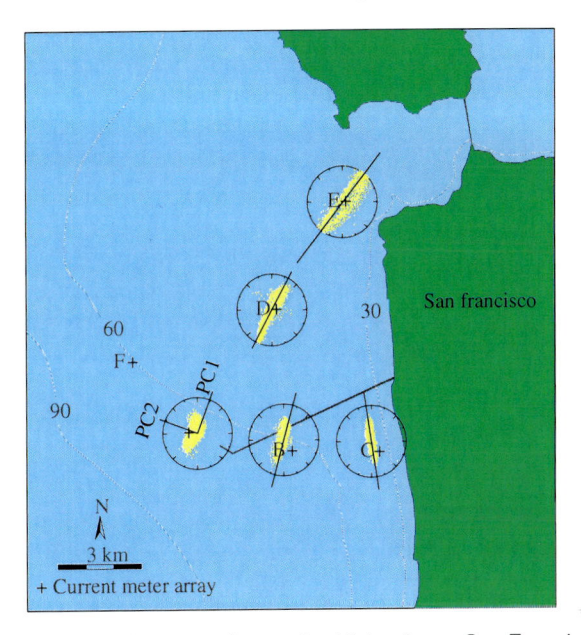

Figure 5.4 Polar scatter diagrams of currents obtained near San Francisco, California

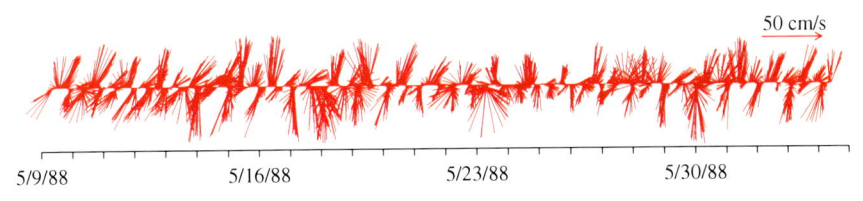

Figure 5.5 Feather plot of currents obtained near San Francisco, California

Current data are also used directly as input into wastefield simulation models. Examples, to predict time series of near field dilution were discussed in Section 4.7.2, and prediction of far field characteristics such as visitation frequency and bacterial concentrations in Sections 3.3.3.3 and 4.7.2.

Surface currents can have important consequences. If the plume surfaces, it may be transported by the surface current, and floatables such as grease and oil can also move with the surface current and wind. It is usually assumed that the wind can induce a surface current whose magnitude is 1% to 3% of the wind speed.

ADCPs do not measure all the way to the surface so surface currents are best measured by drogues or floats (Figure 5.6). A recent development is low-cost transmitting GPS units (around $200) that can record the drogue's position and trajectory. These drifters can be set at different depths and are particularly useful to obtain Lagrangian water motions (as opposed to the Eulerian data at a point that are obtained by recording current meters). This enables better prediction of travel from a potential diffuser location to a point of particular significance such as a water intake or beach. Drifter data can be visualized as progressive vector plots (sequential "tail to head" plot) such as shown in Figure 5.7. ADCP studies should be supplemented by periodic drogue releases at the proposed discharge sites. Their positions should be recorded continuously for one to five days depending on study area characteristics.

Figure 5.6 Surface drifters (Right, Courtesy Ocean Surveys)

Another important recent development is the use of surface current radar that can continually map out surface currents over a wide area. Such data have been used in delineating mixing zone areas for ocean outfalls (Chin *et al.*, 1997).

Field surveys and data requirements 145

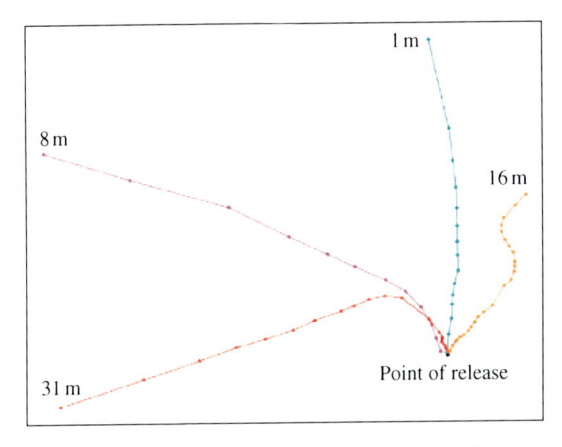

Figure 5.7 Trajectories of drogues released at different depths (courtesy Ocean Surveys)

5.2.3 Density stratification

Coastal waters deeper than about 20 m are usually density stratified, i.e. the density varies with depth. For the water column to be hydrodynamically stable, density must always increase downwards. As discussed in Chapter 3, this stratification plays a critical role in the plume dynamics, particularly in determining plume rise height and dilution. By submerging the plume, it can prevent shoreward plume impaction (see Figure 3.4). It also affects how the wastefield is transported to distant locations, as currents at different depths can move differently. Measurement of stratification and its variability is therefore essential. Even very small density differences, as small as 1 σ_t (1 kg/m^3), over the water column can have profound effects on the plume behavior, so density must be measured to an absolute precision of around 0.1 σ_t.

Seawater density depends on salinity and temperature. The stratification is often primarily thermal, but a nearby freshwater source such as an estuary can dramatically increase stratification. Density is not measured directly, it is calculated from measured salinity and temperature, and salinity is usually computed from conductivity.

Traditionally, stratification data has been obtained by an instrument package traversed up and down through the water column from a boat (Figure 5.8). These packages typically measure conductivity, temperature, and depth (CTD) and transfer the data to a recorder on the ship deck. They can also be attached to a "V-fin" (Figure 5.8) from a moving boat. This is "flown" though the water and winched up and down so it follows a "tow-yo" pattern, or it can be winched up and down with the boat stationary to profile at a fixed location. The data are typically recorded in a deck acquisition system at, say, 1 Hz. Other sensors that measure parameters, such as dissolved oxygen and pH, can also be used. These instrument packages are manufactured by several companies.

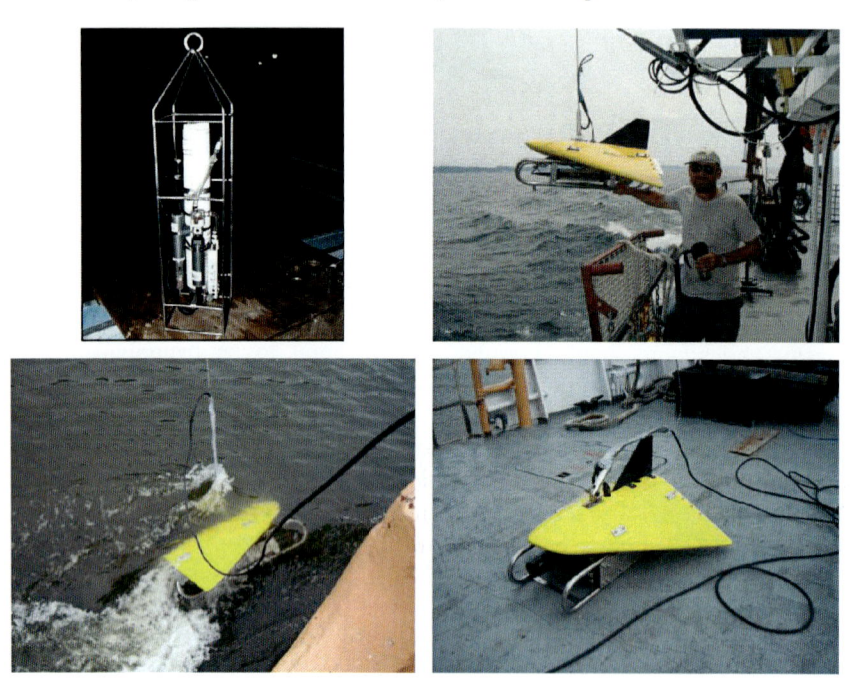

Figure 5.8 Instrument package and V-fin for CTD profiling

Another option is to measure stratification continuously by a string of moored instruments (Figure 5.9). The strings usually consist of thermistors to measure temperature, and conductivity probes to measure salinity. The probes are set at prescribed depths through the water column and their data are saved on a recorder at the bottom. A considerable advantage of instrument strings is that they can record long time series of density stratification at prescribed time intervals by simultaneously measuring densities at multiple depths. The sampling interval and time should be equal to and synchronized

Field surveys and data requirements 147

with any current meter recordings. The long time series of density are especially useful for predicting time series of near field plume behavior. An example of this procedure is given in Section 4.7.2. Figure 4.15 shows some sequential density profiles measured by a thermistor string off the coast of Oahu, Hawaii, and Figure 5.10 shows vertical profiles of temperature, salinity, and density obtained at two hour intervals on the north coast of the Dominican Republic by a temperature-conductivity string moored in water 150 m deep.

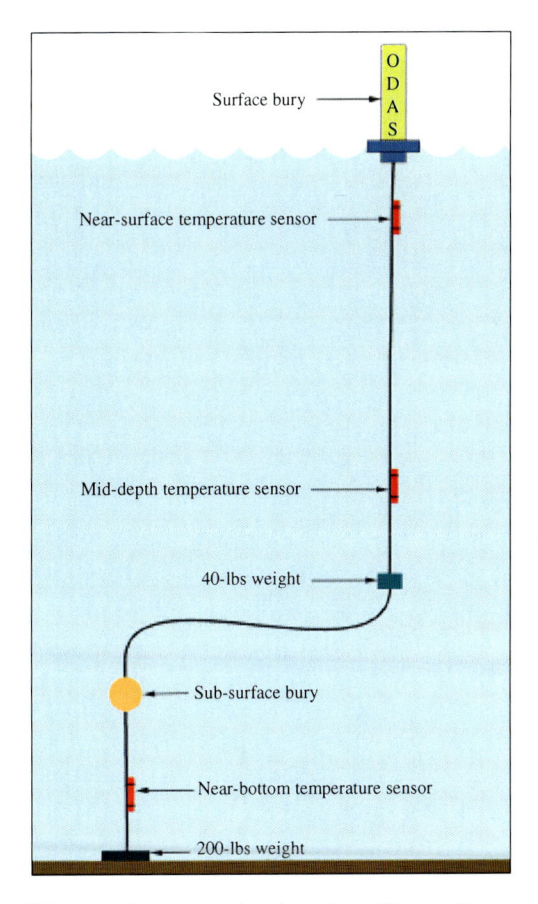

Figure 5.9 Long term mooring (courtesy Ocean Surveys, Inc.)

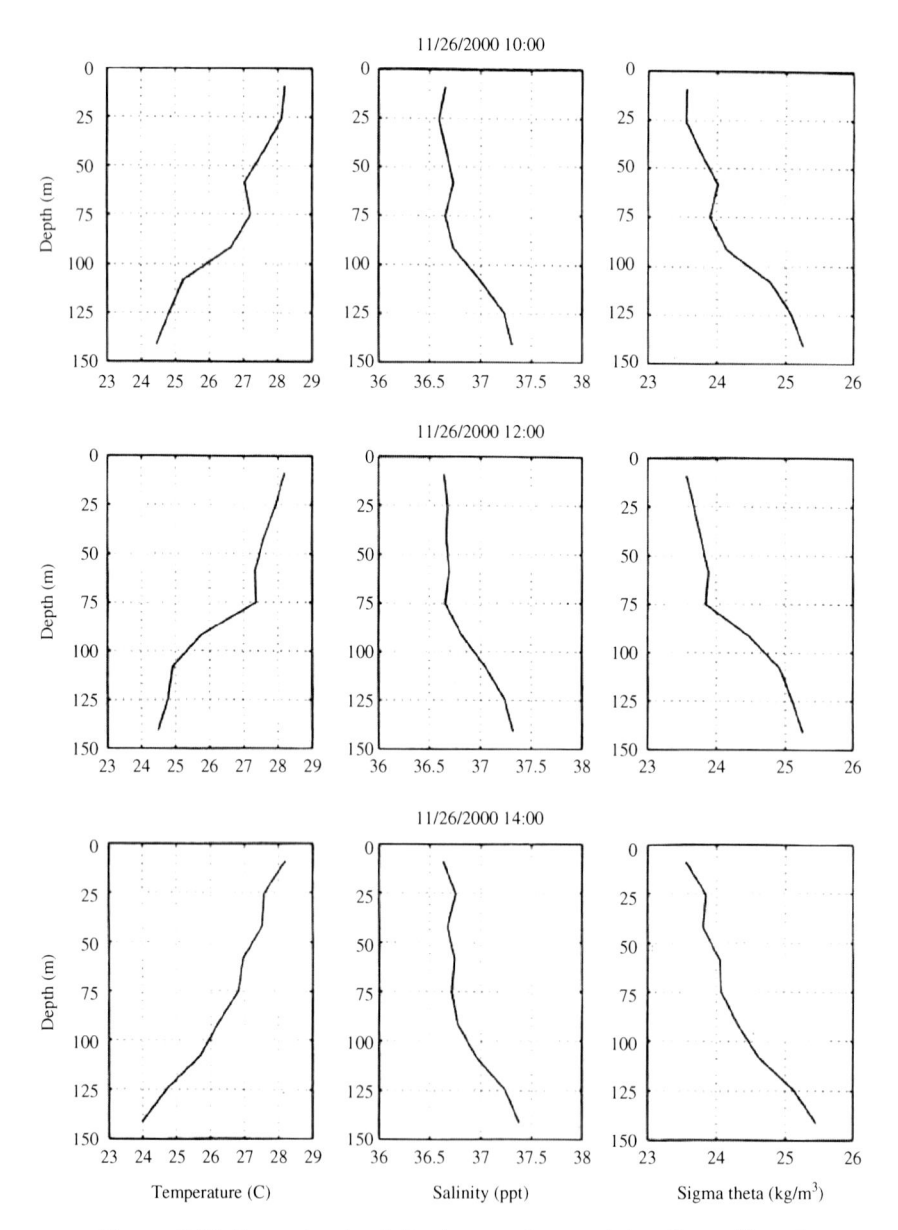

148   Marine Wastewater Outfalls and Treatment Systems

Figure 5.10 Example of salinity, temperature and density profiles at two hour intervals

As with currents, stratification varies over a wide range of time scales from minutes to years. Ideally, stratification should be measured continually (say at 15 to 30 minute intervals) over one year to capture seasonal variations. The instrument strings described above now make this possible. If this is not feasible, stratification should be measured several times during the major seasons over one year and during representative tidal conditions by a profiling instrument deployed from a boat (Figure 5.8). After processing, the data are usually presented in tabular and graphical forms for use in plume analysis.

5.2.4 Waves and tides

Many failures of unburied submarine pipelines are related to surface waves. They cause hydrodynamic forces on the pipe, differential erosion of the seabed or bedding material on which the pipe is laid, mass movement of seafloor materials, deposition of material over the diffuser, and liquefaction of the supporting seabed. Knowledge of waves and the forces, accelerations, and velocities beneath them is therefore necessary to enable design of an outfall that precludes such failures.

This requires determination of the most severe wave that is likely to impact the outfall during its lifetime, referred to as the deepwater design wave. Determination of the height, period, and direction of travel of the wave front of the design wave is described in Chapter 8.

Information on tides is important for several reasons, in addition to their influence on currents. Bathymetric maps are usually referenced to mean low low water level (MLLW). In areas with large tides, the elevation difference between high and low tides can be significant and must be accounted for when analyzing the internal outfall hydraulics. This can be especially important when the outfall operates on gravity head rather than pumps.

For low slope seabeds with large tides, the surf zone, where waves break, can move dramatically during the tidal cycle. The area of extra or special outfall protection in the surf zone must be adjusted accordingly.

Note that the capacity to simultaneously measure water depth and wave characteristics can be added to ADCPs.

5.2.5 Meteorology

Meteorological information, particularly wind speed and direction, is important to many aspects of an outfall project. Although historical data are often available, such as from nearby airports, additional measurements are usually needed. The data are needed to predict surface wind-induced currents, as input to hydrodynamic coastal circulation models (Chapter 4), and are essential for planning and carrying out construction, especially in timing the outfall flotation and submergence. It is highly desirable to conduct these activities during the time of the year and time of day when wind speeds are low.

If there is no permanent meteorological station nearby, a temporary meteorological station should be set up as close to the diffuser area as possible. The data should be correlated with that obtained at the nearest permanent station to enable projection of winds based on long term historical data.

The meteorological station can be shore-based (Figure 5.11a), or on a floating buoy (Figure 5.11b). Either can be equipped with telemetry to transmit data to a central location in real time. A floating buoy can be equipped to measure surface waves, air temperature, wind speed and direction, and can also serve to telemeter data from a submerged ADCP. The advantage of a floating buoy is that winds can be significantly affected by local topography, and measuring wind directly over the water is more representative of actual conditions near the proposed diffuser.

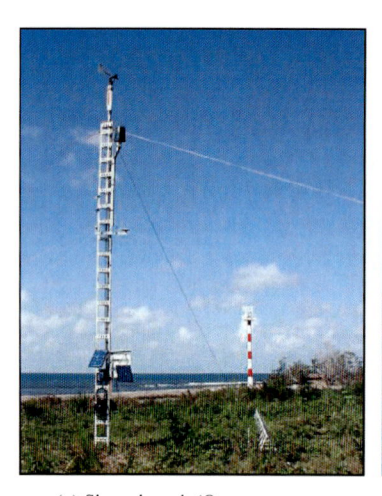

(a) Shore-based (Ocean surveys, Inc.)   (b) Floating

Figure 5.11 Meteorological stations

Wind speed and direction should be measured at least hourly in summer and winter. They should be recorded concurrent with the hydrodynamic surveys. Wind data should be correlated with current measurements to determine how they affect them. This is especially important if the discharge plume is expected to surface as the wind may induce a surface drift current as previously discussed.

Wind data can be presented in many forms. One of the most useful for initial planning purposes is the wind rose, which is a radial plot of the frequency with which the wind blows in each of the major 16 compass directions. The plots are divided into speed ranges to provide a histogram of speed and direction. This enables an estimate of the frequency with which a surfaced plume might reach a particular area of interest and the time for it to get there.

Wind conditions are also important to divers working on outfall construction tasks such as route surveys, drilling, blasting, and guiding the outfall into its final seabed location. Wind can also decrease visibility by stirring up the seabed, especially when it consists of fine sand, silt, or clay.

Heavy rains during various months of the year can also be important because runoff can greatly increase the turbidity of streams discharging into the ocean, reducing visibility near shore. This makes it difficult, if not impossible, to make videos and visually inspect the seabed along the outfall route. Poor visibility can also seriously complicate outfall construction during the process of sinking, as well as hinder inspection of the constructed outfalls.

Historical information on tropical storms and hurricanes is especially important for the estimation of the design wave.

5.3 BATHYMETRY AND GEOPHYSICAL STUDIES

5.3.1 Bathymetry

Because bathymetric information is essential for the planning, design, and construction of an outfall, obtaining it should be one of the first tasks. This usually begins with a search for existing information such as navigational and nautical charts, topographic maps of coastal areas, harbor maps, oceanographic maps, and sounding records. Sometimes sufficient bathymetric information is available to conduct a feasibility study or preliminary design, but additional studies will usually be needed to obtain additional or more detailed information. The savings in time and money associated with the final design, construction, and monitoring of an ocean outfall will usually justify the investment.

The precision and level of detail of the bathymetric chart should be compatible with the magnitude and complexity of the outfall project. It is now possible, with modern multibeam echosounders linked to GPS navigational equipment, to cover an area of 250,000 square meters in one day. The raw data are then entered into computer programs with graphics capabilities to produce a detailed bathymetric chart with isobath intervals of 1 m and horizontal precision less than 1 m. Examples are shown in Figures 5.12 and 5.13.

Accurate bathymetric charts provide information on the physical relief of the seabed. This is necessary for selecting the optimum outfall alignment and diffuser location, especially where the seabed is particularly convoluted, has abrupt escarpments, numerous reefs or other obstacles to be avoided.

Even with this accuracy, however, experience indicates that these surveys can miss anomalies that may affect the outfall design and construction. These can be disclosed by direct observation, so it is recommended that the route be visually inspected by divers before finalizing the design. Diver investigation of the seabed materials and classification of reefs and other anomalies can supplement a bathymetric chart with sufficient geomorphologic information to enable planning and design. An ROV can also be used for this purpose, but if visibility is poor its view can be narrow and limited, increasing the possibility of overlooking important features. Terms of reference for the divers could typically consist of:

- Place a permanent concrete marker (anchor) on the seabed at the diffuser site and attach a temporary buoy to it;
- Dive between the onshore entrance point and the diffuser site and determine the most direct route that is free of obstructions and escarpments;
- Set concrete markers along the route at specified intervals of 50 or 100 m. The markers should be color coded and/or numbered and the depth at each marker should be recorded;
- Make a video of the route following the markers that include close-ups and details of any anomalies. Use artificial light when needed;
- Record bottom conditions in relation to the markers, such as seabed material, reefs, escarpments, obstacles, and marine life attached to the seabed, if any;
- Note direction of bottom currents.

For depths beyond diver limitations, if greater precision is needed for identification of anomalies not indicated on the bathymetric chart or identification of seabed materials and characteristics, sidescan sonar (Figure 5.14) should be used.

Nearshore bathymetry should be checked at different seasons of the year and compared to each other as well as to historical data to determine if wave action and currents cause sufficient fluctuations in the profile of the seabed to warrant increased depth of burial or other special considerations for the outfall in the surf zone.

Field surveys and data requirements 153

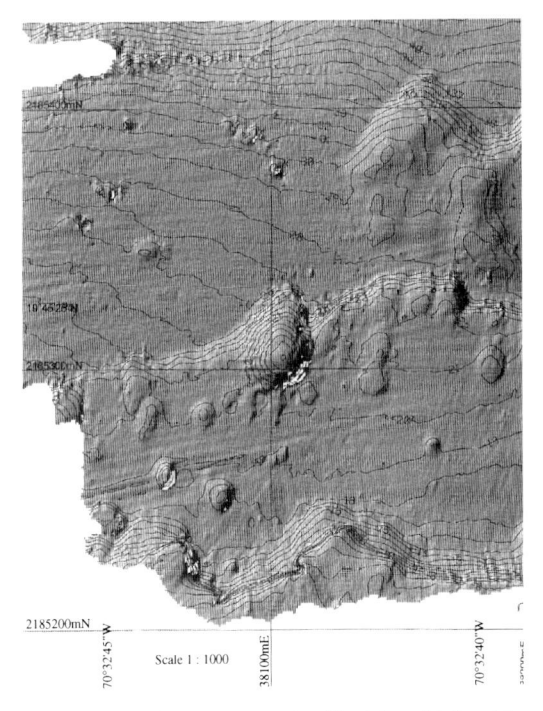

Figure 5.12 Bathymetric chart prepared with data obtained by a multibeam echosounder

For outfall diameters less than 300 mm the cost of multibeam echosounders for bathymetry and mapping can be very high relative to the cost of outfall construction and may not be justified. If the diffuser depth is less than about 30 m, an adequate alternative is to use a team of divers trained for outfall site exploration to identify suitable routes, measure distances along the routes, and determine depths with sufficient accuracy to construct bathymetric profiles for planning and preliminary design.

5.3.2 Geophysical

Geophysical information is essential to selecting the best route for the pipeline, estimating construction costs, designing facilities to stabilize the outfall on the seabed, and evaluating the need for a trench and its design.

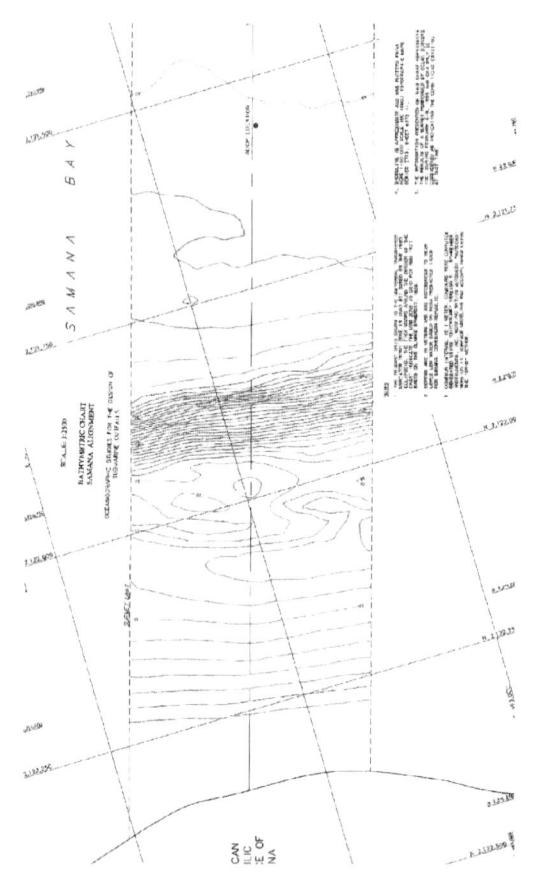

154   Marine Wastewater Outfalls and Treatment Systems

Figure 5.13 Bathymetry chart prepared with data obtained by a single beam echosounder

Knowledge of seabed geomorphology is needed to determine the best diffuser location and select the optimal alignment. In addition to the surface features of the seabed it is also necessary to know its composition. For example, whether it is soft clay, silt, fine sand, gravel, or stone, boulders, or solid rock affects the system used to fasten the outfall to the seabed and the method of trench excavation.

If a trench is needed, the seabed material should be determined preferably before final design and definitely before construction. Sub-bottom profilers are being increasingly used to detect the depth and type of different density layers. These are basically devices that direct a narrow beam of very high energy pulses of sound, usually 3–4 kHz, through the water into the seabed. At each change in density some sound will be reflected and some will pass through to the next layer. The reflected returns from each layer are recorded, analyzed, and an image of the various strata below the seabed is plotted. Interpretation of sidescan sonar data can also provide a general classification of the seabed material. Figure 5.14 shows an imaging system that is specially designed for sidescan sonar imaging and subsurface profiling.

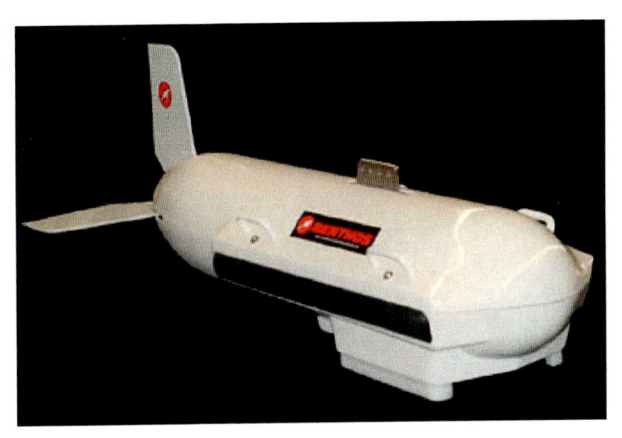

Figure 5.14 Teledyne Benthos SIS-1625 sidescan seafloor imaging system

When the seabed is comprised of sedimentary deposits, samples should be taken along the outfall route and classified according to the Udden-Wentworth grain-size scale. If construction in the surf zone may require piles or sheet piles, test borings in the seabed and penetration tests are usually recommended. Seismic evaluations are required in earthquake zones to determine vulnerability to liquefaction of the seabed and potential displacement at fault zones. If microtunneling or directional drilling is to be used (see Chapter 11), detailed geological information is essential.

5.4 WATER QUALITY

Background water quality surveys during the design and planning phases around the disposal area and at nearby beaches with significant recreational use are usually imperative to enable pre- and post- construction comparisons of water quality. These data can also serve as input to any required Environmental Impact Assessment (EIA). Background water quality monitoring as well as post-construction monitoring to verify the outfall's performance is discussed in Chapter 13.

5.5 BACTERIAL DECAY (T_{90})

As discussed in Chapter 3, an important parameter for outfall design is the rate at which indicator bacteria decay as they travel in the ocean. This is usually quantified by T_{90}, the time required for 90% mortality (one order of magnitude mortality or disappearance) of the bacteria. The use of T_{90} assumes a simple first-order decay (Eq. 3.45):

$$\frac{c}{c_o} = 10^{-\frac{t}{T_{90}}}$$

where c is the indicator bacterial concentration after time t, and c_o is the initial indicator bacterial concentration, usually at the end of the near field. As discussed in Chapter 3, this equation, combined with detailed current data and estimates of oceanic diffusion are key to determining the outfall length needed to comply with bacterial water quality standards.

Although models are available for estimating T_{90} (such as in the EPA model Visual Plumes), its value depends on local conditions. T_{90} also varies diurnally; for an example of modeling with a variable T_{90}, see Carvalho et al. (2004). Some typical values that have been observed in tropical and semi-tropical waters during daylight hours are summarized in Table 5.1. Because of the wide variability of T_{90} and its dependence on local conditions it must usually be measured *in-situ*. Three of the many techniques that have been used to measure T_{90} are presented below.

Table 5.1 Indicator bacterial decay rates in tropical waters (Ludwig, 1988)

Location	T_{90} (hours)
Honolulu, Hawaii	≤0.75
Mayaguez Bay, Puerto Rico	0.7
Rio de Janeiro, Brazil	1.0
Nice, France	1.1
Accra, Ghana	1.3
Montevideo, Uruguay	1.5
Santos, Brazil	0.8–1.7
Fortalleza, Brazil	1.3±0.2
Maceio, Brazil	1.35±0.15

5.5.1 On-site measurement in artificial plume

In this method, some wastewater is mixed with a suitable conservative tracer, such as Rhodamine WT or a radioactive substance, and the mixture discharged into the ocean as a slug release. Samples are taken in the center of the diffusing cloud and the concentrations of dye and indicator organisms (e.g. enterococci and/or fecal coliforms) are measured in subsequent laboratory analyses. An on-board fluorometer can be used to provide rapid dye measurements to locate the peak dye concentrations where subsequent samples should be collected. Pearson (1971) suggests that samples be drawn at depths of 0.6 to 0.9 m and about 3 m at the initial drop point and then at times of 10, 20, 30, 45, 90, 120, 180 and 240 minutes. Judgment in the field is required, but the key factor is that samples be collected at the location of the peak dye concentrations.

The data are corrected for physical dilution by multiplying the observed bacterial concentration by the ratio of the initial dye concentration to the observed dye concentration. T_{90} is then obtained from a best-fit line drawn on a semi-log plot of corrected indicator bacterial concentration versus time. This procedure may be expensive due to the large volumes of wastewater and dye required.

5.5.2 *In-situ* measurement in existing wastewater discharge

If there is already a wastewater discharge, an alternative procedure is to release floats and/or dye, and collect samples as above. The samples are analyzed for the indicator organism and some conservative wastewater constituent. Although not strictly conservative, dissolved orthophosphate has been used by ENCIBRA (1969) with the justification that its decay rate is much slower than that of the indicator organisms. The data are analyzed as described above. Alternatively, dye can be directly added to the wastewater as a tracer.

5.5.3 Bottle method

Transparent bags or large bottles are filled with raw sewage diluted with ambient ocean water at levels expected after initial dilution, e.g. 1:100. They are placed at fixed locations nearshore just below the surface to maintain ambient temperature and light conditions. Samples are extracted every 15 minutes. This procedure substantially reduces cost and equipment by eliminating the use of a conservative tracer and reducing required waste volumes. The results are plotted versus time on semi-log paper to obtain T_{90}. Simultaneous use of opaque receptacles, (e.g. painted black) to simulate nocturnal conditions can provide information on diurnal variations.

Although this procedure has been criticized by some investigators, (Ludwig, 1976), satisfactory results were obtained in the study of Guanabara Bay in Rio de Janeiro (Hydroscience, 1977) and the coastal areas of Mar del Plata (INCYTH, 1984). It is noted that Acra *et al.* (1990) reports that ordinary glass transmits

90% of incident ultraviolet radiation, which is considered to be the most important germicidal component of sunlight. Plastic transparent materials, such as Lucite and Plexiglas, and translucent materials such as polyethylene can also transmit the germicidal components of sunlight. This method has the advantages of simplicity and low costs. Other techniques for measuring T_{90} are presented by Britto (1979).

6
Wastewater management and treatment

6.1 INTRODUCTION

About 50% of the world's population resides in or near coastal areas, and in the developing world much of this population lacks access to safe sanitation. Rapid population growth is outstripping capabilities to provide basic sanitation to much of the world's population beyond the meager limits that already exist in too many places. For developing countries, where resources for treatment and disposal are limited, low cost, reliable solutions are essential. The most probable receiving bodies of the sewage generated by coastal communities are the oceans and seas, so management, treatment, and ocean disposal of sewage are very important issues.

The level of treatment required before discharge is often a controversial and contentious issue. Unfortunately, in many countries, the tremendous capability of the oceans to assimilate wastewater without harm when discharged through a

© 2010 IWA Publishing. *Marine Wastewater Outfalls and Treatment Systems.* By Philip JW Roberts, Henry J Salas, Fred M Reiff, Menahem Libhaber, Alejandro Labbe, and James C Thomson. ISBN: 9781843391890. Published by IWA Publishing, London, UK.

well-designed outfall is not recognized in the effluent quality requirements. Frequently, treatment to arbitrarily specified levels is legislated that are the same as for discharges to inland waters. The National Research Council of the US National Academy in its major study *Managing Wastewater in Coastal Urban Areas* (NRC, 1993) specifically recommended against this "one size fits all" treatment approach, stating:

> *"Coastal wastewater and stormwater management strategies should be tailored to the characteristics, values, and uses of the particular receiving environment based on a determination of what combination of control measures can effectively achieve water and sediment quality objectives,"* and
>
> *"Coastal municipal wastewater treatment requirements should be established through an integrated process on the basis of environmental quality as described, for example, by water and sediment quality criteria and standards, rather than by technology-based regulations."*

Treatment to arbitrary levels beyond these can waste scarce financial resources, require excessive energy use, and generate large quantities of sludge that must be disposed of on land.

Further guidance on appropriate treatment is provided by the World Health Organization (WHO). In October 2003, they published a report (WHO, 2003) that contained guidelines for the protection of recreational water quality. They considered the relative risks to human health of exposure to effluent from various combinations of sewage treatment and disposal; their results were shown in Table 1.1. They assumed three principal types of discharge: (i) directly onto the beach; (ii) through a "short" outfall, where sewage contamination of recreational waters is likely; and (iii) through an "effective" outfall, where the effluent is efficiently diluted and dispersed and does not contaminate recreational waters.

The WHO concluded that, if an effective outfall is provided, the level of treatment has minimal bearing on the risk to human health. The risk from effluent discharged through an effective outfall is low regardless of whether the treatment is secondary, primary, or even preliminary. On the other hand, if a short (ineffective) outfall is provided, even secondary treatment does not reduce the risk to safe levels. These findings are very important for developing countries, since they imply that only preliminary treatment is required with an effective outfall. Preliminary treatment is relatively inexpensive and is especially appropriate where financial resources are scarce. The investment and O&M cost is about one-tenth that of conventional secondary treatment.

The WHO also concluded that stabilization lagoons that discharge directly on the beach or via a short outfall constitute a high health risk. This is important, as stabilization lagoons are commonly used in developing countries, where it is wrongly believed that their effluent does not constitute a health risk and can be safely discharged directly on the beach.

For coastal cities, preliminary treatment followed by discharge through an effective outfall is therefore an affordable, effective, and reliable solution. It is simple to operate and free of negative health and environmental impacts. Many outfalls of this type are successfully functioning and they have a proven track record in many coastal cities all over the world. A monitoring program should be initiated prior to and continuing after the discharge commences in order to verify the performance of the outfall and to determine if higher treatment levels are needed. This prevents unnecessary investment in expensive treatment plants, or, even worse, advanced treatment that is unaffordable and results in "no action," and continued contamination of beaches. Overly stringent treatment requirements (and misplaced faith in the efficacy of treatment) can also result in shoreline discharges with considerable risk to public health.

An outfall system (see Figure 1.1), including the outfall and near field, could in fact be considered a treatment plant. It provides a high level of treatment that is much superior to that which any conventional land-based plant can achieve. Land-based plants can, in extreme cases, remove up to 95%, of BOD and TSS. If the effluent is not disinfected, they remove 50–80% of pathogenic organisms, leaving it with practically the same health risk as raw sewage, although stabilization lagoons can achieve up to 99% pathogens removal with a long detention time. But a well designed outfall system reduces the concentrations of all contaminants by more than 99% by physical dilution. For pathogenic organisms even this may be insufficient, but taking into account their mortality in the marine environment, a well designed outfall can ensure levels below that which could cause health risks on beaches.

The efficacy of preliminary treatment combined with an effective outfall has been demonstrated in many field studies. Their impacts are small and contained in a limited area. For example, five years of observations of two outfalls in Chile, each consisting of preliminary treatment followed by an effective outfall, are discussed in Chapter 14. The results show that the concentrations at 100 m from the diffusers for all parameters, except fecal and total coliforms, are similar to background values. Although the concentrations of coliforms in the effluents are extremely high, their concentrations are reduced to levels that meet stringent standards about 100 m from the diffusers, and with mortality they quickly fall to background levels beyond this. The study concludes that: (i) The effect of the outfalls on the ocean water quality is insignificant; (ii) Heavy metals and micropollutants were below detectable limits around the outfall; (iii) There were no negative effects on benthic communities.

In spite of these and other similar findings around the world and the recommendations of the WHO and the NRC, many industrialized countries require a high level of wastewater treatment, often secondary, prior to discharge. Developing countries tend to copy industrialized nations, so standards in many developing countries are quite stringent and practically require secondary treatment. An exception is the effluent quality standards for outfalls that were adopted in 2000 in Colombia. They recognize the importance of outfalls as a legitimate and capable method of wastewater disposal. It requires that the level of treatment, in combination with initial dilution, dispersion, assimilation and decay, will guarantee compliance with the quality objectives of the receiving body. In other words, the combination of treatment and the outfall must comply with the quality standards, not just the treatment alone. Other wastewater treatment standards in developing countries are discussed below.

In this chapter we discuss wastewater management and appropriate treatment for outfall discharges. We address the implications of population growth and propose guidelines for good practice in wastewater management. The staging of construction of the treatment installations is also discussed. Treatment technologies are compared based on their environmental impacts, cost, and practicality.

It is concluded that discharge through an effective outfall often requires only preliminary treatment. Rotating fine screens, discussed in Section x.x.x, are especially appropriate. Because some developing, and certainly industrialized, countries require higher levels of treatment, we also present an overview of treatment methods with a particular focus on appropriate processes for developing countries. Preliminary and physico-chemical treatment processes are discussed in some detail as they can provide high-quality effluent that is well-suited to ocean discharge at much lower cost than conventional advanced processes. Either will provide an effective and reliable means of wastewater management and disposal with minimal environmental and health risks.

We advocate preliminary treatment for outfall discharges and argue that advanced levels of treatment are often counterproductive. Although the emphasis is on developing countries, much of the discussion, especially of preliminary and physico-chemical treatment for ocean discharges, is equally applicable to developed countries. Water quality standards can usually be met with only preliminary treatment if an effective outfall is used.

6.2 GLOBAL WATER SUPPLY AND WASTEWATER DISPOSAL

Most of the World's population lacks access to safe sanitation. At the United Nations Conference on Sustainable Development held in Monterrey, Mexico in

March 2002, the goal was set (among other Millennium Development Goals, MGD) of reducing, by 2015, the number of people lacking access to safe water by 50%. The Johannesburg World Summit, held in August 2002, identified deficient water and sanitation as a priority issue affecting developing countries and drew up an action plan aimed at increasing access to water and sanitation services. These goals were ratified in the Third World Water Forum held in Kyoto, Japan in March 2003.

UN and World Bank studies show that: about 1 billion people (1 out of 6 inhabitants of the planet) mostly in low and middle income countries lack satisfactory access to safe water supplies; about 2 billion people (2 out of 6 inhabitants of the planet) mostly in low and middle income countries lack access to sanitation services; about 4 billion people (4 out of 6 inhabitants of the planet) are not connected to wastewater treatment systems; about 30,000 children die daily from water borne diseases; and mortality is high but morbidity is even higher.

A forecast of the world's population growth for developing and high-income countries is shown in Figure 6.1. It indicates that almost all the growth will take place in developing countries, and a strong migration from rural to urban areas will take place. These pose enormous challenges to the water sector. For example, meeting the MGD sanitation goals would require construction of sanitation facilities for populations of around 500,000 every day for the next twenty years! The consensus from the Second World Water Forum held in the Hague in March 2000 was that the annual investment required by 2025 to meet the world's needs for drinking water and sanitation, irrigation, industry, and environmental management will increase to US$180 billion from the current level of about US$75 billion.

The trends of population growth and urbanization have major implications for the water sector. They include: increased urban water demand; increased generation of municipal wastewater; and increased demand of water for agriculture. Future water resources management strategies must therefore be based on: efficient irrigation; water conservation; small scale solutions for better use of all available resources; reuse of municipal effluents for irrigation; effluent recycling for non-potable industrial and municipal use; application of adequate tariff policies that cover costs; application of effective regulations; and desalination.

A major deficiency in developing countries is inadequate disposal of wastewater. The percentage of wastewater that undergoes any type of treatment is often very low, typically less than 10%. Nevertheless, we consider that for most developing countries the highest priorities should be increasing the safe drinking water supply and increasing the area of sewerage coverage rather than sewage treatment.

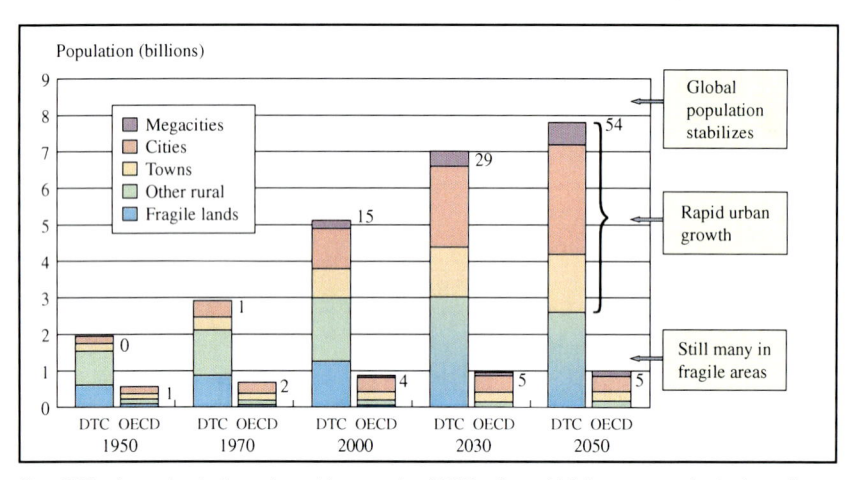

Figure 6.1 World population growth forecast (World Bank, 2003)

Because of expected population growth and prevailing priorities, it seems unlikely that the percentage of wastewater that is treated will increase substantially in the future, unless innovative affordable treatment options become available. Physico-chemical treatment may provide this, as discussed in Section 6.7.3.

6.3 PROPOSED GUIDELINES FOR GOOD WASTEWATER MANAGEMENT PRACTICE IN DEVELOPING COUNTRIES

6.3.1 Overview

Many cities in developing countries suffer major deficiencies in water and sanitation infrastructure, especially in wastewater management. Suggested guidelines for correcting these deficiencies are:

- Ordering investment priorities as:
  - water supply;
  - sewerage (i.e. expanding wastewater collection);
  - wastewater treatment.

- The greatest health benefits result from removal of wastewater from neighborhoods, not from treatment. Therefore, wastewater should be conveyed downstream of the city limits for treatment and disposal. Construction of multiple treatment plants within city limits should be avoided (microtunnels can be used for wastewater conveyance in congested areas; see Chapter 16);
- Realistic standards for effluent quality should be adopted. They should be flexible in terms of quality and timing, and take into account the assimilation capacity of the receiving water bodies;
- The type of treatment must be adjusted to the receiving water body. Wastewater contains many types of pollutants, whose relative impact on the environment depends on the characteristics of the receiving water body. Emphasis should be given to removal of critical pollutants, and unnecessary removal of pollutants that cause insignificant impacts on the environment should be avoided);
- Appropriate technologies for wastewater treatment should be used. They should be based on simplified processes which are cheaper than conventional processes, simple to operate and maintain, and can provide the required effluent quality;
- Wastewater management schemes should be developed in stages in accordance with financing availability. They should account for the system's construction time to enable more stringent ultimate effluent standards to be met while maintaining acceptable environmental standards in the early stages. This can often be achieved by taking into account the assimilation capacity of the receiving body;
- Involve, from the start of project planning, all stakeholders through participatory processes to ensure support and mitigate opposition;
- Secure subsidies for developing countries. Contrary to the position held by many, users in developing countries cannot pay the full cost of wastewater treatment, so it is necessary to secure subsidies for these users. Such subsidies are well justified.

6.3.2 The utility perspective

Steps that water utilities can follow to achieve these goals are:

- Develop the wastewater management scheme in stages, taking into account the assimilation capacity of the receiving body (using water quality simulation models), to ensure that even during the first stages no environmental problems are caused. This is usually a feasible task;

- Apply a comprehensive water quality monitoring program prior to and following the construction of the treatment installations;
- If monitoring results demonstrate no environmental impacts after the first stage, delay or eliminate construction of subsequent stages, to prevent waste of resources;
- If the monitoring results demonstrate environmental problems, implement the subsequent management stages.

6.3.3 The government perspective

Governmental subsidies are usually needed to solve wastewater management problems. Since financial resources are scarce in developing countries, their governments must develop sound strategies of wastewater management. High priority and support should be given to municipalities whose wastewaters cause significant public health or environmental problems such as contamination of critical water bodies or water resources that are used as downstream water supplies. Cities whose wastewater discharges do not cause severe problems, for example cities that discharge to waters with high assimilation capacity, are lower priority. The priority cities must be identified prior to beginning the support program.

Governments must also ensure that the utilities which receive support have a sustainable institutional structure to ensure that the treatment installations are put into operation. Institutional improvements should be included in projects when necessary. Finally, governments must ensure that its resources are used for appropriate, simple, and low cost technology, and not on costly and difficult to operate processes. The appropriate technologies that each government will support should be identified prior to initiating the support program.

To achieve these goals, governments in developing countries should provide subsidies (grants) for wastewater projects to communities that meet the following conditions (eligibility criteria):

- High priority communities whose wastewater causes significant environmental or human health threats (these communities should be identified upfront);
- Communities that agree to apply the appropriate technologies as defined by the government;
- Communities that agree to undertake strict institutional strengthening to ensure sustainability, preferably through participation of specialized private operators.

The amount of the grants should be based on a financial analysis of the utility and on the institutional improvement commitment that the utility is willing to undertake. The issue of establishing levels of subsidies is beyond the scope of this book.

The international community can help to promote good wastewater management practice in developing countries by setting up and contributing to an international fund (similar to the Global Environmental Facility, GEF). This will enable transfer of funds to governments, in the form of grants, loans or a combination of the two, for financing national projects based on the strategies and principles presented above.

6.4 EFFLUENT STANDARDS

6.4.1 Standards in industrialized countries

We consider now specific issues relating to wastewater treatment for marine outfalls. Defining the level of treatment and selecting the unit processes often depends on effluent quality standards prescribed by law rather than on actual impacts on the receiving waters. Industrialized countries often have stringent wastewater standards (which are becoming more stringent with time) that require almost complete elimination of pollutants prior to discharge.

Secondary treatment is often mandated in industrialized countries for ocean discharges, the same as required for inland waterways. This is the case in the US and the EU, although not all cities in industrialized countries comply with this standard. The level of treatment is often primary, although the tendency is for more stringent enforcement of the regulations. In the US, effluents discharged through ocean outfalls are usually treated to the secondary level. A few cities, particularly San Diego, California (Union Tribune, 2002) and Honolulu, Hawaii (Grigg and Dollar, 1995) provide only Chemically Enhanced Primary Treatment (CEPT, see Section 6.7.3.2) prior to ocean discharge. These two cities succeeded in proving that upgrading to secondary would not have any perceptible benefits to the environment, so they received a waiver from the secondary treatment requirement under the US EPA 301(h) provision of the Clean Water Act (USEPA, 1994) that allows less than secondary treatment if the discharge is shown to not adversely affect the marine environment.

Application of such stringent discharge standards may be reasonable for industrialized countries, where users can pay for high levels of treatment and governments usually provide significant subsidies for construction of treatment plants. These standards evolved over many years, however, and were initially less stringent, since the need for phased development was well understood.

6.4.2 Standards in developing countries

In developing countries, wastewater treatment and reuse standards are often based on those in industrialized countries, especially the US and the EU. They do not account for financing and construction time limitations, however. The standards often prescribe an effluent of such quality that secondary treatment or higher is required to achieve it. Since the cost of complying with these standards is very high, they are usually beyond the reach of developing countries, resulting in "No Action" with devastating public health and environmental impacts.

For example, Colombia used to require removal of 80% BOD and 80% Total Suspended Solids (TSS) prior to discharge to any receiving body, and applied discharge fees to violators that multiplied every six months. Only 8% of the country's wastewater is treated, however, mostly to lower levels. Recently, the discharge fee approach was changed (the fees were reduced, since most violators did not pay them) and a major modification for outfall discharges was made in 2000 (see below). Another example is Costa Rica which requires an effluent quality of 50 mg/l BOD and 50 mg/l TSS for all municipal wastewater, even though only 4% of the country's wastewater is treated. Costa Rica has also recently modified this requirement (see below).

The populations of developing countries deserve the same rights as those of industrialized countries, so their effluent quality standards should ultimately approach those of industrialized countries. Such high levels cannot be achieved immediately, however. They must be achieved in stages in accordance with their financial capabilities and actual impacts on the receiving waters.

Unfortunately, the concept of achieving quality standards in stages is not widespread in developing countries and is almost completely absent from their legislations. An exception was recently adopted in Costa Rica, where Supreme Decree 32133-S was enacted on December 7, 2004. It allows compliance with prescribed effluent standards by implementing the required investment program in stages. It allows the first stage to be directed to priority investments in wastewater collection and its conveyance downstream of the city limit, without treatment, to focus on improving public health within the urban areas.

Another interesting approach to setting effluent quality standards for ocean outfall discharges was adopted in 2000 in Colombia. The Technical Norm of the Water and Sanitation Sector in Colombia – RAS (Resolution 1096 of November 17, 2000 del Reglamento de Agua y Saneamiento) in Article 180, which was amended in 2000, states that: *"The level of the required pretreatment prior to wastewater discharge via submarine outfalls is such that, in combination with the processes of initial dilution, dispersion, assimilation and decay it will guarantee compliance with the quality objectives of the receiving body, indicated by the environmental norm in effect and with other provisions which modify, expand or substitute it."* This legislation is notable in that it recognizes the importance of outfalls as a disposal method and the fact that the outfall is part of the treatment system. So the combination of treatment and outfall must comply with the quality standards, not just the treatment alone. Articles 177–179 of the RAS specify procedures for design and construction of outfalls, the studies required prior to design, and the mathematical modeling required.

Despite the requirement in many developing countries of secondary treatment prior to discharge, in most cases the level of treatment (if any) is only preliminary. Experience has shown that this is satisfactory (with effective outfalls) so there is little demand to upgrade treatment.

6.5 KEY POLLUTANTS IN WASTEWATER

Typical municipal wastewater consists of about 99.9% water and 0.1% pollutants. About 60 to 80% of the pollutants are dissolved and the rest are suspended matter.

The pollutants include mineral and organic matter, suspended solids, oil and grease, detergents, nitrogen (in various forms such as ammonia, nitrate, and organic nitrogen), phosphorous in various forms, sulfur, phenols, and heavy metals. The organic matter consists of many compounds, and since it is practically impossible to identify and quantify all of them, it is usually represented by the Biochemical Oxygen Demand (BOD), which is the amount of oxygen required to biologically decompose the organic matter, or by Chemical Oxygen Demand (COD), which is the amount of oxygen required to chemically decompose the organic matter. Municipal wastewaters also contain large amounts of bacteria and viruses, some of them pathogenic. Total bacteria counts in raw wastewater are typically about 10^7–10^8 MPN/100 ml. As with organic matter, since it is difficult to identify and quantify all organisms present in a sample of wastewater, organisms are usually represented by indicator organisms. The most commonly used indicator organism is fecal coliforms (although others are now being used, such as intestinal enterococci). Fecal coliform counts in raw municipal wastewater are typically in the range 10^6–10^7 MPN/100 ml.

The environmental impacts of the many pollutants depend on the outfall and diffuser design and the characteristics of the receiving body. The treatment should therefore be tailored to the particular conditions. BOD is a key pollutant for wastewaters discharged to rivers, especially low flow rivers, since it may result in oxygen depletion. Nutrients (nitrogen and phosphorus) are important in wastewaters discharged to lakes, lagoons, or enclosed bays, as they can lead to eutrophication.

The impacts of pollutants discharged to open coastal waters are different from inland waters. The National Research Council (1993) prioritized the pollutants of main concern for coastal waters as shown in Table 6.1 (see also Section 2.2). Note that most traditional contaminants are of low importance. For example, BOD is low priority due to the high dilution and large reaeration capacity that can be achieved with long outfalls equipped with effective diffusers.

Table 6.1 Constituents of concern for coastal wastewater discharges (after National Research Council, 1993)

Priority	Pollutant groups	Examples	Notes
High	Nutrients	Nitrogen, phosphorus	An excess of nitrogen and phosphorus may cause eutrophication
	Pathogens	Enteric viruses	Organisms harmful to humans, cause many illnesses and deaths in developing countries
	Toxic organic chemicals	PAHs	
Intermediate	Trace metals	Lead	
	Other hazardous materials	Oil, chlorine	
	Plastics and floatables	Beach trash, oil, and grease	Fats, oil, and grease (FOG) may float on the surface of receiving waters, interfering with natural re-aeration, and can create an unsightly film on the water surface
Low	Biochemical oxygen demand (BOD)		A measure of the oxygen demanding materials. High BOD levels in natural waters cause a drop in dissolved oxygen (DO) concentration.
	Solids		High levels of total suspended solids (TSS) can damage benthic habitats and cause anaerobic conditions on the bottoms of lakes, rivers, and seas

The impacts of these constituents can be controlled by suitable outfall design and wastewater treatment. Near field dilution alone will rapidly reduce the concentrations of all pollutants, including toxic materials, by factors of at least 100 rendering their contamination risk insignificant. Nutrients will usually not cause eutrophication due to the high initial dilution and flushing capabilities of most coastal waters (see Chapter 3). The constituents of main importance are therefore microbial contaminants, which, due to their high concentration in the wastewater can remain in high levels even after near field dilution and cause potential health risks. The combination of treatment and outfall design is intended to minimize health risks by ensuring that diluted wastewater does not reach areas of human usage. To accomplish this, the outfall should be long enough and preferably in deep water so that currents dilute and transport the wastes offshore. Ocean currents should be measured and modeling performed (see Chapters 3, 4, and 5) to ensure that water quality standards are met and the wastes are not transported to bathing beaches or other sensitive areas. If a short outfall is unavoidable, disinfection may be needed to ensure compliance with bathing water standards. Floatables may also be of concern since they can contain bacteria and may be driven onshore by local winds.

Preliminary or primary treatment will therefore often suffice because dilution will quickly reduce contaminant concentrations below those required by water quality standards. Even in sensitive areas such as estuaries or near coral reefs, the diluting capacity of the ocean can lower pollutant concentrations enough to prevent harm. In deep water the plume may be trapped by the stratification (Chapter 3) and not impinge on coral reefs.

6.6 PRINCIPALS AND PROCESSES OF WASTEWATER TREATMENT AND DISPOSAL

6.6.1 Introduction

The treatment process chosen must be tailored to the proposed disposal method, which may include discharge to an ocean, river, lake, or groundwater, reuse for irrigation, or recycling for other purposes. The size and design of wastewater treatment plants depends on the expected wastewater flowrate and contaminant loads (ton/day).

Reliable forecasts of wastewater flows and contaminant loads are important since treatment plants are designed to handle conditions typically 10 to 20 years in the future. Plants are sometimes constructed in a modular form, with new modules added as the flow increases. Predictions of flows and contaminants are based on forecasts of population growth, changes of water consumption, contribution of contaminants per capita, and evolution of the sewerage coverage area. Flow forecasts should be prepared with care, since the flow may not always grow much with time, especially in developing countries. Changes of tariff policy may induce more efficient water use and moderate the evolution of water consumption and generation of wastewater. An example is Bogotá, Colombia, where water consumption has been constant over the past 15 years in spite of significant population growth. This resulted from: (i) periods of severe failures in the water supply, with consequent severe rationing that brought people to consume less, and (ii) significant tariff increases which decreased consumption.

In this section, we give an overview of the main processes used for wastewater treatment. For more details, see specialized texts such as Tchobanoglous *et al.* (2003).

6.6.2 Treatment processes

The main wastewater treatment processes are shown in Figure 6.2. Treatment is conventionally divided into preliminary, primary, secondary, and tertiary processes. Preliminary and primary treatments are based on physical processes, secondary on biological processes, and tertiary usually on physico-chemical processes, although it sometimes includes biological processes. For each of these four unit processes there are several alternatives that can achieve similar results, so the total number of combinations is large. Many treatment processes also generate sludge as a byproduct. There are several alternatives for sludge treatment and disposal (see Section 6.6.3) so the number of combinations of treatment methods is myriad; one can find treatment plants with completely different unit processes that achieve similar results with widely differing costs.

The main conventional processes are:

Preliminary Treatment: All treatment processes, with the exception of septic tanks and household systems, require some sort of preliminary or screening process to remove large and floatable objects. Most widely used are screening and grit removal, which also removes floatable material and rags. Because of the importance of preliminary treatment for ocean discharges and as an appropriate treatment for developing countries, it is discussed in detail in Section 6.7.2.

Primary Treatment: Primary treatment often precedes biological treatment in conventional secondary wastewater treatment facilities; its main purpose is to reduce BOD and suspended solids. Sedimentation tanks are the most common form for domestic sewage. The suspended solids settle on the bottom, are scraped to a central point and then removed by a sludge pump. Scum, primarily oil and grease, floats to the surface where it is collected by a mechanical arm and periodically drawn off. A dissolved air flotation (DAF) process can remove oil and grease in less space. The wastewater and air are pressurized and released into a tank open to the atmosphere. Small bubbles float to the top and become enmeshed in the light solids and oils and bring them to the surface where a skimmer collects them. The clarified liquid continues to downstream processes. DAF is more commonly used for industrial wastewater.

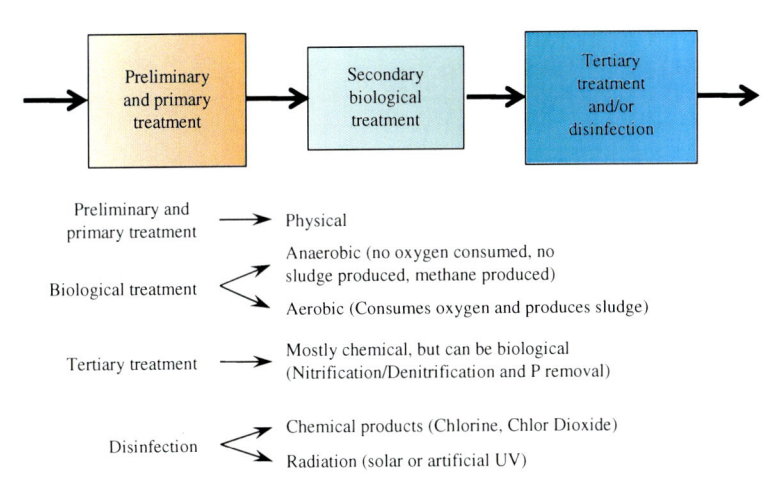

Figure 6.2 Main wastewater treatment processes

A conventional sedimentation tank removes 25–40% of BOD, 40–60% of total suspended solids, and about 50% of the bacterial load. The main residuals collected are solids, scums, and oils. For a medium-strength wastewater, the amount of sludge generated is about 0.10 to 0.17 kg/m^3 of wastewater (about half that generated in full secondary treatment).

The cost of primary treatment is about half that of full secondary treatment. Although primary treatment mechanical processes are relatively simple, some routine maintenance is necessary. This mostly consists of upkeep of pumps, sludge scrapers, scum collectors, and motors. The Operation and Maintenance (O&M) costs of primary treatment are mainly due to treatment and disposal of sludge. The capital cost of the sludge treatment unit is a significant part of primary treatment.

Secondary Treatment: The main original purpose of secondary treatment was BOD removal. Aerobic, anoxic, and anaerobic bacteria feed on organic matter, transforming the BOD to bacterial mass. Aerobic processes are by far the most common for large treatment facilities. In suspended-growth systems, aerobic bacteria need sufficient oxygen to metabolize the organic material. Oxygen is provided by aeration through mechanical aerators, a diffuser, or some other process. The bacteria must be subsequently separated from the wastewater. Most secondary processes have a separate secondary sedimentation tank to settle the flocculated cell mass. The effluent continues to discharge or to downstream processes. The sludge is returned from the sedimentation tank to the aeration tank, which maintains a viable concentration of bacteria to metabolize the incoming organic material. This is called return activated sludge, or RAS. A portion of the sludge must be removed and not returned to the aeration tank in order to prevent sludge concentrations increasing to levels that would be too high to sustain sedimentation. The removed sludge is called excess activated sludge or Waste Activated Sludge (WAS). Sludge return is not necessary for fixed film processes or the Sequencing Batch Reactor (SBR) process.

Lagoons are often used in developing countries. They are natural systems that can provide secondary treatment. In aerobic lagoons the oxygen is provided by algae that develop in the lagoon; in anaerobic lagoons oxygen is not required since the metabolism of the organic matter is by anaerobic bacteria that do not consume oxygen.

Common high-rate secondary treatment processes that require less land than lagoons and wetlands are:

- The **activated sludge** process (Figure 6.3), where effluent is brought into an aeration tank, air is bubbled into the wastewater mixture (mixed liquor), and aerobic bacteria metabolize the dissolved and suspended organic material. From the aeration basin, the effluent flows into a secondary sedimentation tank, where the cell mass settles out.
- The **oxidation ditch** process is an activated sludge process where wastewater flows into a ring-shaped channel. Oxygen is not evenly mixed throughout the ditch as in a conventional activated sludge process. This provides zones of varying reaction, allowing more operational control. Cell mass is settled out in a secondary sedimentation tank and recycled back to the oxidation ditch.
- In a **trickling filter**, primary effluent is evenly distributed over a circular bed of fist-sized stones 900 to 1800 mm deep. Bacteria, fungi, and algae grow on the rock surface. As wastewater flows between the rocks, aerobic bacteria metabolize the organic material in the wastewater. As the biomass grows, the influent wastewater flow sloughs off the excess, which settles out in a secondary sedimentation tank.

- In **sequencing batch reactors (SBRs)**, all the treatment steps take place in a single completely mixed tank to which influent is intermittently directed. The treatment consists of discrete, timed processes: fill, mix/aerate, settle, withdraw effluent, and withdraw sludge. Historically, SBRs have been used for only small treatment facilities. There has been a recent resurgence of interest in the SBR process because it eliminates the need for secondary sedimentation and pumps.

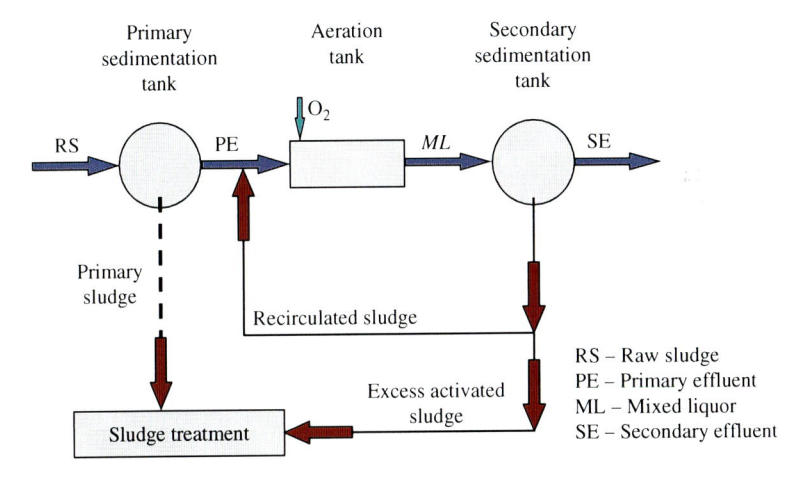

Figure 6.3 The activated sludge process

These secondary treatment processes are usually best for large, high population density communities because of their high cost and the high skill levels required for O&M. In industrialized countries, they are also used for medium and small communities. They can produce good quality effluent if operated and maintained properly, but very poor effluent if operated improperly. They generate 0.10 to 0.15 kg of sludge per day per m^3 of wastewater, whose treatment and disposal is complicated and expensive. It is generally high in volatile solids and can become septic quickly, producing offensive odors if not treated or disposed immediately. Sludge treatment installations form part of the treatment plant and should not be omitted (as sometimes happens). Flies can be a serious nuisance with trickling filters, as they live and breed within the filter medium. Energy costs can be substantial and a standby generator must be provided, especially in developing countries, in case of power failures lasting longer than a few hours.

Anaerobic processes present an interesting option for secondary treatment in hot climates. This is because they produce little sludge, so sludge treatment costs are low, allow energy recovery from the methane they produce, and have low capital and operation costs. They are not in widespread use, however, and have been used only for small communities. This because they are highly temperature sensitive and function well only in hot climates, so are not of interest in industrialized countries, most of which have cold climates and where most new technologies are developed; they do not require sophisticated mechanical equipment and are therefore not interesting to equipment manufacturers and do not enjoy their aggressive marketing of aerobic processes. In recent years the use of Upflow Anaerobic Sludge Blanket (UASB) reactors has spread in Latin America, especially Brazil and Colombia. This process does not produce a high quality effluent, however, and when such an effluent is necessary, the UASB effluent needs a further polishing process. UASB in Brazil is commonly followed by trickling filters.

Tertiary Treatment for Nutrient Removal: Conventional secondary treatment removes some nitrogen and phosphorus; specialized processes are needed to remove more. These include nitrification and denitrification for nitrogen removal, partial phosphorus removal by biological processes, and chemical precipitation for complete phosphorus removal. Nutrient removal processes are more complex and expensive than secondary treatment. They are rarely required in developing countries, except for discharges into enclosed water bodies where eutrophication may occur such as lakes, estuaries, or bays. Capital costs are high and include construction of additional tanks, pipes, and recirculation pumps. O&M costs increase significantly when nutrient removal is included. It includes maintenance of the aeration systems, pipes, and pumps. The processes are complex and require skilled labor for efficient operation. Chemical costs for phosphorus precipitation can substantially increase O&M expenses.

Disinfection: removes pathogens and other bacteria from treated wastewater effluent. Common processes include chlorination, ultraviolet (UV) radiation, ozonation, and pond disinfection. Chlorine and ozone are strong oxidizing agents, which oxidize organic and inorganic matter and quickly kill all the pathogens they contact. Chlorine can be added to wastewater in gas, liquid, or tablet form; ozone is added as a gas only. UV radiation sterilizes pathogens and is applied through low-pressure mercury lamps. Pond disinfection is a natural process that occurs in successive stabilization ponds due to visible light and ultraviolet radiation from the sun, sedimentation, and natural die-off. Wastewater effluent discharged below ground generally undergoes significant pathogen and bacteria removal as it travels through the soil, so a properly designed discharge of effluent to groundwater can also achieve disinfection.

Typical maintenance of disinfection installations includes replacing chemicals, adjusting feed rates, and maintaining mechanical components. Most chlorine systems are designed for minimal maintenance; UV radiation requires little maintenance other than regular cleaning and lamp replacement. Ozone generating and feeder equipment uses large amounts of electricity and is complicated. The USEPA estimates that 8 to 10 kW-hours are used for each pound of ozone generated.

Chlorination is the most widely used disinfection process around the world. UV radiation performs less well with effluents that are high in turbidity or suspended solids so sand filtration prior to radiation is common. Ozonation disinfects more powerfully than chlorine, with no harmful by-products. It is usually used to disinfect highly treated secondary or filtered effluent. Ozone must be generated on-site, which can be costly and requires a reliable power supply. Pond disinfection is a simple technology and maintenance-free but requires a large land area. Chlorination produces many undesirable organic compounds that are toxic to humans and aquatic life. For this reason, it is desirable to remove as much of the organic material as possible before adding chlorine. Dechlorination may be needed to lower the residual chlorine concentration in the effluent. Chlorine gas is very hazardous, and robust safety features must be employed where it is stored. Because of these safety concerns, liquid sodium hypochlorite is now commonly used instead of chlorine.

6.6.3 Sludge treatment and disposal

The solids, or sludge, generated by treatment processes must be disposed of safely and economically. There are various processes for this including sludge thickening, sludge stabilization, sludge dewatering, and drying and cold digestion/drying lagoons. These processes are described below.

Sludge thickening removes water from sludge to reduce the cost of subsequent treatment or sludge disposal as a concentrated liquid. Typical processes include thickening by gravity, lagoon, gravity belt, and centrifuge. Gravity thickening feeds liquid sludge to a tank. The effluent is discharged over a weir and the thickened sludge is pumped from the tank bottom to be digested or disposed as a liquid sludge. Lagoon thickening is gravity thickening in an earthen basin. Gravity belt thickening (GBT) uses the gravity zone of a belt filter press for sludge thickening. Centrifuge thickening involves pumping the sludge to a solid bowl centrifuge rotating at up to 3,000 rpm. Lagoon thickening is appropriate for many applications in low to medium population density communities; it is simple and inexpensive but requires a large land area. Gravity thickening uses less land area, but requires more operator attention and equipment maintenance. GBT and centrifuge thickening are appropriate for high population density communities and industrial use. GBT thickening requires higher operator attention and regular maintenance by qualified technicians. Centrifuge thickening has high power requirements. Maintenance can require highly skilled maintenance workers and expensive shipment of spare parts from outside the country of use.

Sludge stabilization is performed on thickened waste solids to reduce the volatile solids and pathogen content so they can be safely disposed or used for land application. It also reduces the solids volume. Typical processes include: anaerobic digestion; aerobic digestion; composting; lime stabilization and air drying. Anaerobic digestion is oxidation in the absence of free oxygen in closed tanks; aerobic digestion is oxidation in aerobic conditions. The stabilized sludge is drawn off the bottom or from the mixing tank. Composting is a process where aerobic organisms degrade and disinfect already thickened sludge. The sludge is mixed with bulking material, such as wood chips, to provide porosity for aeration and then laid over a network of porous piping and aerated. The stabilized sludge can be used as fertilizer. Lime stabilization is the addition of alkaline compounds to raise the pH of the sludge mixture. Equipment and O&M costs can be very high, and trained operators are needed for proper operation. Air drying beds are shallow paved or earthen basins where thickened waste sludge is allowed to naturally dry. Composting and air drying require large land areas and large quantities of organic materials such as wood chips or waste plant material as a bulking agent. Air drying is easiest to operate, but may not be suited to rainy areas.

Sludge dewatering removes water from sludge to reduce the cost of subsequent treatment processes or prior to sludge disposal as concentrated solids. The processes are similar to thickening processes, but higher solids concentrations are achieved. Typical processes include belt filter press dewatering, centrifuge dewatering, screw press dewatering, and plate and frame dewatering. Belt filters press the sludge with belts that apply pressure and squeeze out the liquids. In centrifuge dewatering, sludge is pumped to a solid rotating bowl. In screw press dewatering the sludge is pumped into a perforated cylinder surrounding a rotating screw. The screw forces the sludge toward the end of the container and progressively dewaters it by the pressure of the screw against the sludge. Plate and frame presses are an established, but high maintenance, and high cost dewatering processes. Belt filter press, centrifuge, and screw pump operations require close operator attention for control of loading rate. The equipment requires regular maintenance and may require periodic import of maintenance parts from outside the country of use.

Cold digestion/drying (CDD) lagoons are a low-technology alternative that incorporates thickening, stabilization, dewatering, and storage in a series of earthen basins. Digestion and stabilization takes place at ambient temperatures. Two lagoons are needed, one for fill and one for maturation. At the end of the one-year filling period the first lagoon is isolated and allowed to dry for up to one year and sludge fill is directed to the second lagoon. Rooted aquatic plants grow on the surface during the maturation period and assist in sludge drying by evapotranspiration. They are limited to hot climates with a prolonged dry season. CDD lagoons require a large land area but little operation or maintenance during filling.

6.6.4 Summary and costs

The following observations can be made about the treatment processes outlined above:

(1) Physical treatment has limited efficiency (which in certain cases may be sufficient), it does not remove dissolved matter, is low cost;
(2) Chemical treatment increases sludge quantities, is more costly than physical treatment but can reach a high level of treatment at much lower costs than that of biological processes;
(3) Biological treatment by lagoons is simple to operate but requires large areas so is not adequate for large communities. It is not easily controlled and therefore needs to be located far from cities. The cost of the lagoon system is low, but the cost of transporting the raw wastewater to the lagoon can be high because of the long distance;
(4) Aerobic biological treatment is efficient and can produce a high quality effluent. It removes soluble organic matter but produces high quantities of sludge. It has a high and costly consumption of energy and needs extensive electromechanical equipment. It has high investment and O&M costs, is complicated to operate, and so is not recommended for developing countries;
(5) Anaerobic biological treatment is simple to construct and operate but functions well only in hot climates. It requires limited area, can reach relatively high efficiencies, and produces low quantities of stabilized sludge (less than 30% of that of aerobic processes). It does not consume, but rather produces, energy and it requires low investment and O&M costs.

A summary of the relative costs of typical treatment processes and their removal efficiencies is given in Table 6.2, in which the processes are presented in increasing order of sophistication. This table includes some other processes, such as milliscreening and CEPT, which are discussed in Section 6.7. Some of the treatment processes are unit processes that would be combined with others to constitute a treatment plant. For example, primary treatment may include screening, grit chambers, and sedimentation tanks. On-site treatment systems commonly use subsurface disposal, especially septic tanks. It can be used as a high-density treatment system if the soil is permeable enough and there is no significant risk of groundwater contamination.

Capital costs vary widely for different plants of the same type and in different countries for two main reasons. First, each process can be designed over a wide range of different criteria. The choice of design criteria affects the effluent quality and investment costs. For example, secondary treatment by activated sludge can be designed as high rate activated sludge with a detention time of 2–4 hours in the biological reactor, or as an extended aeration activated sludge with a detention of 10–24 hours. The cost of an extended aeration reactor is much higher than that of a high rate reactor. Second, labor, material, and equipment costs, land costs as well as taxes and duties, vary in different countries. O&M costs also vary due to different design criteria, and costs of labor, energy, chemicals and spare parts. Only typical values and not ranges are presented in Table 6.2 for O&M costs.

The costs rise very rapidly as the level of treatment (and contaminant removal) increases. This is shown by the estimated annual costs to treat 100 mgd (4 m^3/s) of raw wastewater in Figure 6.4. The level of treatment sophistication is expressed by the percentage BOD removed. These costs include recovery of investment plus O&M costs.

6.7 APPROPRIATE TREATMENT TECHNOLOGY

6.7.1 Introduction

Developing countries cannot afford expensive wastewater treatment plants such as discussed above that are difficult and costly to operate. Experience shows that mechanical-biological processes are usually quickly out of operation due to high operating costs or inability to properly operate them.

Table 6.2 Costs and percent removal capabilities of typical treatment processes[a]

Constituent	Milliscreening without chemicals	Stabilization lagoons	Conventional primary	CEPT[b] Low dose	CEPT[b] High dose	Secondary, plus: Conventional primary	Secondary, plus: CEPT	Secondary, plus: Nutrient removal
Suspended solids, mg/l TSS	5–50[c]	70–85[d]	41–69	60–82	86–98	89–97	88–98	94
BOD, mg/l O_2	5–30[c]	75–85[d]	14–36	45–65	67–89	86–98	91–99	94
Nutrients:								
As mg/l TN		10–25	2–28	26–48	NA	0–60	NA	80–88
As mg/l TP		10–25	19–57	44–82	90–96	10–56	83–91	95–99
Treatment cost (US$/m³[c])	0.04	0.05	0.13	0.16	0.18	0.29	0.32	0.37
Capital cost (US$/Person)	5–20	20–40	50–100	50–100	50–120	100–200	100–200	150–250

[a]After NRC (1993) and other experience
[b]Chemically enhanced primary (CEPT)
[c]Depending on screen opening
[d]Depending on design criteria

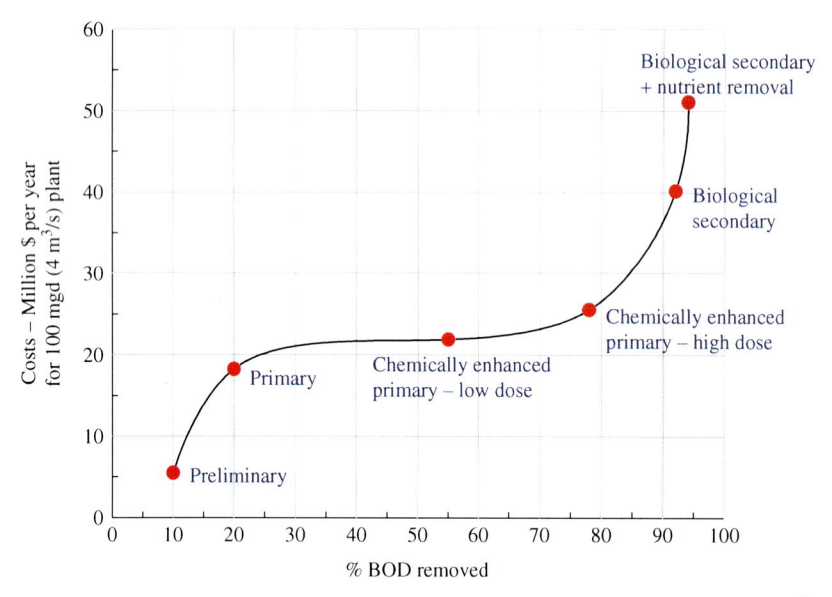

Figure 6.4 Annual costs of wastewater treatment for a flow of 100 mgd (4 m³/s)

Simple treatment processes, otherwise known as utilization of appropriate technology, are therefore recommended. Treatment plants based on appropriate technology are less expensive than high technology plants in investment and O&M. The issue of O&M costs is very significant in developing countries. Although they can often obtain soft loans and even donations to finance the construction of high technology treatment plants, they are often quickly abandoned when their owners find they cannot pay the high O&M costs.

Some appropriate technology processes are: Preliminary treatment; physico-chemical treatment; lagoon (anaerobic, facultative and aerated); anaerobic treatment; reuse for irrigation; and combinations of the above.

Some are often impractical. Lagoon treatment is inappropriate for large cities because of its large land requirements. Upflow Anaerobic Sludge Blanket (UASB) reactors have become popular in Latin America, especially Brazil and Colombia. Their treatment capacity is limited, however, and when a higher effluent quality is required, subsequent treatment must be added. Simpler anaerobic processes are mainly used for small cities. Anaerobic treatment is highly temperature dependent and does not function in cold climates, so its use is limited to hot climates. Reuse for irrigation requires high institutional capacity and a high level of institutional coordination. Anaerobic treatment, lagoons, and reuse for irrigation are not discussed further, mainly because they are not adequate and are not recommended as treatment prior to ocean outfall discharge. Information on these processes can be found elsewhere.

The most promising appropriate technology processes, especially for large communities, are therefore preliminary treatment (when a low level of treatment is sufficient) and physico-chemical treatment. These two processes are discussed in more detail below because they are particularly well better suited to, and recommended for, ocean outfall discharges.

6.7.2 Preliminary treatment

For ocean discharges, the WHO considers (Table 1.1) that preliminary treatment followed by an effective outfall imposes low risk to human health, and further treatment does not provide additional benefits. This is extremely important for developing countries and emphasizes the importance of preliminary treatment combined with an effective outfall for wastewater disposal.

The most widely used preliminary treatment method includes screening and grit removal. The screens also remove floatable material and rags. Grit removal removes inert solids and sands that could damage pumps and other mechanical equipment. Removal of gross settleable and floating materials are very import for marine outfalls. Grit may settle and accumulate on the seabed near the diffuser and may also settle in the outfall and clog it. Removal of floatables is important for two reasons: (i) large floating particles and materials (like plastic bags) cause aesthetic nuisance; (ii) oil and grease can cause a surface oil film. This may be an aesthetic problem and may prohibit oxygen transfer into the water resulting in oxygen depletion.

Conventional preliminary treatment usually consists of coarse screens followed by a grit chamber. The course bar screens usually consist of two rows of screens, the first with an opening of about 1", and the second around ½". There are many different types of grit removal processes, but most include a chamber through which wastewater flows, so that heavy, inert solids settle to the bottom. Grit removal installation usually includes also some kind of oil and grease removal. Aerated grit chambers are commonly used, in which the aeration causes the oil and grease to float and be skimmed off. After being washed, drained, and compacted, the preliminary treatment residuals are usually disposed in sanitary landfills. The screenings and grit can be removed mechanically or manually. This type of preliminary treatment requires extensive manual work and is inefficient.

New types of preliminary treatment equipment have been under development for many years. These include disk form screens, revolving tray screens, and cylindrical revolving screens. The first generations of these types of equipment were not very effective, were mechanically complicated and unreliable. More recently, innovative, automatic, efficient, and reliable equipment for preliminary treatment has been developed. The coarse bar screens and the first generation rotating screens are being replaced by rotating drum shaped fine screens (or microscreens) and the conventional grit chambers by compact vortex grit chambers. These are discussed below.

Rotating Fine Screens. In modern rotating fine screen systems, the effluent is passed through a filter screen with an opening from 0.2 to 6 mm. Various configurations are used, but the rotating drum type seems to be the most common (Huber *et al.*, 1995).

A typical rotating fine screen is shown in Figure 6.5 and its operating principles in Figure 6.6. The effluent enters the rotating drum from the inside and passes to the outside while floating and suspended material is retained. The drum is cleaned with a water or effluent spray. As the screen rotates, the captured screenings are transported upwards by a screw conveyor and dropped into a compactor that compresses them. They are then transported out by a conveyer screw or belt conveyor into a central hopper and hauled for final disposal in a solid wastes disposal site such as a sanitary landfill.

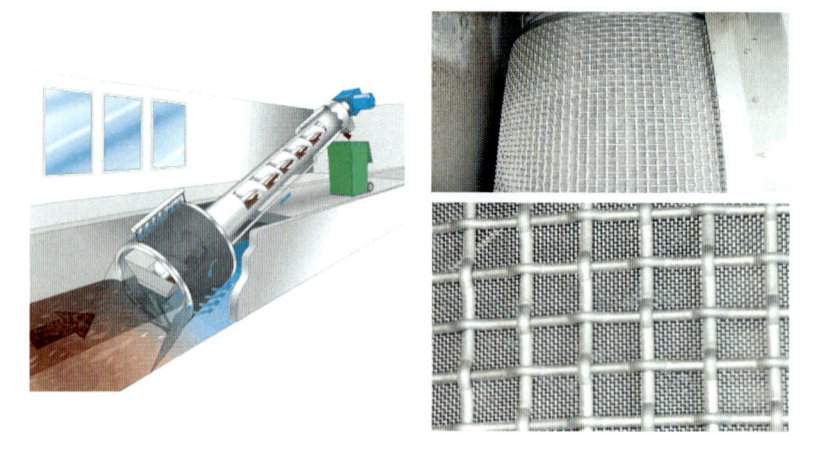

Figure 6.5 Preliminary treatment by rotating fine screens
(Courtesy Huber Technology)

Screen openings are available in the range of 0.2 to 6 mm, and drums in diameters from 600 to 2600 mm. The units are made completely of stainless steel. The wastewater flow depends on the screen opening and manufacturers provide tables or curves that give the flow capacity as a function of screen opening.

For example, a 2600 mm diameter rotating screen with an opening of 0.5 mm can handle a wastewater flow about 0.5 m³/s, while the same diameter drum and screen opening of 6 mm can handle a wastewater flow of about 1.8 m³/s. Larger flows are accommodated by arranging several machines in parallel.

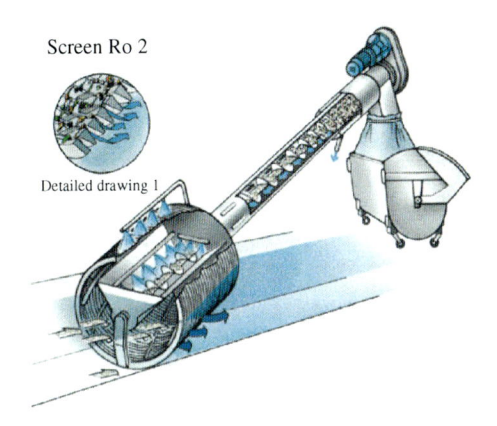

Figure 6.6 Preliminary treatment by rotating fine screens
(Courtesy Huber Technology)

Preliminary treatment by rotating fine screen is now a well-proven technology with robust design. It has reduced investment cost, is simple and reliable to operate, and can be adjusted to particular hydraulic inflow conditions. Land and maintenance requirements are low, and packaged plants are available.

Removal of suspended solids and organic matter depends on the screen opening. According to information from the Huber Group, for a 1 mm screen opening, removal efficiencies of up to 20% of BOD, up to 90% of floatable solids, and up to 70% of floatable oil and grease can be achieved. Koppl and Frommann (2004) report removal of 20–30% BOD and up to 50% filterable solids (suspended solids) in rotating screens with screen opening of 0.2 to 0.3 mm. These high removals of suspended solids result in large quantities of sludge, similar to that of primary treatment, and would require installation of sludge treatment. For wastewater discharge via an effective outfall, screen openings less than 0.5 mm are not required; openings in the range of 0.7 to 1.5 mm are more appropriate.

Photographs of a preliminary treatment plant in La Plata, Argentina, are shown in Figure 6.7. This plant consists of three drums of 2,600 mm diameter. A drawing of a rotating fine screen installation is shown in Figure 6.8.

Figure 6.7 A preliminary treatment plant in La Plata, Argentina

Wastewater management and treatment

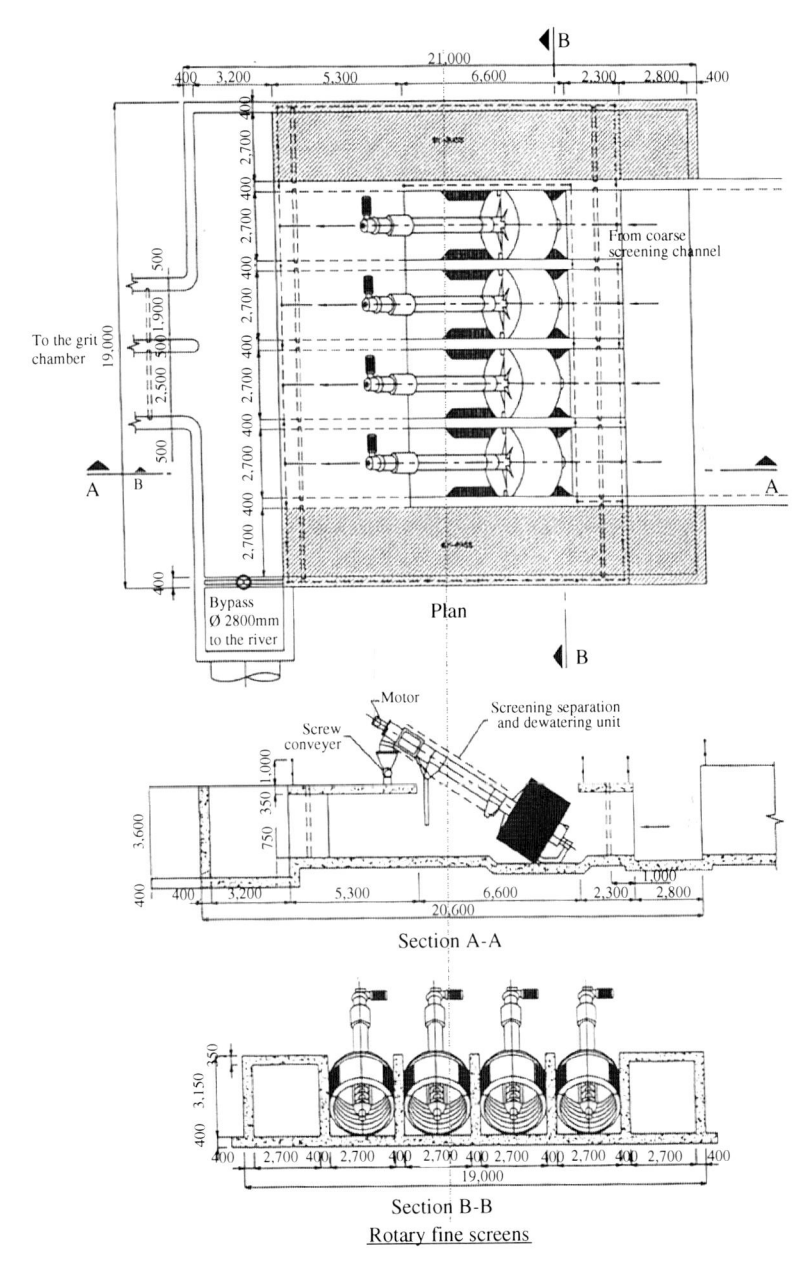

Figure 6.8 Typical plan and section of a rotating fine screen

Details of a preliminary treatment plant designed for Cartagena, Colombia, for effluent discharge to the Caribbean Sea through a marine outfall is presented in Figure 6.9. The design capacity is 4 m^3/s and it consists of eight rotating drums, each of 4 m diameter, with 1.5 mm screen openings. The plant also includes a vortex type grit chamber (see below).

It is advisable to include facilities for effluent disinfection. Although a well-designed outfall would not normally require disinfection, the capability for disinfection is helpful in gaining the acceptance and confidence of the public for preliminary treatment. It is also useful for emergencies, such as pipe leaks, until the outfall can be repaired.

Vortex Grit Chambers In this type of grit removal, a free vortex is generated by the flow that enters tangentially at the top. Effluent exits the upper part of the unit cylinder. Grit settles by gravity at the bottom, where it is removed by a belt conveyor or lift station, and organic solids exit with the effluent. Another type of vortex grit chamber (Figure 6.10) uses a rotating turbine to maintain constant angular flow velocity and to promote separation of organics from the grit. The grit settles by gravity into a hopper, from which it is removed by a grit pump or lift pump. Vortex grit chambers are more compact than conventional grit chambers, and are efficient and reliable. Because of their smaller size they are also less expensive.

Preliminary treatment consisting of rotating fine screens and vortex grit chambers are less costly than conventional preliminary treatment that uses screens and aerated grit chambers, more reliable, require less land and maintenance, and remove more solids. The investment and O&M cost are about one-tenth that of conventional activated sludge.

Preliminary treatment will often suffice as the only treatment required for discharges to seas or large rivers. Even for small rivers, if the budget is limited it is better to provide preliminary treatment to several communities than to provide secondary treatment to one community. The rationale for this is explained in Table 6.3.

For a fixed budget of €1 million (about US$1.5 million), preliminary treatment could serve a population of 140,000, eliminating about 3,400 kg/day of COD. But if secondary treatment were mandated, the same budget could serve only a population of about 8,000, eliminating about 1,270 kg/day. Therefore, to protect the river and the environment with a limited budget (which is usually the situation in developing countries), it is better to start with lesser treatment serving a wider population, rather than advanced treatment serving a smaller population.

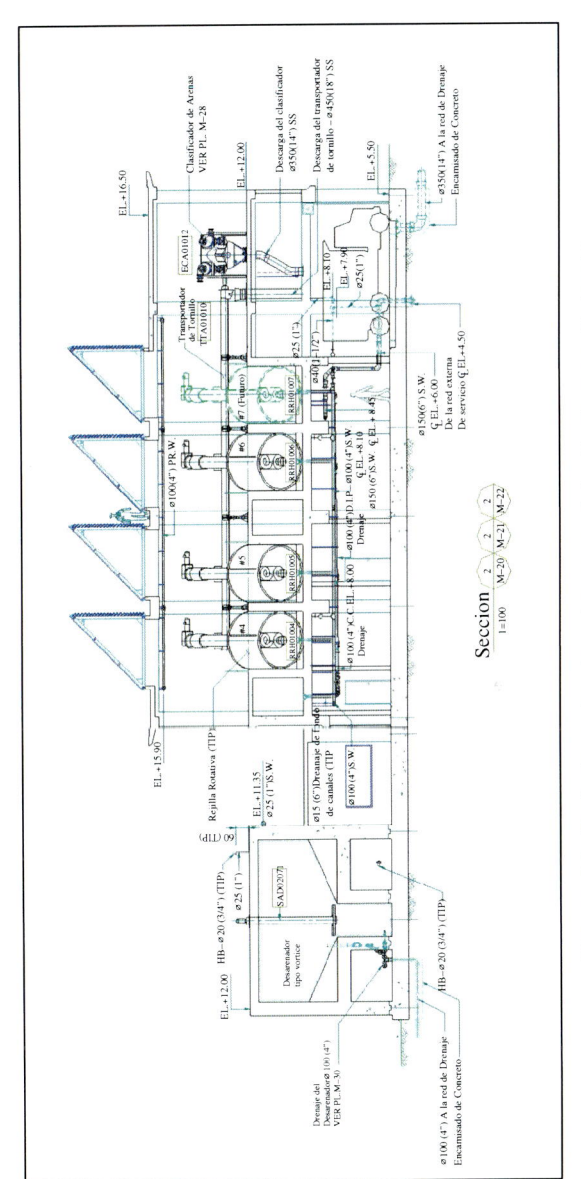

Figure 6.9 Preliminary treatment proposed for Cartagena, Colombia

Figure 6.10 Vortex grit chamber

Table 6.3 Comparisons of rotating fine screens and secondary treatment for a fixed budget of €1 million (courtesy Huber Technology)

Parameter	Unit	Rotamat high removal screen	Secondary biological treatment
Per capita investment	€/Capita	7	125
Connected population	Population	142,000	8,000
Influent COD load	kg O_2/d	17,000	960
Influent N-NH_4 load	kg O_2/d	6,700	380
Removal efficiency of COD	%	20	95
Removal efficiency of N	%	0	95
Eliminated COD load	kg O_2/d	3,400	910
Eliminated N load	kg O_2/d	0	360
Eliminated oxygen consuming substances load	kg O_2/d	3,400	1270
Investment per daily unit of oxygen removed	€/(kg O_2/d)	294	787

6.7.3 Physico-chemical treatment

6.7.3.1 The case for physico-chemical treatment

In view of expected population growths in developing countries and the order of priorities in water and sanitation, it is difficult to envisage them adopting more advanced treatment unless innovative and affordable treatment schemes become available. Conventional biological processes are not feasible since they require high investment and O&M costs and are difficult to operate. Furthermore, the effluent quality they deliver is frequently higher than needed. For large cities and all type of climates the most promising appropriate technologies are preliminary treatment and, when higher effluent quality is needed, physico-chemical treatment.

These treatments can provide the innovative and affordable treatment schemes needed in developing countries. They utilize compact units for liquid-solid separation and can achieve, at modest cost, effluent qualities that immensely improve existing conditions (under which no treatment may exist at all), and solve public health and environmental problems. Moreover, it is possible to build and operate compact systems consisting of fine screens followed by sand filtration with an option to add precipitants like coagulants and flocculants to achieve a higher quality effluent also at relatively modest cost. In the near future, it may be possible to replace sand filtration by microfiltration.

Consider the city of Sao Paulo, Brazil. This is one of the largest cities in Latin America, with a population of about 17 million. About 40% of the city's wastewater currently undergoes secondary treatment in five activated sludge plants. They discharge their high quality effluents, along with the other 60% raw sewage, to the Tiete-Pinheros river system, which conveys mainly this mixture of raw sewage, secondary effluent, and storm water. The large investment for wastewater treatment currently bears little benefit since the combined volume of raw sewage and secondary effluent is practically equivalent to raw sewage. The strategy is to build secondary treatment plants in stages, so as to ultimately achieve close to full secondary treatment of all the wastewater produced in the city. This is not an effective strategy because the huge investments will not improve river water quality or public health for many years.

A better strategy would have been to submit the raw sewage of the city to chemically enhanced primary treatment (CEPT, see the following section). This could achieve BOD and TSS removals of about 70% and 80% for the entire flow at a cost similar to that invested for a portion of the city's flow and with lower O&M costs. Although the conveyance costs to the treatment plants must also be taken into account, first providing CEPT treatment to the entire city's wastewater, and later upgrading to secondary treatment makes more sense than building secondary treatment plants in stages.

This example illustrates the potential of physico-chemical treatment to contribute to rational wastewater treatment strategies in developing countries. Physico-chemical treatment is particularly appropriate for outfall discharges in countries where treatment beyond preliminary is required. It can produce effluent quality similar to that of secondary treatment at lower costs and with lower land requirement. This has been done in Hong Kong, Hawaii (Grigg and Dollar, 1995), and San Diego (Union Tribune, 2002). In Hawaii and San Diego it was shown that upgrading the physico-chemical treatment to secondary would not provide any significant benefit to the marine environment.

6.7.3.2 Chemically Enhanced Primary Treatment (CEPT)

The first process that comes to mind when considering physico-chemical processes is Chemically Enhanced Primary Treatment (CEPT), Figure 6.11. Although this is a well known process that has been used for over a hundred years, it has not yet found the widespread acceptance that would be expected based on its performance. Its main uses are: (i) treatment prior to ocean discharge; (ii) phosphorus reduction; and (iii) as a compact treatment unit. CEPT involves addition of chemicals, typically salts such as ferric chloride or aluminum sulfate, to primary settling tanks. This causes the particles to coagulate and flocculate and settle out, resulting in higher removal rates of TSS and BOD than conventional primary treatment. Removal rates in CEPT are 80–85% TSS and 50–70% BOD as compared to 40–50% TSS and 30–35% BOD in conventional primary tanks. CEPT also has higher performance efficiencies per surface overflow rate (in terms of m^3/d per m^2 of settling tank area) compared to conventional primary tanks. This increased efficiency allows for smaller CEPT settling tanks compared to conventional primary treatment for the same wastewater flow.

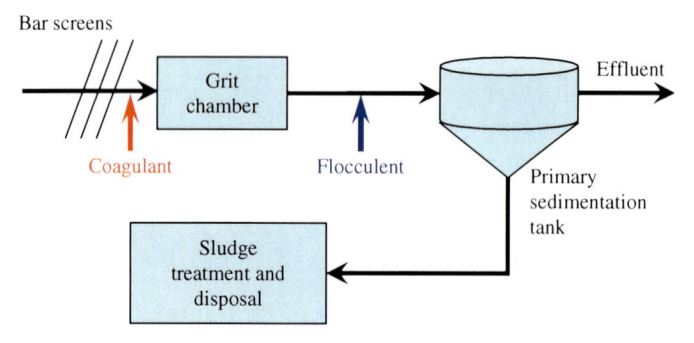

Figure 6.11 Schematic diagram of the CEPT process

To upgrade a conventional primary treatment facility to CEPT, addition of a chemical coagulant and optionally a flocculent is needed. The chemical dosing systems are inexpensive and simple to install. Since the addition of chemicals generates additional quantities of sludge, upgrading the sludge handling and treatment is also required. CEPT relies on small doses of metal salts, usually in the range of 10 to 50 mg/l so the chemicals themselves are only a slight contribution to the sludge. Most of the increased sludge production is due to the increased solids removal, which is precisely the goal of CEPT. The main options for sludge treatment, which must be part of the installation, are lime stabilization and use of drying beds for drying the sludge, anaerobic digestion, and composting (see Section 6.6.3).

If a CEPT unit is designed from scratch, the sedimentation tanks will be smaller and cheaper than primary sedimentation tanks for the same wastewater flow without chemicals addition. In this case, the additional investment required for treating the increased sludge generated is offset by the lower cost of the sedimentation tanks. When a primary treatment unit is upgraded to CEPT, however, this offset does not exist and additional investment for sludge treatment is required.

The quality of CEPT effluent approaches that of a biological secondary effluent in terms of TSS and BOD removal. The CEPT effluent can be effectively disinfected, which is important for developing countries where high levels of morbidity result from waterborne diseases.

CEPT is much cheaper than conventional activated sludge. According to Harleman *et al.* (1996, 1998, 2001) the capital cost of a conventional activated sludge process is about four times that of CEPT, and O&M costs (both including disinfection) are about twice as expensive.

Removal of 70% BOD by CEPT will often suffice for discharge to rivers with adequate assimilation capacity. Phosphorus removal by CEPT may also suffice for lake discharges.

CEPT treatment does not preclude additional treatment such as secondary or tertiary. It reduces the size and cost of these subsequent treatments due to increased removal of suspended solids and organic matter. CEPT is a relatively simple technology, which provides low cost, easy to implement and effective treatment.

6.7.3.3 CEPT followed by filtration and disinfection

If a higher effluent quality than can be produced by CEPT alone is required, it can be followed by filtration and UV disinfection as shown in Figure 6.12. The filtration can be a multimedia sand filter or membrane microfilter. Microfiltration is a relatively new technology and its use in that combination requires investigation.

According to Cooper-Smith (2001) the combination of CEPT followed by sand filtration and disinfection can produce a high quality effluent (averaging

20 mg/l BOD, 15 mg/l TSS) using a Tetra Deep Bed Filter at a filtration rate of 5 m/h and a coagulant (Kemira PAX XL60, a polyaluminium silicate) dose of 40 mg/l. Further, UV disinfection of the filtered effluent complied with the required Bathing Water Directive standard, and net present value calculations showed this option to be considerably cheaper than secondary biological treatment. Sludge treatment in this system is identical to that of CEPT sludge.

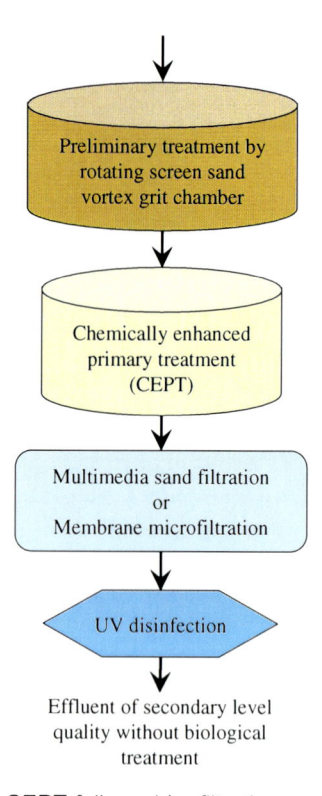

Figure 6.12 CEPT followed by filtration and disinfection

CEPT followed by filtration can produce an effluent comparable to secondary effluent at lower cost, reduced power needs, and ease of operation. The investment and O&M costs for CEPT followed by multimedia sand filtration are about one-half of those for conventional activated sludge (both including disinfection).

6.7.3.4 Chemically enhanced solids separation by rotating fine screens (CERFS)

Multiple-stage screening and the addition to flocculants can significantly increase contaminant removal rates. The rotating fine screen devices discussed in Section 6.7.2 can remove up to 50% of particulate COD and up to 30% of total COD with screen openings around 0.2 mm. Recent investigations (Huber, 2003) indicate that addition of organic coagulant to the raw sewage entering a fine rotating screen of 0.2 to 0.3 mm opening achieves up to 60% reduction of total COD and BOD, and TSS reduction from 50% to 95%. This process is still under development and seems to be based (Figure 6.13) on two-stage screening where the first screen has an opening of about 6 mm and the second is a fine screen with an opening of 0.2 mm.

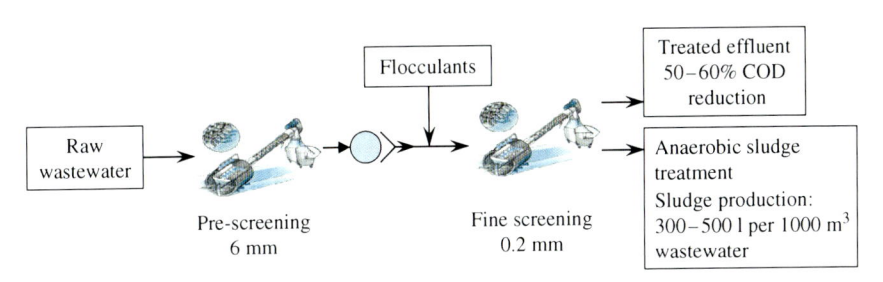

Figure 6.13 Chemically enhanced solids separation by rotating fine screens (CERFS)

This process is called chemically enhanced rotating fine screens (CERFS). Although it may not quite reach the level of solids and BOD removed by CEPT, it is less costly and simpler to operate. The organic matter removal achieved in CERFS can prevent environmental problems if the effluent is discharged to a high flow river or other receiving bodies with high assimilation capacity. CERFS may be a better first stage goal than CEPT, especially in developing countries.

The high removal of solids in CERFS generates larger quantities of sludge than preliminary treatment without chemicals addition, so sludge treatment facilities must be provided. Sludge treatment is identical to that of CEPT, i.e. the main treatment options are lime stabilization and drying beds, anaerobic digestion, and composting.

The costs of CERFS are not yet available since it is a new concept. Rough estimates are that the investment cost would be about one-sixth that of conventional activated sludge, and O&M costs (both including disinfection) about one-half. More research is required to better establish the performance and cost of CERFS.

6.7.3.5 CERFS followed by filtration and disinfection

Similar to the concept of improving CEPT effluent quality by filtration and disinfection, the CERFS effluent might undergo identical processes to achieve better effluent quality. As for CEPT, UV disinfection and filtration by either multimedia sand filtration or membrane microfiltration could be used. The proposed process flow diagram is presented in Figure 6.14.

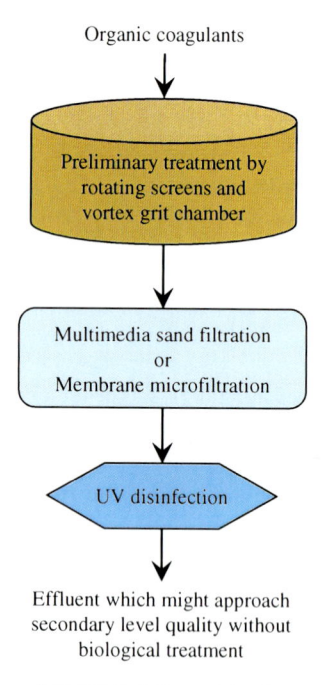

Figure 6.14 Proposed CERFS followed by filtration and disinfection

The quality of CERFS effluent followed by filtration is not yet known. If it approaches the levels of secondary effluent this process would be very attractive for developing countries as it could provide a high quality effluent at much lower cost, reduced power needs, and easy operation. Sludge treatment should be identical to that of CEPT sludge.

The costs of CERFS followed by filtration and disinfection are not yet available. Rough estimates suggest that the ratio of investment cost in conventional activated sludge to that of CERFS followed by filtration and disinfection is about 2.5:1, and the ratio of O&M costs (both including disinfection) is about 2:1. Note that the investment cost of CERFS followed by filtration and disinfection is not much lower than that of CEPT followed by filtration and disinfection. The reason is that a large portion of the cost for both systems is in filtration and sludge treatment which are similar in both. The difference between these systems is the cost of the rotating screens in CERFS, which is lower than that of the primary settlers in CEPT. Future improvements in microfiltration might reduce the cost of these systems, and sale of the sludge for agricultural or any other beneficial use could improve their overall economic performance.

Since CERFS followed by filtration and disinfection has not yet been implemented, research is needed to generate information about its performance and cost.

6.7.4 Management and disposal of solid wastes

The appropriate technologies described above generate two types of solid wastes. Rotating screens with coarse openings larger than 0.7 mm usually produce residual solids which are mainly mineral, consisting of sand, grit, coarse particles and large materials like plastics, bottles, cans and other floatable material. This material is defined as solid wastes from preliminary treatment. Rotating fine screens with openings less than 0.7 mm (and all the other mentioned process, like CEPT with or without filtration and CERFS with or without filtration), produce residual solids which consist of a combination of mineral and organic matter, defined as sludge. These two types of solids, solid wastes and sludge, are handled and disposed of differently, as discussed below.

6.7.4.1 Solid wastes from preliminary treatment processes

Disposal of screenings (solid wastes collected in preliminary treatment plants) may include: (1) removal by hauling to disposal areas (sanitary landfill), (2) disposal by burial on the plant site (small installations only), (3) incineration alone or in combination with sludge and grit (large installations only), (4) disposal with solid wastes, or (5) discharge to grinders or macerators where they are ground and returned to the wastewater.

Removal to a sanitary landfill is the most common disposal method. In a rotating fine screen plant the solids are compacted prior to disposal. A schematic presentation of the solids collection process is presented in Figure 6.6 and a photograph of typical solid wastes of a preliminary treatment plant is presented in Figure 6.15. The solid wastes container fills automatically without manual intervention. When the container is full it is put on a removal truck by a crane built onto the truck. The same truck leaves an empty container at the plant for filling and then travels to the landfill site to dispose of the solid wastes.

Figure 6.15 Solid residues from a preliminary treatment plant

The quantities of solid wastes generated in preliminary treatment plants are generally smaller than in primary and secondary plants. Estimated volumes of screened material under hot climate conditions, based on data from preliminary treatment plants in Florida, USA, are presented in Table 6.4.

Consider, for example, the quantities of solid waste that will be generated in the preliminary treatment plant of Cartagena, Colombia (see Chapter 14). For the current population of about 1 million, about 14 m^3/d of solid waste will be generated, growing to about 24 m^3/d after 20 years. This would require one truck-trip per day initially and two in the future. The materials will be disposed of by the same method as for solids retained in wastewater pumping stations at the La Paz Regional Landfill, located in a rural area 17 km from Cartagena. This landfill has an environmental license issued by the local environmental authorities and is expected to be operative for 15 years, with a capacity of more than 600 tons per day of solid waste. The average daily weight of the solids wastes from the preliminary treatment plant will be about 10 ton/day, around 1.6% of the capacity of the landfill.

Table 6.4 Volume of screened material for various screen openings

Type of screen	Screen opening (mm)	Volume of screened material (m^3 per m^3 wastewater)	(ft^3 per MG wastewater)
Mechanical bar	19	0.90×10^{-5}	1.2
Mechanical bar	13	1.4×10^{-5}	1.8
Mechanical bar	8	1.7×10^{-5}	2.3
Step	6	2.2×10^{-5}	3.0
Step	2	3.4×10^{-5}	4.5
Rotating fine	1.5	4.7×10^{-5}	6.3
Rotating fine	1.0	8.2×10^{-5}	11.0
Rotating fine	0.5	13.2×10^{-5}	17.7

MG: Million U.S. gallons

According to Colombian legislation, the solid wastes generated in Cartagena's preliminary plant are not classified as toxic but rather as municipal solid wastes. This was confirmed by laboratory analysis. The material collected at the screening machines has a low liquid content. It will be stored and transported in hermetic containers to the regional landfill by a contractor. This ensures that there will be no leaks during transport and that the community will not be disturbed, all in compliance with the Colombia's norms and regulations. The dry solid waste will be deposited at a specially prepared area of the landfill.

Each month the waste management company will certify the services carried out, specifying collection dates, type and quantity, and the procedures undertaken. This will serve for keeping records on the company's environmental management system and also to prove to the environmental and sanitation authorities that the operation is being properly managed.

The quantity of screened material will not be a major problem in Cartagena. In the first stage, the volume is equivalent to the solid wastes generated by a population of about 7,000, and in the final stage by a population of about 12,000. The quantity of solid wastes generated in the treatment plant will constitute only about 0.7% of the total solid wastes generated by the city.

6.7.4.2 Sludge management

The treatment and disposal of sludge generated by rotating fine screens with openings less than 0.7 mm (and all the other mentioned processes such as CEPT and CERFS with or without filtration) is basically identical to that generated by conventional secondary treatment.

The main options for treating this type of sludge are shown in Figure 6.16. They are anaerobic digestion, lime stabilization drying beds, and composting. Details on the unit treatment process within each option, such as thickening, stabilization, dewatering and drying were discussed in Section 6.6.3. Composted and lime stabilized sludge are used as organic fertilizers in agriculture. Anaerobic digested sludge has also been used as an organic fertilizer, in liquid or dewatered form, for many years. It is still used in this manner in many parts of the world, although its use has recently been prohibited in some countries due to public health risks from pathogenic organisms. Anaerobic digested sludge should therefore be disinfected prior to agricultural use, disposed of in a landfill, or reused for other purposes, such as a supplement to construction materials.

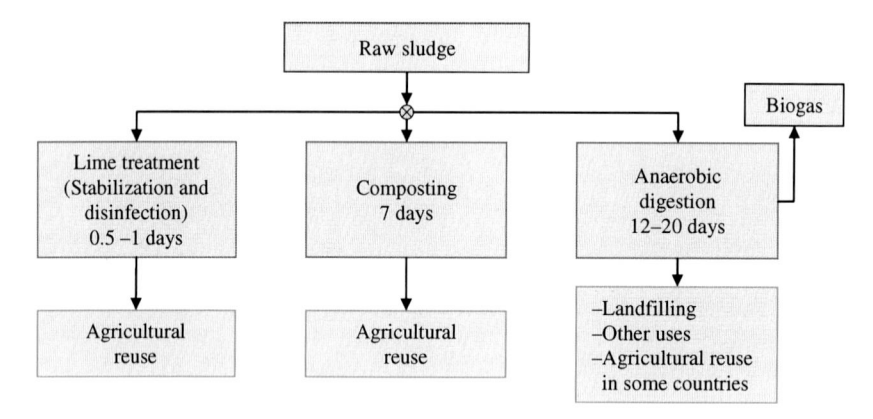

Figure 6.16 Main sludge management options

6.7.5 Treatment costs

Costs of the preliminary treatment process in rotating fine screens, CERFS, CERFS followed by filtration, CEPT, and CEPT followed by filtration are presented in Table 6.5. Removal efficiencies of BOD and total suspended solids (TSS) are also shown.

As for Table 6.2, investment and O&M costs vary widely for different plants of the same type and in different countries and for the same reasons as discussed there.

Table 6.5 Costs and removal capabilities for CERFS and CEPT processes

Constituent	Rotating fine screens	CERFS[a]	CERFS followed by filtration	CEPT[b] Low dose	CEPT[b] High dose	CEPT[b] followed by filtration	Conventional secondary treatment
Suspended solids, (TSS, mg/l)	5–50[c]	50–95[d]	70–95	60–82	86–98	85–95	89–97
BOD (mg/l O_2)	5–30[c]	40–60[d]	60–75	45–65	67–89	75–90	86–98
Treatment cost (US$/m^3)	0.03	0.10–0.15	0.10–0.15	0.16	0.18	0.18	0.29
Capital cost (US$/Person)	5–20	18–36	40–80	50–100	50–120	60–120	100–200

[a] Chemically enhanced rotating fine screens
[b] Chemically enhanced primary treatment
[c] Depending on screen opening
[d] Depending on type of chemicals and dose

6.8 WASTEWATER TREATMENT AND GLOBAL WARMING

The impact of wastewater treatment on global warming depends on the type of treatment. Aerobic treatment is neutral, in spite of releasing carbon dioxide to the atmosphere. This is because, in producing the organic matter contained in the wastewater, the same amount of carbon dioxide is removed from the atmosphere as is released in the treatment process. However, supply of oxygen for the aerobic process, usually through aeration, consumes energy and the production of this energy from fuel generates carbon dioxide. Consequently, aerobic treatment processes contribute a net amount of carbon dioxide to the atmosphere and thus have a negative impact from the standpoint of global warming. Anaerobic treatment produces methane gas and carbon dioxide. Methane is a strong greenhouse gas, 22 times stronger than carbon dioxide, so its contribution to global warming can be significant.

Combining treatment with discharge through a marine outfall changes the picture somewhat. Consider three scenarios:

(1) If the treatment is aerobic, the system generates carbon dioxide, as explained above, and has a negative impact of contributing to global warming.
(2) If the treatment is anaerobic, the system has a significant negative impact on global warming, as explained above.
(3) If the treatment is only preliminary, as recommended, the system has a significant positive impact on global warming. There is no on-land release of carbon dioxide (or methane) to the atmosphere. The organic matter is diluted in the sea and decomposes aerobically, however, the carbon dioxide released is dissolved in the seawater, so the balance is positive. At some future time the carbon dioxide might be released to the atmosphere, and then the impact becomes neutral.

For developing countries the positive impact of preliminary treatment and marine discharge is even greater. This is because, if a marine outfall is not used, raw wastewater is usually discharged to small rivers or creeks where it undergoes anaerobic processes that releases methane. A system that combines preliminary treatment with an effective outfall prevents this release of methane. Other advantages of preliminary treatment compared to advanced treatment is that secondary effects such as greenhouse gas production from the increased energy usage and materials transportation are also avoided. It should be noted that wastewater treatment plant process-related greenhouse gas emissions are about 1% of the total emissions in developed countries.

6.9 CONCLUSIONS

The following conclusions refer particularly to developing countries, with emphasis on appropriate treatment for ocean discharges through effective outfalls. The observations on appropriate treatment for ocean discharges also apply to developed countries, however.

In most developing countries the percentage of wastewater that undergoes any type of treatment is very low. They can also anticipate rapid population growth that will impede the expansion of wastewater treatment. Authorities must understand that the needs of developing countries are different from those of industrialized countries. Realistic standards for effluent quality should be adopted that are flexible and permit staged development. Developing countries should use appropriate treatment technology based on simplified processes that are less expensive than conventional processes in investment and O&M, are simple to operate, and can produce effluent of the required quality.

Wastewaters contain many types of pollutants, whose relative impact on the environment depends on the receiving water body and outfall design. The treatment should be chosen accordingly and not arbitrarily specified.

Wastewater management in developing countries should be developed in stages. It should account for the assimilation capacity of the receiving water body (using water quality simulation models) to ensure that even during the initial stages any environmental impacts are minimal. This is usually feasible. To support this strategy, a comprehensive monitoring program should begin prior to and following construction of the first stage. If this indicates no public health or environmental problems, consideration should be given to delay or even eliminate construction of subsequent stages so as to prevent unnecessary waste of resources. If monitoring indicates environmental problems, the subsequent stages should be implemented. This is a logical strategy that would prevent unnecessary investment in costly treatment installations that are unaffordable in developing countries.

For ocean disposal, the WHO (Table 1.1), and extensive experience, indicates that preliminary treatment with discharge through an effective outfall poses a low risk to human health. And more advanced levels of treatment do not further lower this risk. Preliminary treatment and discharge to the sea by an effective outfall is therefore often the best wastewater management strategy for coastal cities in developing countries. It is affordable, effective, and reliable, simple to operate and with little risk to public health and the environment. Many ocean outfalls are successfully functioning all over the world and have a proven track record. The most reliable and inexpensive preliminary treatment consists of rotating fine screens and vortex grit chambers. The capability of disinfection by chlorination or other means should be included for use in emergencies.

Unfortunately, many developing countries treat wastewater in lagoons and then discharge it either directly on to the beach or through a short outfall. The WHO (Table 1.1), indicates that this practice poses a high risk to human health and should be discontinued.

Developing countries that have adopted the stringent standards for ocean discharges of industrialized countries, such as secondary treatment, must adopt and understood the principle of staging. This allows for construction of a first stage with preliminary treatment and an effective outfall. Insistence on a first stage consisting of treatment above and beyond preliminary would render most projects in developing countries financially non-viable and will prevent improvements in wastewater disposal. Avoiding the problem of wastewater disposal is the worst option of all, and usually leaves the most vulnerable population (the poor) under the worst conditions.

If advanced treatment is required or legislated, physico-chemical treatment is a good option. It has the potential to become an innovative and affordable technology that can boost wastewater treatment in developing countries. For example, the CEPT process and its derivatives, such as CEPT followed by filtration and disinfection, are proven technologies that can yield an effluent comparable to that of secondary biological treatment at much lower cost. This observation applies to industrialized countries as well as developing countries.

Other technologies based on liquid-solid separation by rotating fine screens preceded by dosing with coagulants and flocculants and followed by filtration or microfiltration membranes also have considerable potential. They can yield high quality effluent at costs even lower than that of CEPT and its derivatives, thereby presenting an excellent solution to many large cities in the developing world that need advanced treatment but cannot afford secondary biological treatment.

7
Materials for small and medium diameter outfalls

7.1 GENERAL CONSIDERATIONS

Up to the year 2005, most ocean outfalls with diameters larger than two meters were constructed from concrete pipe. Outfalls with smaller diameters have used many materials including the thermoplastics polyethylene (PE), polypropylene (PP), and polyvinyl chloride (PVC); GRP (fiberglass); reinforced concrete and prestressed concrete; and the ferrous metals steel, ductile iron, and cast iron. The diameters of pipes made of GRP, polyethylene, and polypropylene have steadily increased over the years, and they are finding increased application to large outfalls. By the year 2004 solid wall polyethylene pipe was available in diameters up to two meters and GRP up to 2.4 meters. Polyethylene has now become the predominant material for outfalls with diameters of about one meter or less.

© 2010 IWA Publishing. *Marine Wastewater Outfalls and Treatment Systems.* By Philip JW Roberts, Henry J Salas, Fred M Reiff, Menahem Libhaber, Alejandro Labbe, and James C Thomson. ISBN: 9781843391890. Published by IWA Publishing, London, UK.

Each material has advantages and disadvantages and the ultimate success of an outfall depends greatly on selecting that most appropriate for the specific oceanographic and site conditions, adequate mitigation of any weaknesses of the material chosen, the capabilities of the entity responsible for outfall construction, and the capabilities of the agency responsible for operation and maintenance.

Factors that enter into the material selection include:

- Cost of the pipe and ancillary materials:
  - Shipping and transportation
  - Import and/or export duties
  - Construction
  - Outfall operation
  - Outfall maintenance
- Quantity of effluent discharged
- Characteristics of the effluent
- Corrosive effects of the marine environment
- Hydrodynamic effects of the marine environment
- Geomorphology of the seabed
- Ocean traffic conditions
- Availability of necessary skills and equipment.

The most important characteristics of the most common materials used for marine outfalls and their relative costs are summarized and compared in Table 7.1.

The cost of the pipe delivered to the site usually ranges from 8% to 20% of the total outfall construction cost. However, the pipe with the lowest cost may entail more expensive construction methods, more expensive protection against physical oceanic forces, more expensive monitoring and surveillance, and/or more expensive maintenance and repair. Different kinds of pipe often require different ancillary facilities, for example steel pipe requires cathodic protection but polyethylene and GRP pipe do not. A polyethylene outfall requires concrete ballast weights to prevent it from floating whereas ductile iron pipe usually does not. It is important to consider the total life cycle costs associated with all components of an outfall when selecting the type of pipe to be used.

Flowrate is important. For example, small flows that require pipe diameters less than 400 mm would usually be limited to pipes that can be installed by flotation and submergence because the cost of pipes that necessitate submarine assembly is much greater. High flows may necessitate diameters that are so large that installation by flotation and submergence is not possible or is not cost effective.

Materials for small and medium diameter outfalls

Table 7.1 Characteristics and relative costs of common outfall pipe materials

Type of pipe	Resistance against				Axial flexibility		Relative costs	
	Internal pressure	External impact	Corrosion		Pipe	Joint	Material	Construction
Steel	Excellent	Excellent	Poor		Good	Excellent	Low	Low
Cement coated steel	Excellent	Good	Fair		Poor	Poor	Moderate	Moderate
Ductile iron	Excellent	Very good	Poor		None	Fair	Low to moderate	Moderate
Reinforced concrete	Good	Good	Good		None	None	Moderate	High to very high
Prestressed concrete	Very good	Good	Good		None	None	Moderate to high	High to very high
Polyethylene	Fair	Good	Excellent		Very good	Very good	Moderate to high	Low
Polypropylene	Fair	Good	Excellent		Very good	Very good	Moderate to high	Low
GRP	Good	Fair	Very good		Poor	Poor	Moderate to high	Moderate to high
PVC	Fair	Poor	Excellent		Fair	Poor	Low	Low

The effluent characteristics also affect the choice of pipe. If it contains significant quantities of abrasive sediment it is prudent to select a material that is resistant to abrasion. If the effluent is likely to produce hydrogen sulfide then the material should be resistant to its corrosive effects.

Impact strength is extremely important if the outfall is located in an area where boat anchorage is likely to occur or where storm driven submerged objects may strike the outfall.

An important consideration is the capacity and/or experience of the entities who will construct the outfall. The type of pipe should be compatible with their specialized skills and the equipment that is available for construction.

Maintenance and repair requirements differ for the various types of pipe materials. The material selected should be compatible with the capabilities of the entities responsible for these operations.

The most important characteristics of the most common materials used for marine outfalls and their relative costs are summarized and compared in Table 7.1.

7.2 SPECIFIC CONSIDERATIONS

7.2.1 Pipe connections

Compressed O-ring gasket joints are used in concrete, ductile iron, GRP, and PVC pipe. If this type of joint is subjected to pulling forces the pipes can separate. This can also occur if there is lateral pipe movement caused by hydrodynamic forces or snagging by a boat anchor. It can also occur when there has been vertical movement due to differential settlement of the sea floor due to liquefaction during storms or earthquakes, or undermining of the pipe by scour due to waves or currents.

There are joints available that can prevent this separation. For ductile iron pipe, restrained joints that incorporate locking devices to prevent pullout and allow articulation of the joint while maintaining its integrity are used and restraint harnesses are available for PVC pipe. Properly made welds in steel pipe will usually withstand axial flexure almost as well as the pipe itself. Concrete pipe and GRP pipe with standard compressed O-ring couplings must rely on both an adequate foundation for the pipe and adequate pipe stabilization against hydrodynamic and other forces to prevent pullout. GRP couplings that prevent pullout are available.

There are many types of material and shapes of O-ring type gaskets. Some materials are vulnerable to attack by marine organisms and others deteriorate prematurely in the marine environment. It is important to specify that the gasket material be suited to the marine environment.

Welded joints in steel pipe are strong but can be subject to corrosion because of dissimilar metals in the welding rod and the steel pipe. This can be mitigated by protective coatings and galvanic corrosion protection. Bolted mechanical joints should be avoided. If necessary, however, all joint components should be made of materials that resist marine corrosion, and the bolts and nuts should be protected by sacrificial anodes.

Heat fused joints in HDPE pipe and polypropylene pipe are almost as strong as the pipe itself. When properly carried out they have proven to be one of the most trouble free joints in the marine environment.

7.2.2 Internal pressure

The internal pressure in an operating ocean outfall will be greatest at the design peak flow, which will usually occur near the end of its useful life. The highest pressure will usually occur at or near the point where the outfall enters the ocean. The pressure head at that point is the sum of the friction loss in the outfall pipe and fittings, head losses in the diffuser, and the density head between the seawater and effluent. The pressure head depends on the outfall length and diameter, the effluent flowrate, the diffuser depth, and the design of the diffuser ports, but for many outfalls is often in the range of 10 to 25 m. For very deep outfalls that are installed by flotation and submergence the maximum internal pressure will probably occur during installation.

The resistance to rupture from internal pressure is proportional to the tensile strength of the pipe material and the wall thickness. Of the materials commonly used for ocean outfalls, steel has the greatest tensile strength. Concrete pipe has reinforcing steel that gives it high burst strength. Ductile iron is next strongest, followed in order by GRP, PVC, polypropylene, and polyethylene. All of these materials are strong enough to withstand the internal pressures usually encountered. Polyethylene has the lowest tensile strength, but with adequate wall thickness can readily withstand the pressures usually encountered.

7.2.3 Resistance to externally imposed forces

Resistance to externally imposed forces must be considered when choosing the pipe material. These are commonly caused by hydrodynamic forces due to waves and currents, impacts from boat anchors, snagged fishing nets, and impact from submerged logs and other debris during storms and hurricanes.

External stress on an outfall pipe also occurs during loading, transporting, and handling of the pipe as well as during construction and installation. The pipe must be designed to withstand buckling that could occur if the external pressure on the pipe is greater than the internal pressure. It must also resist buckling when the pipe is subjected to bending. Ductile iron concrete and standard wall steel pipe are stiff enough to easily resist buckling. But for GRP, PVC, polyethylene, polypropylene, and thin wall steel pipe the potential for buckling and collapse of the pipe must be considered when negative pressure, burial in a trench, or bending during handling and installation may occur.

7.2.4 Flexibility

Flexibility of the pipe and/or the pipe joints is an important factor in selecting the pipe material or the type of joint when:

- There are obstacles on the seabed that must be avoided.
- The material of the ocean floor is structurally weak and there is good possibility of consolidation and subsidence.
- Seasonal transport of the seafloor material may occur.
- There is a possibility of spontaneous liquefaction in high-risk seismic zones.

7.3 TYPES OF PIPE USED IN SUBMARINE OUTFALLS

7.3.1 Concrete pipe

Three basic kinds of concrete pipe have been used for outfalls: Non-cylindrical reinforced concrete pressure pipe, bar-wrapped concrete cylinder, and prestressed concrete cylinder.

Non-cylinder reinforced concrete pressure pipe is manufactured by placing a reinforcing cage in a mold and filling the mold with concrete. The mold is shaped to form a joint to accommodate one or more preformed rubber gasket O-rings of circular cross section as shown in Figure 7.1.

A cross section of the pipe wall and joint of bar-wrapped concrete cylinder pipe is shown in Figure 7.2. This pipe is constructed around a steel cylinder with steel bell and spigot O-ring joints welded to each end. A variation of this joint is a double O-ring joint with the rings spaced apart to enable pressure testing of the joint after the spigot has been inserted into the bell.

For bar-wrapped pipe the concrete is centrifugally cast in place on the interior of the cylinder. The exterior is then wrapped with mild steel reinforcing rods in a helical pattern, and a slurry of Portland cement is applied followed by an exterior mortar coating. The pipe sections are joined when they are placed in

Materials for small and medium diameter outfalls 211

their final location. This is accomplished by placing an O-ring in the groove on the spigot end of one section of pipe and inserting it into the bell end of another. For outfalls, two grooves and O-rings are sometimes used on each joint. The O-ring is compressed between the bell and spigot forming a watertight seal. After joining, cement mortar is placed inside and outside the joint to protect the steel bell and spigot from corrosion.

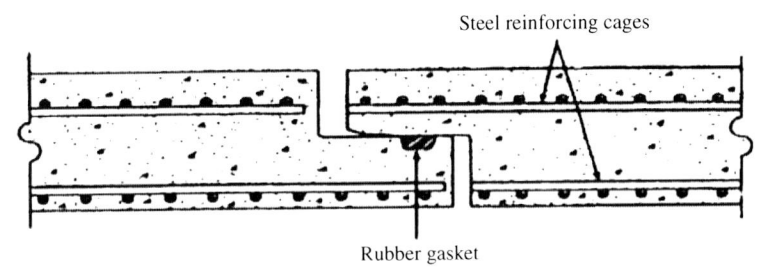

B. AWWA C302-type pipe with concrete joint

Figure 7.1 Non-cylinder reinforced concrete pipe with concrete joint (courtesy American Concrete Pressure Pipe Association)

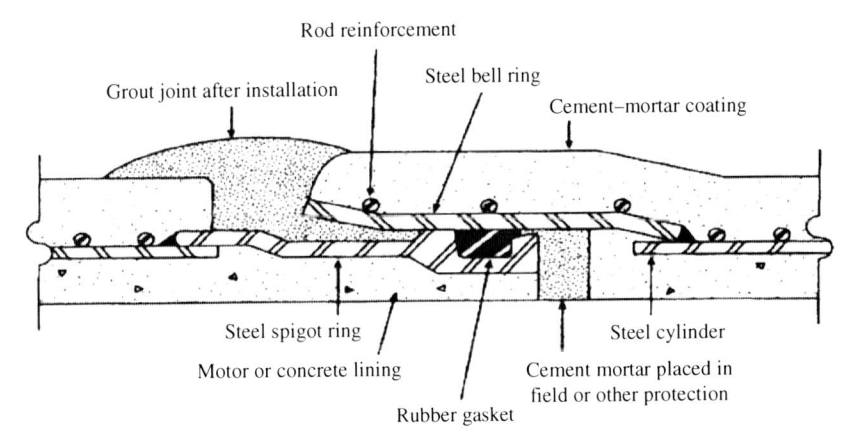

Figure 7.2 Cross section of wall and joint of a bar-wrapped concrete cylinder pipe (courtesy American Concrete Pressure Pipe Association)

Prestressed concrete cylinder pipe is illustrated in Figure 7.3. This pipe is also constructed around a steel cylinder with steel bell and spigot O-ring joints welded to each end. The concrete core is placed in one of three ways: the centrifugal process, radial compaction, or vertical casting. After the core is cured, the pipe is wrapped helically with high strength wire that is pretensioned as it is wrapped to produce a residual compression in the concrete core. The wire is then coated with a thick cement slurry followed by a dense mortar coating. This type of pipe is joined in the same manner as the bar wrapped cylinder pipe.

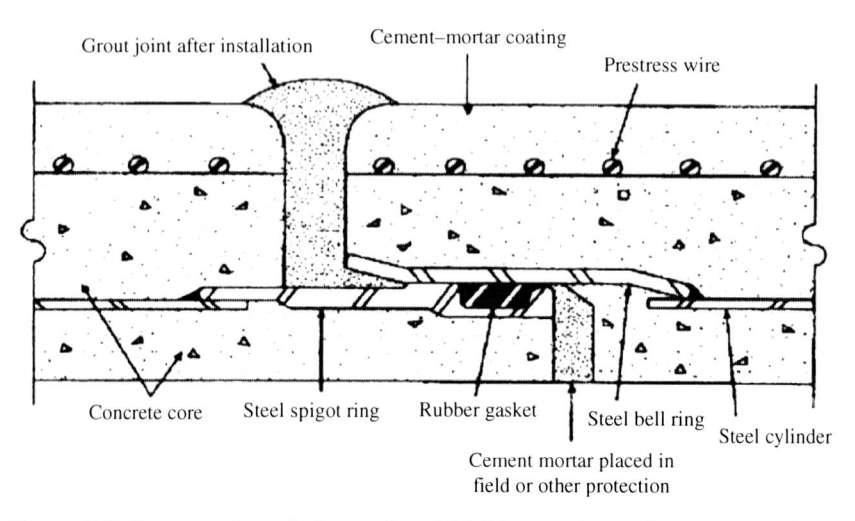

Figure 7.3 Cross section of pipe wall and joint for prestressed concrete cylinder pipe (courtesy American Concrete Pressure Pipe Association)

The construction method for an outfall using concrete pipe requires preparation of a stable bed (foundation) along the ocean floor. The bed should have as uniform a slope as possible because the pipe joints can only tolerate very small angular offsets. Any areas with weak or unstable seabed materials must be stabilized by filling with suitable material or by attaching the pipe to piles driven into the seabed.

The pipe sections are connected underwater on the prepared bed by use of a laying platform or barge and an alignment frame or "horse" with divers directing the connections. The sections are connected by first applying a lubricant to the O-ring gasket(s) on the spigot end of the pipe and then inserting that end into the bell end of the next section. This compresses the O-ring making a watertight seal. After each section is connected the joint must be pressure grouted with cement mortar inside and outside the pipe to protect any metal surfaces from corrosion.

Concrete pipe has good resistance to marine corrosion, much better than ductile iron and steel, but not as good as polyethylene or GRP pipe.

Welded joints are used when the bottom pull method is used to install steel cylinder concrete pipe. In this case the steel cylinder at the ends of the pipe sections are not coated with prestressed or bar wrapped concrete. The bare ends are welded together, checked by radiography or another suitable method. The exposed portion of the steel cylinder is then provided with a protective coating of bar, wire, or mesh wrapped reinforced concrete and the exposed internal portion also recoated with concrete.

7.3.2 Steel pipe

Steel pipe is very strong, has very good resistance to external impact, and enough ductility that it can conform to moderate changes in seabed slope. It is quite inexpensive compared to other pipe materials. It is widely used in the marine environment by the oil and gas industries and has also found use in submarine outfalls.

Steel's major disadvantage for ocean outfalls is that it is very susceptible to corrosion from seawater and sewage effluent. Various measures to protect against corrosion have been used. Cement mortar coating can protect against seawater corrosion but will spall off or crack if the pipe is bent. Cement mortar lining on the pipe interior can be susceptible to attack by hydrogen sulfide. Asphalt and coal tar coatings for steel pipe have been abandoned in favor of epoxy, polyethylene, PVC, and polyurethane coatings that have proven effective at reducing corrosion of the internal pipe wall. Plastic coatings have also been used for external corrosion protection.

Steel pipe usually has slightly negative buoyancy when filled with effluent so it does not float. It frequently requires additional ballasting or mechanical anchors, however, to prevent displacement by ocean currents and waves. A common way to provide extra ballast weight is to coat the outside of the pipe with concrete or cement mortar that is applied along with galvanized steel wire reinforcing. This serves the dual purpose of providing additional weight and protection against marine corrosion but it has the limitation that the pipe cannot be bent in the installation process.

Steel pipe sections are usually joined by welding. After the welds are made and tested (by radiography or ultrasound) external and internal protective coatings are applied to the weld. Flanged connections are usually limited to the extreme end where diffusers are connected, or entrance hatches for cleaning are attached. Where bottom and/or oceanic conditions require the joints to be articulated, special ball-type joints have been used to allow movement without breaking the seal or damaging the protective coatings.

Cathodic corrosion protection is sometimes used in addition to the protective coatings, especially around the joints, to provide protection in the event that the protective coating is damaged or loses its integrity. Cathodic protection can also be used for steel pipe instead of protective coatings but will require considerable surveillance, upkeep, maintenance, and expense. See Section 7.4.

7.3.3 Ductile iron pipe

Ductile iron pipe and gray cast iron pipe have been used for marine outfalls. Gray cast iron pipe is no longer manufactured, however, and has been replaced by ductile iron because of its superior physical strength that approaches that of steel. Restrained articulating joints such as the ball joints shown in Figure 7.4 are usually required for the majority of ocean outfalls. Such joints are available for pipe diameters up to 54 inches.

Flex-Lok (4–24)

Ball and socket (6–36)

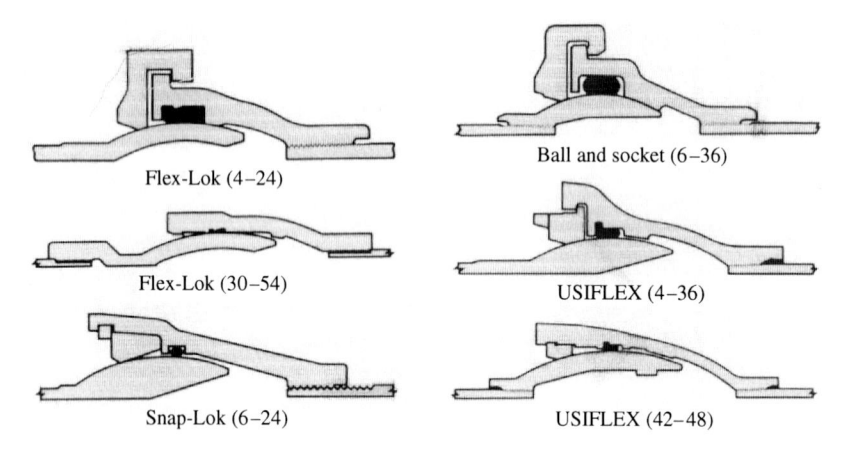

Flex-Lok (30–54)

USIFLEX (4–36)

Snap-Lok (6–24)

USIFLEX (42–48)

Figure 7.4 Typical ductile iron pipe ball and socket joints used in outfall applications (courtesy Ductile Iron Pipe Research Association)

The excellent physical strength of ductile iron pipe can be advantageous for ocean outfalls that are subject to intense physical abuse. It has excellent resistance against impact from submerged objects driven by waves and currents as well as ship anchors and commercial fishing gear. For this reason it has found application in areas such as Alaska where extremely hard solid rock seabed make entrenchment very difficult and expensive but storms can generate extremely turbulent sea conditions that can move submerged logs and large rocks.

By using supplemental flotation (Figure 7.5), ductile iron pipe has been installed in Alaska (and possibly elsewhere) by flotation and submergence. This was found to be the least expensive method. It can also be installed by bottom pull if the site and ocean floor conditions are suitable. This is usually the second least expensive method of installation. Both of these methods require the use of restrained articulating joints. Another method that can be employed for ductile iron pipe is a lay barge with an articulated stinger. This is usually considerably more expensive than the previous two methods.

Figure 7.5 Installation of a ductile iron submarine outfall in Alaska by flotation and submergence

Although ductile iron is somewhat more resistant to marine corrosion than steel pipe it definitely requires corrosion protection. Internal protection is usually by means of an epoxy or polyethylene coating; external protection is usually by cathodic corrosion protection (see Section 7.4).

7.3.4 GRP pipe

Glass reinforced plastic (GRP) pipe is increasingly being used for ocean outfalls in diameters up to 2400 mm. It comes in a standard length of 12 m for diameters greater than 300 mm, but lengths of 6 and 18 m are also available. For diameters smaller than 300 mm the standard length is 6 m. ASTM currently has standards for GRP sewer pressure pipe and ISO is finalizing their proposed standards.

A major advantage of GRP pipe is that it is almost immune to corrosion due to the marine environment as well as the effluent conducted by the pipe. Another is that it is lighter than steel, concrete, and ductile iron pipe but is more dense than water so sinks in seawater. This light weight facilitates handling the pipe on the laying barge or platform as well as joining the sections underwater.

New GRP pipe has a very smooth interior that results in a Colebrook-White absolute roughness height of 0.000029 m and a Hazen-Williams coefficient of $C = 150$. After a few of years of conveying sewage effluent, however, the roughness increases and the Hazen-Williams coefficient approaches $C = 140$ because of accumulation on the pipe walls.

There are several types of joints used for GRP outfalls, Figure 7.6. Most common is the standard compressed O-ring in Figure 7.6(a). Individual pipe sections are joined by inserting a lubricated pipe stub end into the coupling thereby compressing the O-ring and providing a watertight seal. This type of joint will permit a small deflection (between about 0.5 and 3.0 degrees depending upon pipe diameter) and still be watertight. Its disadvantage is that it can pull out if the outfall is subjected to axial force or if the permissible deflection is exceeded. To prevent pullout during construction, marine harness lugs can be fixed to the pipe barrel on each end by fiberglass-polyester bands and then bolting the lugs together. The bolts are usually removed before backfilling or placement of armor rock.

A similar joint that is sometimes used is the lock coupling joint in Figure 7.6(b). After the pipe spigot has been inserted, a locking key is driven through openings in the coupling into matched locking keyways in the coupling and pipe spigot. This joint will resist pullout due to moderate axial force that could occur during installation or by soil movement or hydrodynamic forces.

A third type of joint is the butt wrapped joint. It is used when the joint will be subjected to strong axial tensile force such as installation by bottom pull or

Materials for small and medium diameter outfalls 217

flotation and submergence or where conditions might result in pipe displacement subsequent to installation. It is made by bringing two pipe sections together and joining them by wrapping with glass fiber reinforcement and polyester resin. It is limited to onshore fabrication under clean and dry conditions. It is sometimes used to join several sections of pipe onshore for subsequent transport offshore where they are connected underwater by standard O-ring couplings to the previously installed pipe.

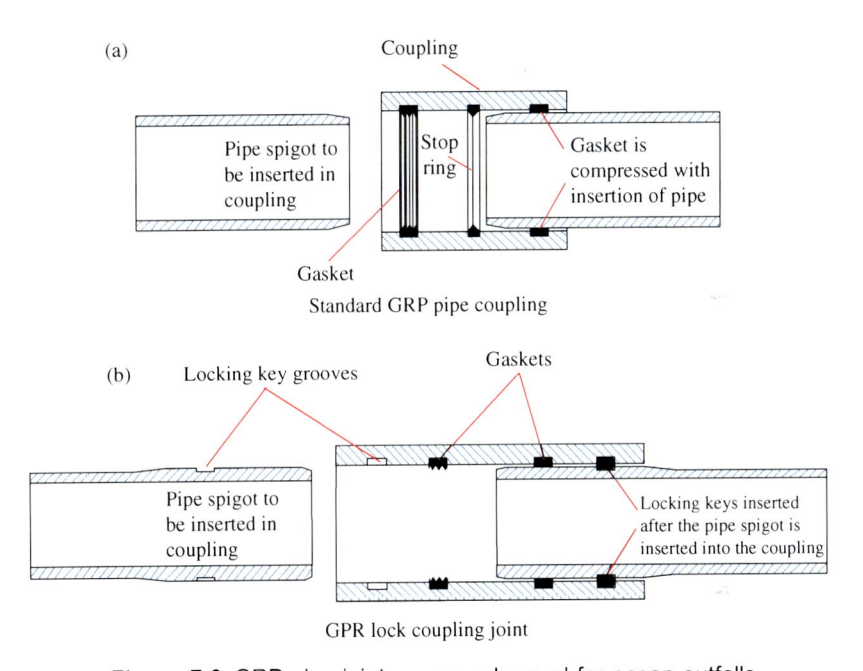

Figure 7.6 GRP pipe joints commonly used for ocean outfalls

Flanged joints are frequently used for GRP marine outfalls for special situations such as connection to different materials. Flexible stainless steel couplings are sometimes used to make underwater connections to elbows, wyes, or other fittings as well as to make underwater closure for repairs.

7.3.5 PVC Pipe

PVC pipe has found limited use for ocean outfalls. Although it has excellent resistance to marine corrosion, there have been incidents of marine organisms attacking PVC pipe such as in Rio de Janeiro where marine mollusks drilled through the pipe wall and rendered an outfall inoperative in less than one year (Grace, 1978). In other situations, however, such as the San Blas Islands of Panama, and Playa Dorada in the Dominican Republic, PVC has operated in the ocean for more than 15 years with no such attack.

PVC pipe is available in diameters up to 24 inches and it usually is available in sections 20 feet (6 m) long. Two types of joints are commonly used: the bell and spigot with a compressed O-ring, and the solvent weld joint. The bell and spigot joint has virtually no resistance to pull out under axial tensile force on the pipe. The solvent weld is only applicable for diameters 200 mm and smaller, but when properly made this joint has resistance to pull out almost equal to the tensile strength of the pipe.

New PVC has pipe has a very smooth wall and an absolute roughness height of 0.000029 m can be used in calculating the Colebrook-White friction coefficient, or a Hazen-Williams coefficient $C = 150$. After a few of years of conveying sewage the absolute roughness increases because of accumulation on the pipe walls and the Hazen-Williams coefficient is closer to 140.

PVC has a specific gravity of 1.4, so a PVC pipe will sink when filled with seawater, but when filled with effluent it will have almost neutral buoyancy. Ballast weights or mechanical anchors are therefore required to to stabilize it against movement from forces due to currents and waves.

For the solvent weld type joint it is possible to use the flotation-submergence process with the ballast weights attached. If this is done with O-ring joints, they must have restraint harnesses, as depicted in Figure 7.7, to prevent pullout. In addition, submergence must be carried out very carefully with supplemental flotation attached to the ballast weights to control the radius of bend of the pipe and preclude buckling of the pipe wall. Another method of installation of PVC pipe is for the pipe sections to be connected underwater and the ballast weights attached underwater, usually by means of a laying barge and divers. Both of these installation methods are tedious, time consuming, and costly. This is why PVC pipe has had limited use for marine outfalls.

7.3.6 Polyethylene (HDPE) pipe

Because of its excellent resistance to marine corrosion and the speed with which it can be installed, high density polyethylene (HDPE) pipe is currently the dominant type of pipe for ocean outfalls with diameters less than one meter, and increasingly for diameters up to two meters. Because of the increased use of this type of pipe, Chapter 9 of this book is devoted to designing outfalls with polyethylene pipe.

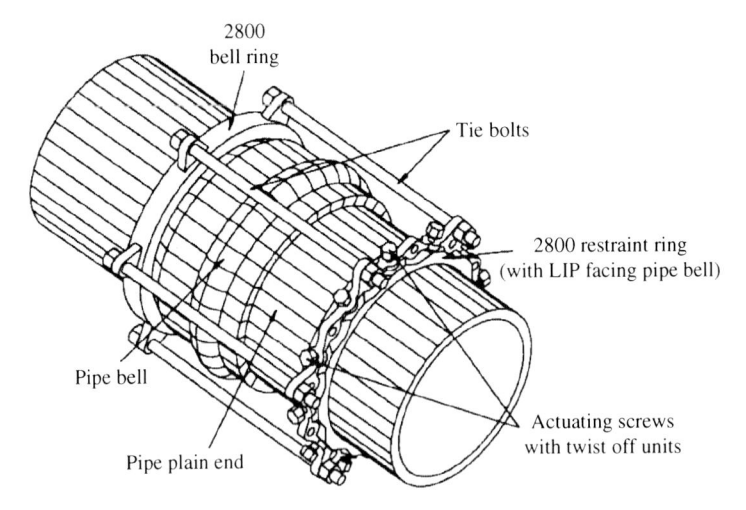

Figure 7.7 Mechanical joint restraint for PVC Unibell pipe (courtesy EBAA Iron Sales, Inc.)

7.4 CORROSION PROTECTION

HDPE, polypropylene, PVC, and GRP do not require corrosion protection for the pipe itself, but ancillary items such as bolts, nuts, washers, mechanical anchors, joint restraints that are made of materials vulnerable to marine corrosion must be protected. Some HDPE pipe installations have been in service for more than 30 years with no evidence of deterioration of the material. A small number of installations suffered severe environmental stress cracking but this was attributed to questionable quality control of the resins and the pipe manufacturing process.

Steel pipe and ductile iron pipe require corrosion protection and reinforced concrete pipe and cement coated steel pipe require some protection. Fittings, bolts, nuts, and washers need protection unless they are made of material resistant to marine corrosion. For ferrous pipes using compressed O-ring gasket joints and provided with cathodic corrosion protection, it is either necessary that each pipe length be connected electrically or separate cathodic protection be provided for each section.

Ferrous metals are subject to two types of corrosion: oxidation and galvanic. Coatings are effective against oxidation corrosion. They are not very effective for galvanic corrosion, however, which is the predominant form of corrosion of

220   Marine Wastewater Outfalls and Treatment Systems

ferrous materials in the marine environment. For this type of corrosion cathodic protection is necessary.

Cathodic corrosion protection converts the surface into a cathode of an electrical cell. This is accomplished either by the galvanic anode system, which entails attaching a sacrificial anode such as aluminum, zinc, magnesium, or a special alloy directly to the surface being protected or by the impressed current anode system which uses low voltage to reverse the flow of electrons that occurs in the galvanic corrosion process. Figure 7.8 illustrates a packaged sacrificial anode attached directly to a ferrous metal pipe. Special sacrificial anode cathodic corrosion protection devices for direct attachment to ferric metal accessories are also available commercially in the form of bolt caps, nuts, and sleeves.

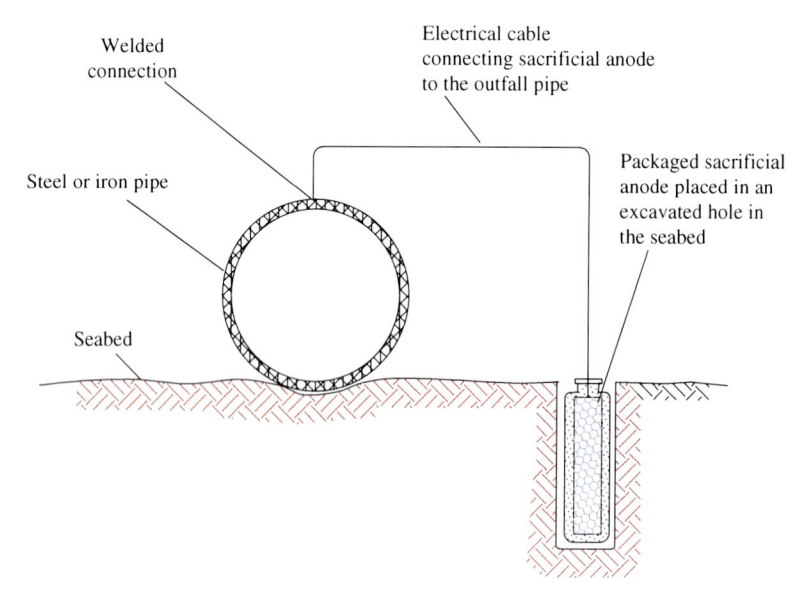

Figure 7.8 Sacrificial anode for corrosion protection of ferrous metal pipe in the marine environment

A schematic of cathodic protection by the impressed current anode system is depicted in Figure 7.9. The anodes are spaced at selected intervals along the pipe but not too close to it so as to maintain a relatively even distribution of current flow from the anode onto the pipe.

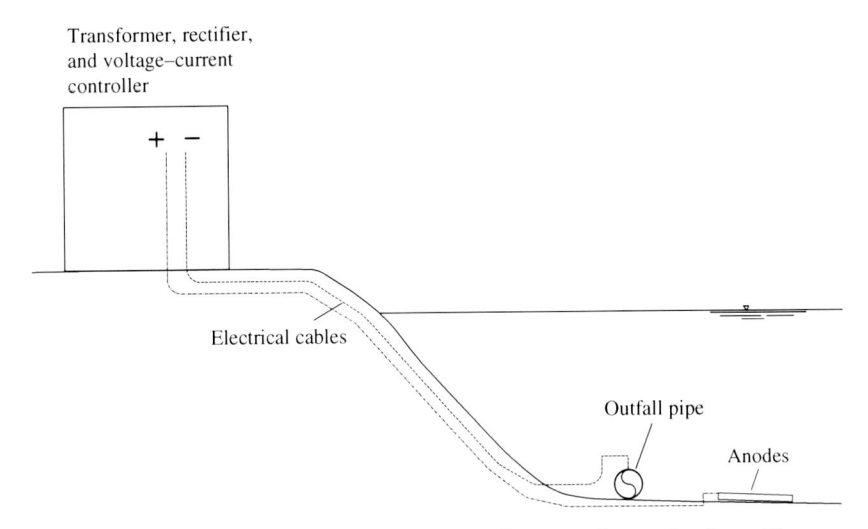

Figure 7.9 Impressed current anode system for corrosion protection of ferrous metal pipe in the marine environment

For bare ferric metals 10.8 milliamps is usually needed for each square meter of surface area to ensure protection in the marine environment. For concrete coated steel pipe, the bare surface area is typically estimated to be 1/10th of the total surface area of the pipe being protected. Anode material, size, and spacing are selected to yield the required current for a period typically ranging from 10 to 25 years, after which the anode is replaced.

8
Oceanic forces on Submarine Outfalls

8.1 INTRODUCTION

A thorough understanding of the physical forces that are prevalent in the ocean, their causes as well as the likelihood and level of forces likely to be experienced over an outfall's useful life is essential for its rational design regardless of the material with which the outfall pipe is made. These forces can threaten the outfall's integrity directly or indirectly through erosion of the seafloor or by undermining it's foundation. They can also cause failure by transport and deposition of sediment, sand, gravel, and even cobbles that can bury the diffuser orifices. All of these factors must be taken into consideration and mitigated in the planning and design process.

For submarine outfalls the physical forces of greatest concern are due to currents, waves, and pressure (depth). They will be addressed in that order.

© 2010 IWA Publishing. *Marine Wastewater Outfalls and Treatment Systems.* By Philip JW Roberts, Henry J Salas, Fred M Reiff, Menahem Libhaber, Alejandro Labbe, and James C Thomson. ISBN: 9781843391890. Published by IWA Publishing, London, UK.

8.2 CURRENTS

It is important to point out that currents can flow at different speeds and directions at different depths of the water column. Strong near-bottom currents are important because they can induce both horizontal and vertical forces on an outfall and therefore must be considered during its design. Currents in the water column are important primarily in determining the initial dilution and direction of plume travel. Surface currents are important primarily for evaluating the travel of a surfacing plume and they can be extremely important during the construction of an outfall by the flotation-submergence method.

The designer should have a good understanding of currents and be able to anticipate the ranges of speeds and directions. The currents (see also Sections 3.2 and 5.2.2) may be wholly or partly due to:

- Geostrophic forces
- Lunar tides
- Winds
- Return water from wave runup.

Geostrophic forces, due to the earth's rotation, drive the large ocean currents such as the South Equatorial Current, the Equatorial Counter Current, The North Equatorial Current, the Humboldt or Peru Current, the Kuroshio Current, as well as the South Subtropical Current in the Pacific Ocean. These forces also drive the Gulf Stream, the North Atlantic Drift, the Brazil Current, the Benguela Current in the Atlantic Ocean and cause the counter clockwise circulation of the Mediterranean Sea. These currents and others are depicted in Figure 8.1. Even though these major oceanic currents may be too far offshore to directly affect the outfall, they can induce significant nearshore currents.

Tides also drive currents. They can be visualized as mounds of water that are traveling around the earth's oceans and seas due to the gravitational forces of the moon and the sun. The high water level of the tide occurs about every 24 hours and 50.5 minutes. Depending on the location on the globe the difference between high and low tides can be less than one meter or more than 15 meters. The difference between the elevations of the high and low tides and the geomorphology of the ocean floor and the coastline configuration determine the velocity of the currents generated. Restrictions in the pathway of the movement of the "mound of water" can result in high velocity currents. For example, narrow waterways between the islands of an archipelago in the open ocean, restricted entrances through an island's fringing reef, and the narrow entrances to an atoll can be subjected to strong tidally-induced currents. Restricted entrances to large fiord or bays where there are tidal differentials in excess of 12 meters can be subjected to current velocities as high as 6 or 7 m/s.

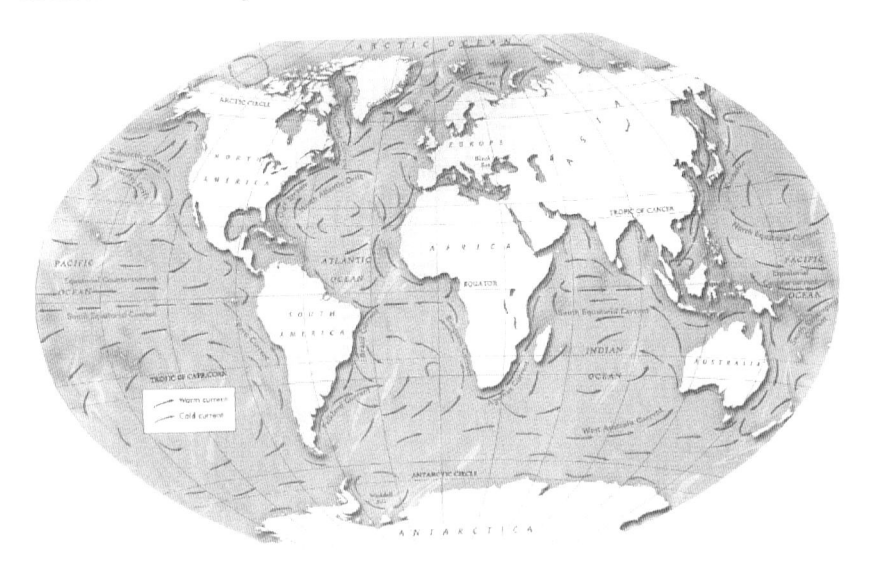

Figure 8.1 Major oceanic currents (courtesy of National Geographic)

Wind-driven currents can also be very important in the design of ocean outfalls. The wind may be set in motion by geostrophic forces, diurnal heating and cooling, and atmospheric pressure gradients often associated with storms and hurricanes. Wind-driven surface currents are frequently the most important factor that determines the direction of a surfaced plume. Wind can also seriously hamper the transport, placement, and sinking of an HDPE outfall so seasonality of strong winds should be considered when scheduling construction.

The seaward return of runup water on the beach from large breaking waves can result in strong littoral currents running parallel to the shore and strong rip currents running more or less perpendicular to the shore. These littoral currents are often perpendicular to the outfall axis and should be considered when analyzing outfall stability and erosion and transport potential of the material in unconsolidated seabeds. Rip currents may be beneficial in transporting discharged effluent away from the shore but can also adversely affect the installation of an ocean outfall. Large waves breaking over an atoll and pouring water into its lagoon can result in strong currents in the lagoon's entrance to the sea. The same can occur through the entrances through a fringing reef of an island or entrances through barrier islands near the mainland.

When planning an outfall, records of peak currents and their direction (when available), tide tables, and navigational charts for the area under consideration should be consulted. Valuable insight can often be obtained by interviewing local naval officers, harbor masters, fishermen, and divers. Where currents are known to be high but there are no records, current measurement by an acoustic Doppler current profiler (ADCP), though expensive, may be justified (see Chapter 5). It is recommended that currents be recorded continuously for a period of one year in the area under consideration. When this is not possible currents should at least be measured at times of the year most representative of peak currents.

8.3 WAVES

Even though waves transport relatively small amounts of water they can transmit enormous amounts of energy and can cause pipeline failure through the hydrodynamic forces they exert on the pipe, differential erosion of the seabed or bedding material on which the pipe is laid, mass movement of seafloor materials, and liquefaction of the supporting seabed.

Wave dynamics induce both horizontal and vertical (lift) forces on ocean outfalls that are resting on the seabed and can also affect the stability of the seabed itself. These forces must be adequately countered by the system used to hold or fasten the outfall to the ocean floor.

The amount of energy that can be imparted by a wave is proportional to the height and the length of the wave. The outfall must be designed for the largest wave that is likely to occur during its useful life.

Among the oceanographic studies, analyses, and calculations that need to be carried out prior to designing a submarine outfall, one of the most important is the determination of the design wave.

8.3.1 Design wave

Many failures of unburied outfalls are due to underestimating the hydrodynamic forces due to large waves. Waves induce both horizontal and vertical (lift) forces on pipelines resting on the seabed. These forces must be countered by a system that holds or fastens the outfall to the seabed or otherwise protect the outfall. Waves can also cause failure through differential erosion of the seabed or bedding material on which the pipe sits, liquefaction of the seabed, and burial of the diffuser ports by sediment. Knowledge of waves and the accelerations, velocities and forces beneath them is necessary to preclude such failures.

There are four major steps in determining the hydrodynamic forces that waves exert:

- Determine the deepwater wave likely to affect the outfall during its useful lifetime;
- Transformation of deepwater waves to shallow water;
- Determine the orbital velocities and inertia forces in shallow water;
- Calculate the forces caused by waves at various outfall depths.

The most severe wave that is likely to impact the outfall during its projected useful life is referred to as the *deep water design wave* and is described by its height, length, period, and the direction of travel of its wave front. A long accurate record of wave characteristics near the proposed outfall site would be very useful in designing an outfall but this information is rarely available, especially in developing countries. Short-term local wave data is sometimes available but its usefulness is limited primarily to determining the preferable time of year (or day) when sea conditions are most amenable for construction and in selecting the construction method to be employed.

To derive a design wave it is first necessary to decide on the return frequency (or recurrence interval) of the wave. Because outfalls commonly have a useful life in the neighborhood of 25 years, a return frequency of 50 years is often selected, but frequencies as high as 100 years have been used when the cost associated with this high level of assurance is justified.

Wind-induced waves are usually most important. However, in some regions, seismic and volcanic activity can produce large destructive waves called tsunamis. They have extremely long wavelengths, very high velocities, and enormous energy and should be considered where there is a history of occurrence. There is presently no reliable means of estimating the magnitude, recurrence interval, and location of tsunamis in a form useable for outfall design. Hence, this subject will not be covered here.

When waves are being generated by wind they are referred to as seas. Seas are composed of waves of many different heights and periods, frequently moving in different directions. The surface oscillations persist long after the wind has quieted and the waves move out of their area of generation. Because longer waves travel faster than shorter waves, the components separate as they travel outside the area of generation and the waves become more uniform and are then known as swells. Seas have a much more disturbed surface than swells appearing as irregular short crested waves and they tend to have a shorter period than swells. Swells are generally more regular than seas and have well defined crests and troughs. Their periods are usually greater than 10 seconds and they propagate in one predominant direction, appearing almost unidirectional when viewed from above. Swells can travel long distances with little loss of energy.

The characteristics of wind waves generated in shallow water are influenced by bottom depth (d) but not in deep water. In the context of ocean engineering and waves, deep water is defined as the depth that exceeds ½ of the wave length. Wind waves in deep water are influenced directly by the speed of the generating wind, the distance that the wind is blowing over the water (the fetch), and the length of time that the wind blows. In deep water, when a given fetch is subjected to a specific wind velocity, the wave height will increase with time of exposure until a maximum height is reached, and additional exposure does not increase this height. The waves are then called fully developed waves. In the open ocean the wave generation process generally is responsive to average winds over a 15 to 30 minute interval.

8.3.2 Prediction of wind-driven waves

The two basic categories of predicting seas (wind waves) and swells are forecasting and hindcasting. Currently, both are being carried out on a global scale by computer-assisted numerical computations. These use one of several spectral methods for which the preparation of meteorological data and the computer processing demand considerable time and experience. Basically, forecasting is prediction of future wave conditions using derived meteorological information. Hindcasting is determination of wave conditions from a past time, utilizing an historical database of storms and hurricanes that pass within a certain distance of the site of interest. The wave conditions that had been generated by each of these events as they traveled along their individual tracks are then computed and the deep water wave height and period are extracted.

Results of deep-water wave analyses are usually presented in a manner similar to that indicated in Table 8.1. This table shows the *significant wave height* ($H_{1/3}$) in meters and the corresponding *wave period* ($T_{1/3}$) in seconds as functions of return period in years. The significant wave height is defined as the average of the highest 1/3 of all wave heights observed over a certain period of time. For engineering purposes this is considered to be a more representative wave height than either the overall average of the wave height or the maximum wave height. The maximum wave height H_{max} can be estimated as approximately 1.8 times the significant wave height. The table usually also shows the equivalent standard deviations, the 95th percentile values of wave heights, and the probability for each return value of significant wave height to be exceeded.

Table 8.1 Typical presentation of wave analyses

Return period (yrs)	Significant wave height $H_{1/3}$ (m)	Standard deviation (m)	95th percentile H (m)	Exceedance probability (%)	Wave period $T_{1/3}$ (s)
5	2.9	0.65	8.3	100	8.1
10	5.0	0.83	9.6	99.5	10.1
25	7.8	1.05	11.1	89.1	12.5
50	10.0	1.21	12.1	62.5	14.3
100	12.5	1.39	13.0	39.5	15.5

A number of computer programs are available to carry out these analyses and they are continually being improved. The ACES computer program of the US Army Corps of Engineers is widely used. The Cooper parametric hurricane-wave model is also popular and is based on the local wind at 20 m elevation (Cooper, 1988).

8.3.3 Goda's simplified procedure for wind wave prediction

Feasibility studies and preliminary designs for small diameter ocean outfalls frequently do not justify the cost, manpower, and time needed for the collection and preparation of meteorological data, numerical computation and/or computer processing to carry out full blown forecasting or hindcasting of wind waves. Frequently it is necessary to make a rough estimate of the significant height and period of the deepwater design wave for these purposes.

A straightforward simplified procedure for predicting the significant height and period of wind waves generated in a uniform fetch in deep water using Wilson's formulas was developed by Goda (2003). The author of this chapter compared the results of the Goda procedure with the results of much more elaborate and expensive methodologies for predicting the height and period of significant waves and found close compatibility. His procedure is summarized in the following paragraphs.

As mentioned previously, the height and period of wind waves are directly influenced by the wind speed, the fetch, and the length of time that the fetch is exposed to the wind. These data are frequently available in meteorological records or can be estimated with reasonable accuracy. When a specific fetch is exposed to a particular wind speed, the height of the waves generated will increase with time until a maximum height is reached and further time of exposure will not result in additional increase in height. Then the waves are categorized as "fully developed." If the time of exposure is less, the waves will not reach this height.

Wilson developed the following empirical formulae to predict the significant wave height and period for wind waves generated by a constant wind of speed U (m/s) over a fetch F (kilometers) for a sufficiently long wind duration that the waves are fully developed:

$$\frac{gH_{1/3}}{U^2} = 0.30\left\{1 - \left[1 + 0.004\left(\frac{gF}{U^2}\right)^{1/2}\right]^{-2}\right\} \tag{8.1}$$

$$\frac{gT_{1/3}}{2\pi U} = 1.37\left\{1 - \left[1 + 0.008\left(\frac{gF}{U^2}\right)^{1/3}\right]^{-5}\right\} \tag{8.2}$$

The minimum time t_{min} required for complete wave development is:

$$\frac{gt_{min}}{U} = \int_0^{gF/U^2} \frac{d(gF/U^2)}{gT_{1/3}/(4\pi U)} \tag{8.3}$$

Goda determined the relationship between the minimum fetch required for completely developed waves as a function of duration and wind speed:

$$F_{min} = 1.0 t^{1.37} U^{0.63} \tag{8.4}$$

where t is the duration (minutes) of the wind, U the wind speed (m/s), and F_{min} the minimum fetch (km) for complete wave development.

The initial step in the Goda procedure is to use this equation to estimate the minimum fetch length (F_{min}) required for the fully developed wave at a specified wind speed. Then if the actual fetch F being considered is greater than F_{min}, the height of the significant wave $H_{1/3}$ would be limited by the duration of the wind, and F_{min} would be used in Wilson's equations to determine $H_{1/3}$ and $T_{1/3}$. Conversely, if F_{min} is less than the actual fetch F, the height of the significant wave would be limited by the duration of the wind so the real fetch F would be used in Wilson's equations to determine $H_{1/3}$ and $T_{1/3}$. Goda developed Figures 8.2 and 8.3 to facilitate a quick evaluation of the significant wave height and period for wind speeds between 5 and 30 m/s.

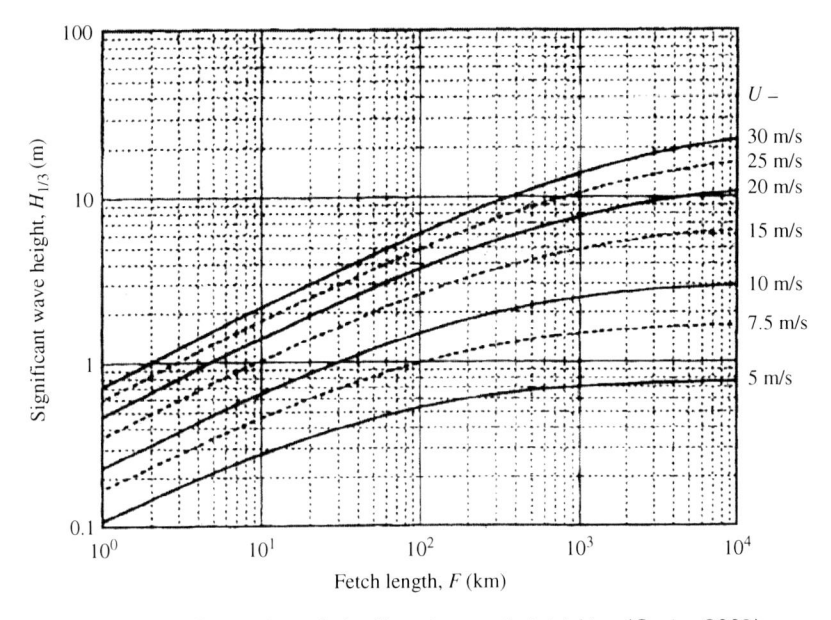

Figure 8.2 Prediction of significant wave height $H_{1/3}$ (Goda, 2003)

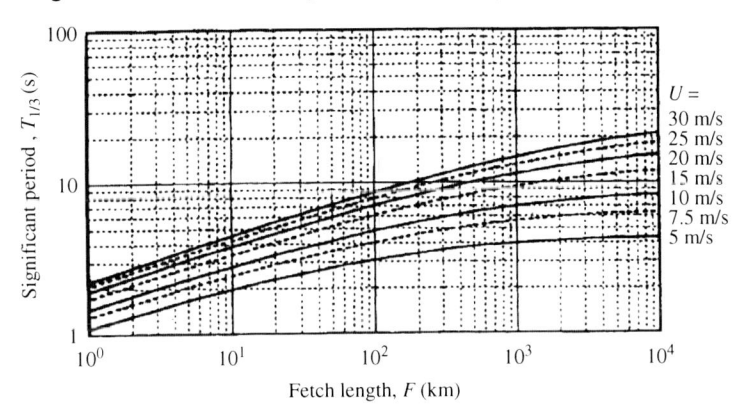

Figure 8.3 Prediction of significant wave period $T_{1/3}$ (Goda, 2003)

8.3.4 Transformation of deep water waves approaching shore

The transformation of waves from deep water to the coast is an important consideration. There are many approaches to analyzing wave transformation that differ in complexity and accuracy. Estimating nearshore waves usually entails simplifying assumptions and approximations, especially where the bathymetry is irregular and complex. The first assumption is that the deepwater wave is characteristic of the design wave that would be propagated to the site under consideration. It is thus important to ensure that there are no sheltering effects or bathymetry anomalies that would negate this assumption.

As a wave propagates from deep into shallow water it can be affected by various processes including, but not limited to, refraction, shoaling, diffraction, dissipation due to friction with the seabed, dissipation due to percolation into the seabed, breaking of the wave, and wave current interaction.

Water depth is the most important physical parameter influencing waves as they propagate towards shore. Waves are thus strongly influenced by bathymetry, and wave interaction with the seabed can cause attenuation. The slope of the seabed, the presence of shoals and reefs, submarine canyons, and submarine escarpments can result in changes in the height, shape, and direction of waves as well as their period. Shoals can more than double the height of waves passing over them. Analyzing wave transformation is further complicated by the fact that water depth at a specific location is usually not constant. It can vary considerably with the tide, storm surges, and seiche.

As waves progress towards shore and enter water that is approximately half as deep as their wavelength, their velocity and wavelength decrease, the wave height increases, and the waves become steeper. The steepness progressively increases with decreasing water depth, and when the water depth is approximately 1.28 times the wave height it breaks. The peak velocity and peak acceleration of the near-bottom water particles increase as a wave progresses into shallower water until just before the wave breaks. The breaking of waves dissipates their energy and induces currents.

There are three general categories for breaking waves. Plunging breakers are characterized by the crest of the wave being thrown forward from the rest of the wave with considerable force giving the appearance of the formation of a tube. These are the waves sought by surfers. They usually occur when deepwater waves encounter a seabed with a moderate slope. Spilling breakers are characterized by foam and turbulent water forming on the wave crest and running down the wave face. They usually occur over flatter slopes. Surging breakers are characterized by the lower portion of the wave being thrown forward. This type of breaker usually occurs on steep bottom slopes, against irregular sea cliffs, and steep escarpments.

The surf zone is defined as the region extending from the shore to the seaward boundary of the breaking waves. The seaward boundary varies considerably with tides and variation in wave height and period. Within the surf zone, wave breaking is the dominant hydrodynamic force. It is common practice to bury a submarine outfall from the shore out through the surf zone to a depth sufficient to protect it from these forces.

8.3.5 Velocity and acceleration under a passing wave

Particle motion under a wave can be viewed from a standpoint of particle trajectories (the Lagrangian approach) or streamlines (the Eulerian approach). Figure 8.4 illustrates both conceptualizations for a sinusoidal water wave progressing over deep, intermediate depth, and shallow waters.

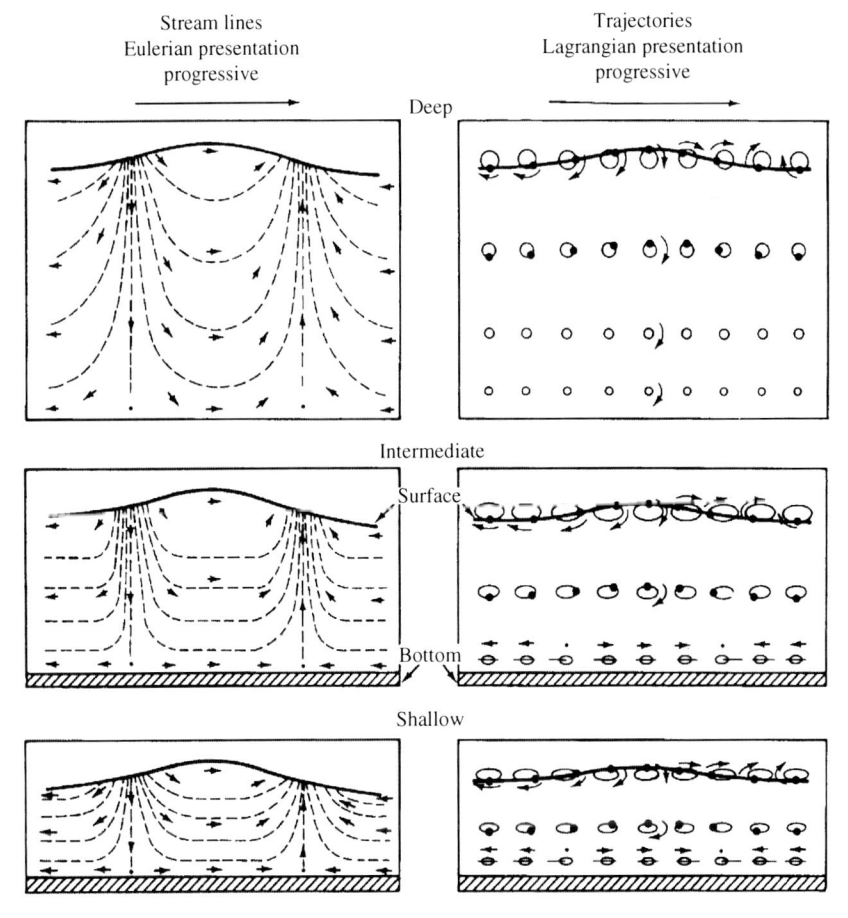

Figure 8.4 Lagrangian and Eulerian views of a sinusoidal wave progressing over deep, medium, and shallow depth water (after Kinsman, 1965)

To calculate the hydrodynamic forces caused by a wave passing over an outfall it is first necessary to determine the wave-induced horizontal velocity and acceleration and the vertical velocity and acceleration at the outfall.

Wave-induced velocities and accelerations should be calculated for various stations (depths) along the outfall route. Three wave theories are frequently utilized for this. They are the linear (Airy), the cnoidal, and Fenton's Fourier series theories.

The ACES computer program of the US Army Corps of Engineers can perform calculations using each of these three methods. It can determine the horizontal and vertical velocity and acceleration components throughout the water column. For outfalls the depth of interest is the outfall pipe centerline. To use ACES for the linear or cnoidal analysis it is necessary to enter the wave height and period, the seabed depth, the depth of interest, and the wavelength fraction passing over the point of interest. Fenton's Fourier series method is somewhat more complex than either linear or cnoidal analysis but it allows input of either the Euler or Stokes wave celerity and it determines energy characteristics as well.

A parameter that is often used to assess the relevance of various wave theories is the Ursell number, U_R:

$$U_R = \left(\frac{L^2 H}{d^3}\right) \tag{8.5}$$

where L is the wave length, H the wave height, and d the water depth.

The linear theory is generally preferred when $U_R < 26$, and the cnoidal theory when $U_R > 26$. The approximate range of the cnoidal theory is for $d/L < 8$ when $U_R > 20$. Fenton's Fourier series is applicable for water deeper than the surf zone and shallower than ½ the wave length.

These theories are based on a two-dimensional water surface. The surface of the linear wave is sinusoidal in time and space and the motion of water particles at the surface are circular. In the cnoidal and Fenton's analysis the motion of a surface particle is ellipsoidal. In all theories the water motion under the wave just above the sea floor is almost parallel to the bottom surface (approximately horizontal); as the distance above the seabed increases the vertical component of the motion also increases.

It is important to point out that when applied to the same wave height and period, each of these theories yield somewhat different results. This is shown in Figure 8.5 that depicts the water surface shapes and the velocities, and accelerations of a water particle one meter above the seabed in a water depth of 20 meters for a wave height of 10 meters with a 14.3 second period as predicted by each theory. It can be seen

Oceanic forces on Submarine Outfalls 235

that the height of the wave crest above mean sea level for the linear theory is equal to the depth of the trough below mean sea level, whereas the crest height predicted by the 2nd order cnoidal theory above mean sea level is considerably greater than the depth of the trough below mean sea level. The height of the cnoidal wave crest is also considerably higher above mean sea level than the linear wave. These differences are even more pronounced for the Fenton's Fourier series wave.

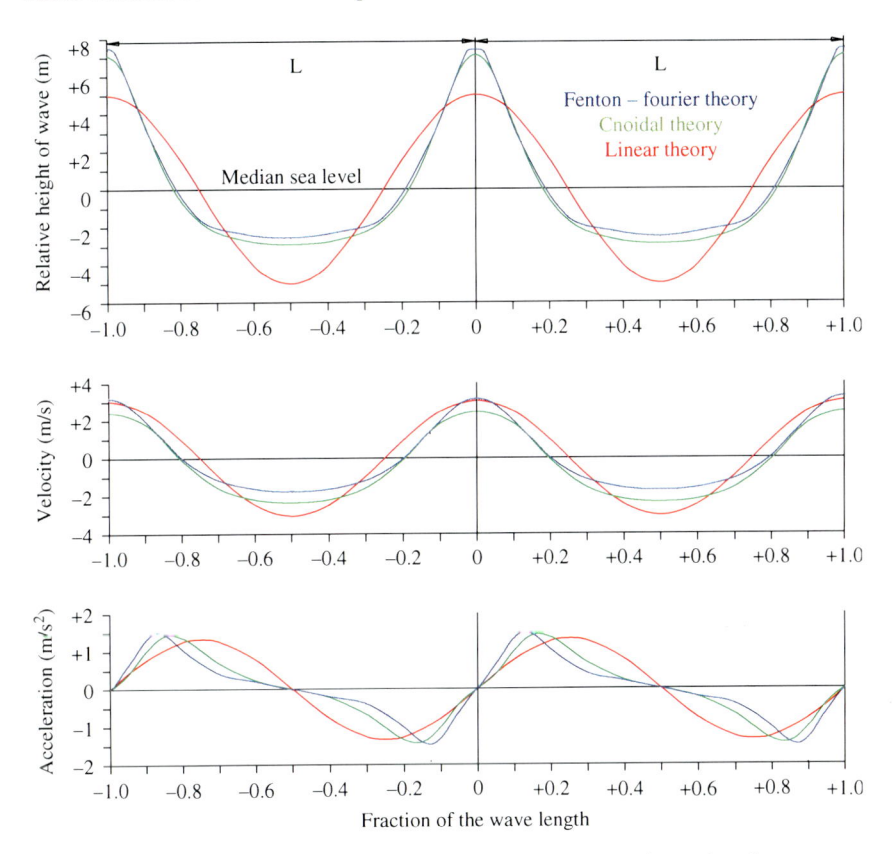

Figure 8.5 Surface elevation and horizontal velocities and accelerations one meter above the seabed as determined by linear (Airy), cnoidal, and Fenton's Fourier series analyses of a wave with a height of 10 m and a period of 14.3 seconds moving through a 20 m water depth

In all three theories the near-bottom water particles reach a peak velocity in the direction of wave propagation as the wave crest passes; as the trough passes, they again reach a peak velocity but in the opposite direction.

In Figure 8.5 it can be seen that for the linear theory the peak velocities under the crest and the trough are equal in magnitude but opposite in direction, whereas the 2nd order cnoidal analysis yields a peak velocity under the crest that is slightly higher than that under the trough; for Fenton's series this difference is even more pronounced. Fenton's Fourier wave has a peak velocity that is slightly higher than the linear wave which is somewhat higher than the cnoidal wave. The peak velocity under the trough of Fenton's wave is less than the cnoidal wave which is less than the linear wave.

For the linear wave, the peak velocity occurs when the acceleration is zero and the peak acceleration occurs when the velocity is zero. In the cnoidal and Fenton waves the peak velocity also occurs when the acceleration is zero but the peak acceleration occurs while the velocity is still significant.

The near-bottom water particles in the linear wave reach peak horizontal acceleration when the point on the wave that is exactly half way between the crest and the trough (1/4 of the wave length) passes over. The peak acceleration of the cnoidal wave is slightly greater than the peak acceleration of the linear theory and it is reached at a distance about 1/5 of the wave length from the crest. The peak acceleration of the Fenton wave is greater than either the linear or the cnoidal wave and is reached at a distance slightly more than 1/10 of the wave length from the crest.

The relative timing of the velocity and acceleration is important in estimating the total horizontal and total vertical forces that are induced by waves. For the linear theory, adding the peak drag force, which is velocity dependent, to the peak horizontal inertia force, which is acceleration dependent, would overestimate the total horizontal force because they do not occur simultaneously. For the cnoidal and Fenton theories, using either the peak drag force or the peak horizontal inertia force, whichever is greater, would underestimate the total force. This is further complicated when the lift force, which is velocity dependent, is taken into consideration. The method used to obtain the combined effect of these forces must be compatible with the theory used to estimate the forces.

It is important to point out that wave theories are approximations and care should be taken to ensure that they are applied when the actual conditions reasonably satisfy the assumptions on which the theories are based.

8.4 FORCES DUE TO A STEADY CURRENT

A steady current running crosswise to a pipeline on the seabed imposes both a horizontal and a vertical force on the pipe. The horizontal force is due to drag, and the vertical force is due to the Bernoulli effect of the difference in velocity of water flowing above and below the pipe. To calculate these forces on an outfall it is necessary to know the velocity and also the direction of the current relative to the outfall pipeline, because the magnitude of their effect varies considerably with the angle of attack.

The maximum horizontal and maximum vertical forces imposed by a steady current both occur when the current is perpendicular to the axis of the pipeline. If the current is parallel to the axis, little if any force will be exerted on the pipe but there will be a drag force on concrete ballast weights attached to the pipe. This longitudinal force on the ballasts is usually insignificant because of the relatively small cross sectional area of the ballast and the fact that the ballasts are firmly affixed to the pipe.

The horizontal drag force on a pipeline on the ocean floor due to a current impinging perpendicular to the pipeline can be expressed by:

$$F_D = C_H \rho A \frac{V^2}{2} \qquad (8.6)$$

where

F_D = drag force (N)
C_H = drag coefficient (after adjustment for the incidence angle)
ρ = seawater density (usually about 1,025 kg/m^3)
A = $D \times l$ (diameter of pipe × length of section considered, m^2)
l = length of pipe (m)
V = horizontal speed (m/s)

The drag coefficient C_H for a pipe resting on or near the seabed with its axis perpendicular to the flow is influenced by the roughness of the pipe wall, turbulence of the flow, and roughness of the seabed, but is independent of the pipe's distance above the seabed. This coefficient depends on the Reynolds number:

$$\text{Re} = \frac{VD}{\nu} \qquad (8.7)$$

where ν = kinematic viscosity of sea water (typically 1.12×10^{-6} m^2/s).

Drag coefficients are plotted in Figure 8.6 as functions of Reynolds number. Two curves are representative of smooth wall polyethylene pipe for relative roughness of the seabed (k/D) of 0.0075 and 0.008–0.02 that are more or less respectively representative of silt-sand and sand-gravel sea beds for typical diameters of polyethylene outfall pipe. A third curve is representative of rougher sea beds and pipe that has been in the ocean for some time and lost smoothness due to accumulated marine organisms. For pipe diameters and oceanic conditions relevant to stability of submarine outfalls the

Reynolds number is frequently above 2×10^5 where the drag coefficient becomes constant.

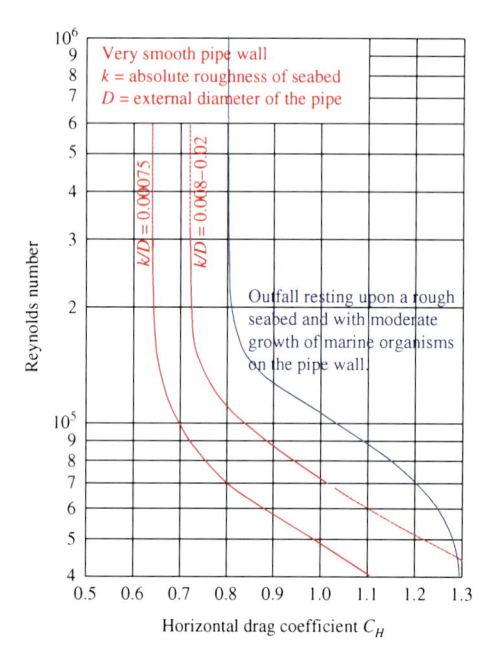

Figure 8.6 Drag coefficient versus Reynolds number for steady flow and three selected seabed conditions

The author of this chapter has inspected numerous outfalls that have significant roughness due to bioaccumulation after only a few years of operation, negating any assumption that initially smooth outfall pipe would remain so. In one extreme case in Alaska, Figure 8.7, the accumulation was sufficient, only eight years after the pipeline's installation, to enlarge the effective diameter more than 100%. Under conditions favorable for marine organisms seeking substrate to attach to, it is recommended that a C_H value of 1.0 be used.

If the current is impinging at an angle less than 90° to the pipe's axis the drag coefficient should be corrected by a factor that can be estimated from Figure 8.8.

Oceanic forces on Submarine Outfalls 239

Figure 8.7 Heavy growths of marine organisms on a submarine pipeline 8 years after installation in Angoon, Alaska (Reiff)

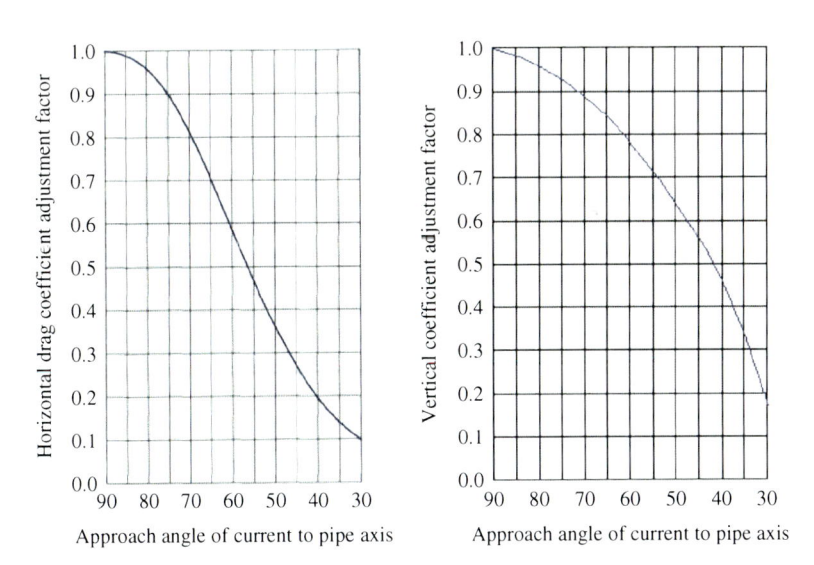

Figure 8.8 Correction factors for drag and lift coefficients for currents impinging on a pipeline at various angles (after Grace, 1978)

The lift force induced by a steady current perpendicular to the pipeline can be determined by:

$$F_L = C_L \frac{\rho}{2} A V^2 \qquad (8.8)$$

where
C_L is the lift coefficient
F_L = Lift force (N)
$A = (D \times l)$ pipe diameter × length of pipe considered (m^2)
V = current speed (m/s)

The lift coefficient C_L for an outfall pipe resting on the seabed in steady state flow perpendicular to the pipe axis can be obtained from Figure 8.9. This shows three curves for a smooth wall pipe resting on seabeds with differing relative seabed roughness. The lift coefficient, like the drag coefficient, stabilizes at a constant value at higher Reynolds number, but unlike the horizontal drag coefficient, the lift coefficient decreases with decreasing roughness of the seabed and it decreases with increasing pipe roughness. A fourth curve, plotted in blue, represents a typical lift coefficient of an outfall pipe with significant accumulation of marine organisms resting directly upon a moderately rough seabed. A lift coefficient value of 1.0 is commonly used for outfalls resting on the seabed to accommodate the uncertainty of changing seabed conditions.

The lift coefficient C_L obtained from Figure 8.9 must be adjusted if the current is flowing at an angle less than 90°. This is done by multiplying it by a factor obtained from Figure 8.8.

Furthermore, the magnitude of the vertical (lift) force due to a horizontal current varies with the height of the pipeline above the seabed. The maximum lift force occurs when the pipe is resting on the ocean floor; as the height increases the lift force decreases. At a clearance of one pipe diameter the lift force is approximately 20% of the value for the same pipe resting on the seabed, and at two pipe diameters the lift force is negligible. The lift coefficient should then be adjusted by an additional factor that is based on the relative clearance of the pipe above the seabed. The relative clearance is the ratio of the distance above the seabed to the diameter of the pipe. This factor can be obtained from Figure 8.10.

It is usually convenient to calculate both the drag force and the lift force per meter of outfall pipe length to facilitate the determination of the optimum spacing of ballast weights or mechanical anchors that are used to stabilize the outfall.

8.5 WAVE-INDUCED FORCES

Even though waves transport relatively small amounts of water they can transmit enormous amounts of energy. Many failures of unburied pipelines are related to waves. They can cause pipeline failure through the hydrodynamic forces on the pipe, differential erosion of the seabed or bedding material on which the pipe is laid, mass movement of seafloor materials, and liquefaction of the supporting seabed.

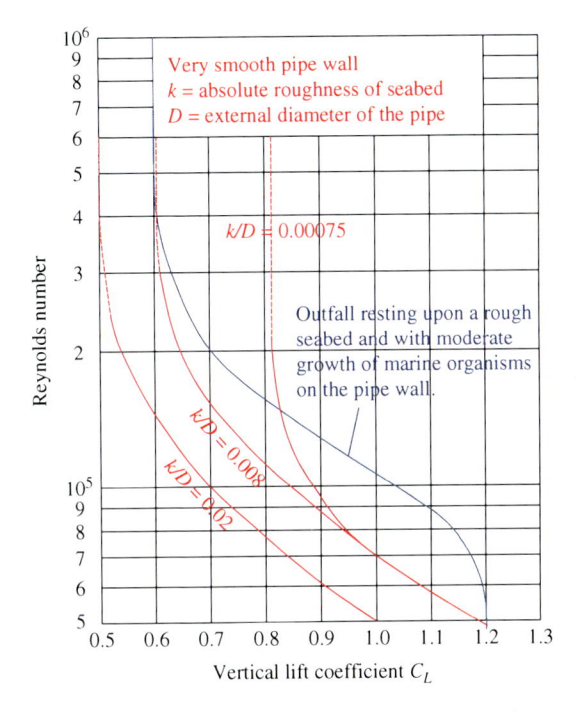

Figure 8.9 Lift coefficient versus Reynolds number for steady current flow and various selected seabed conditions

As previously discussed, waves induce both horizontal and vertical (lift) forces on outfalls that are resting on the seabed. These forces must be adequately countered by the system that is to be used to hold or fasten the outfall to the ocean floor. These forces can be estimated by the following methodology.

8.5.1 Horizontal forces

Horizontal forces induced by waves on pipelines include both drag and inertia forces. These forces are frequently estimated by the Morrison equation:

$$F = F_D + F_I \qquad (8.9)$$

where F is the total horizontal force, F_D the drag force, and F_I the inertia force. For a wave perpendicular to the pipeline, the drag and inertia forces can be calculated using the following equations:

$$F_D = C_H \frac{\rho}{2} A U |U| \qquad (8.10)$$

$$F_I = C_I \rho \frac{\pi D^2}{4} l a_h \qquad (8.11)$$

where:
- F = total horizontal force (N)
- F_D = drag force (N)
- F_I = inertia force (N)
- U = horizontal wave-induced velocity (m/s)
- C_I = inertia coefficient
- a_h = wave induced horizontal acceleration (m/s^2)

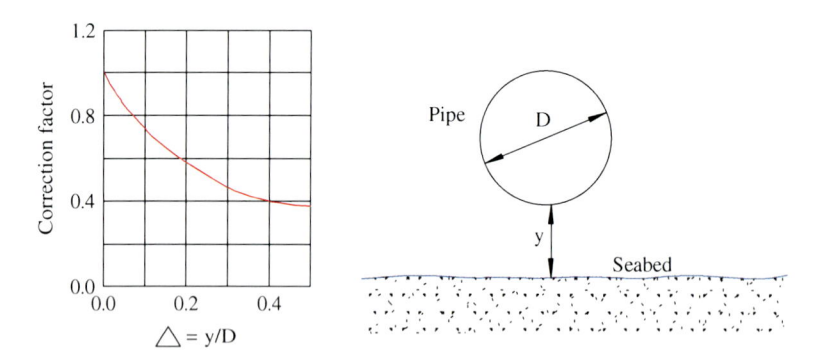

Figure 8.10 Correction factor for the lift coefficient for a pipe at various relative clearances above the seabed (Grace, 1978)

These forces should be calculated for various stations along the outfall using velocities and accelerations determined for the respective depths by means of an appropriate theory such as discussed in Section 8.3.4.

The drag coefficients C_H for wave-induced water movement impinging horizontally at 90° to a submerged pipeline can be estimated from Figure 8.6. The Reynolds Number is based on the maximum horizontal velocity induced by the wave at the specified water depth.

Because outfalls are commonly routed to access deep water as soon as possible, they are usually oriented nearly perpendicular to the contour lines of the seabed. As deep water waves approach shallower water they are refracted by the drag and frictional forces of the ocean floor. Therefore wave fronts and their induced currents frequently intersect an outfall at an angle considerably less than 90°, and the drag coefficient C_H must be multiplied by an angle correction factor (F). This factor, which is not the same as the factor for steady currents, can be obtained from Figure 8.11. The drag force is then calculated using the adjusted drag coefficient from:

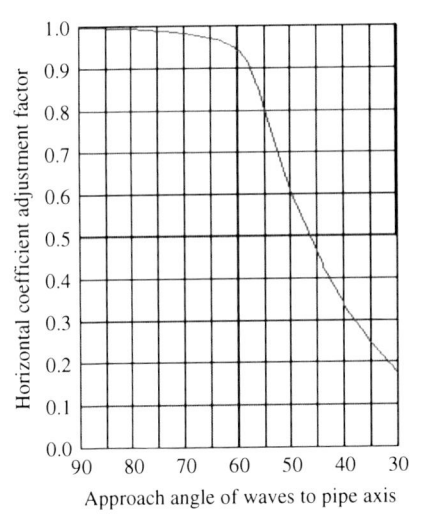

Figure 8.11 Correction factor for drag coefficient C_H for wave fronts approaching a pipeline at various angles (Grace, 1978)

$$F_D = FC_H \frac{\rho}{2} AU|U| \qquad (8.12)$$

The horizontal inertia force is acceleration dependent but independent of the Reynolds number. It is calculated using Eq. (8.11). The coefficient C_I depends on relative clearance of the pipe (Δ) which is the ratio of the clearance of the outfall pipe above the seabed to the outside diameter (D) of the outfall pipe. The maximum value of the inertia coefficient C_I is 3.29 and this occurs when the pipe rests directly on the seabed. It decreases as the pipe is elevated above the seabed, asymptotically approaching 2 when the pipe is remote from the seabed.

The relationship of C_I to its relative clearance above the seabed is presented in Figure 8.12.

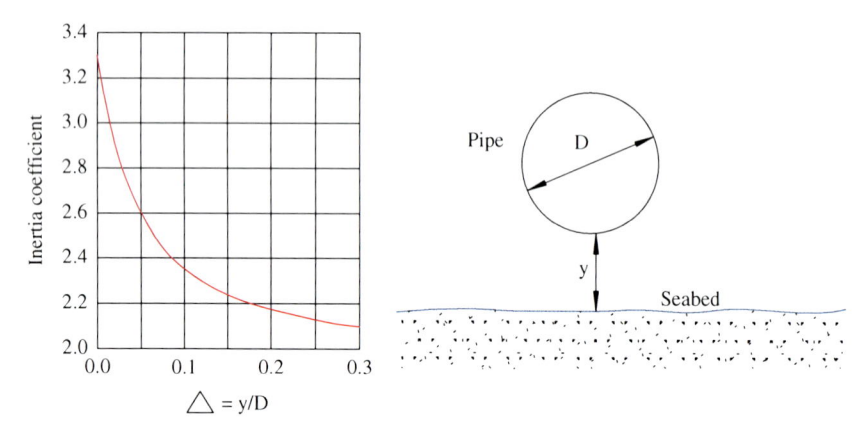

Figure 8.12 Horizontal inertia coefficient for a pipe at various elevations above the seabed (Davis and Ciani, 1976)

8.5.2 Vertical forces

Waves also induce vertical forces on pipelines. This is primarily due to the induced horizontal flow of water across the pipe. There is also a vertical acceleration force of the water mass that is usually negligible because the vertical component of wave-induced water motion near the seabed is relatively small outside the surf zone.

The vertical (lift) force on the pipe that is caused by the horizontal flow due to waves can be determined by the same equation (8.8) used to calculate the lift force due to steady currents:

$$F_L = C_L \frac{\rho}{2} A V^2 \quad (8.13)$$

where V is the maximum wave-induced horizontal velocity.

An initial value for the lift coefficient C_L for water impinging horizontally at 90° to a pipeline can be estimated by use of Figure 8.9 using a Reynolds number based on the maximum horizontal velocity under the wave. When the wave front intersects the axis of the outfall at an angle less than 90°, C_L should be modified by a factor obtained from Figure 8.13.

Oceanic forces on Submarine Outfalls

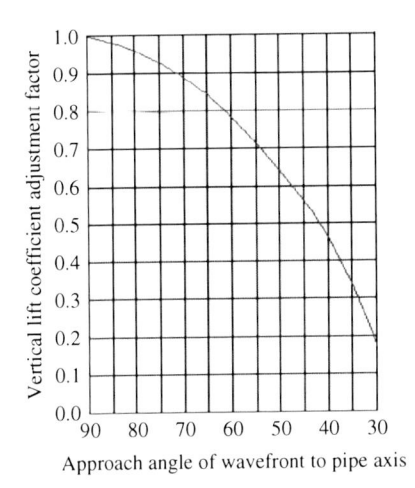

Figure 8.13 Correction for lift coefficient for waves approaching a pipeline at various angles

8.6 HYDROSTATIC PRESSURE FORCES

The hydrostatic pressure in the ocean increases with depth and depends on the average density of the water column. The pressure inside the outfall pipe will be greater than the external pressure by an amount equal to the head losses due to friction. During periods of no flow the internal and external forces are equal which means that the height of the fresh water column inside the pipe will be greater than the seawater depth.

This head differential, due to the difference in density of the effluent and seawater, is called the density head, and is important because it increases the static head at the outfall entrance. The magnitude of the differential, expressed in meters of effluent, is given by:

$$\Delta h = D_S \left(\frac{\gamma_S}{\gamma_E} - 1 \right) \qquad (8.14)$$

where
- Δh = head differential (m)
- D_S = the depth of the outfall port(s) (m)
- γ_S = specific weight of seawater (kN/m^3)
- γ_E = specific weight of effluent (kN/m^3)

For example, an outfall discharging at a depth of 40 m with a specific weight of seawater of 10.05 kN/m^3 and a specific weight of effluent of 9.80 kN/m^3, would have a head difference of 1.0 meters. This static head must be overcome to initiate flow.

During installation of an outfall by flotation and submergence it is important to ensure that the pressure inside the outfall is higher than outside to avoid collapsing the pipe wall. This is especially important when bending places additional stress on the pipe wall.

Hydrostatic pressure also influences ballast weight concrete in that it must contain a high percentage of Portland cement to minimize penetration of the seawater so as to prevent premature corrosion of the reinforcing steel.

9
Design of polyethylene outfalls

9.1 INTRODUCTION

Low density polyethylene (PE) was discovered in 1941 and high density polyethylene (HDPE) was first produced commercially in 1957. Polyethylene pipe has been used to transport water and wastewater for more than 50 years, during which there were many improvements in the resins used and in pipe fabrication techniques. Although there are a number of classes of polyethylene resins, some are not appropriate for marine outfalls. They do not have the high quality needed to ensure that the pipe can resist the forces of construction and installation, the internal and external pressures in service, the hydrodynamic forces of the ocean, and its chemical corrosion and biological aggressiveness.

The design of an ocean outfall commences after the location and orientation of the diffuser is established in accordance with the processes described in Chapters 3 and 4 and the location of the headworks is determined. It includes considerations for internal hydraulics, external hydrodynamic oceanic forces,

© 2010 IWA Publishing. *Marine Wastewater Outfalls and Treatment Systems.* By Philip JW Roberts, Henry J Salas, Fred M Reiff, Menahem Libhaber, Alejandro Labbe, and James C Thomson. ISBN: 9781843391890. Published by IWA Publishing, London, UK.

structural integrity and stability, material suitability, geomorphology of the seabed, competing uses of the ocean, as well as installation and operational methodology. In this chapter the major design issues for outfalls using HDPE pipe are addressed. It is not intended to be a design manual; more detailed specialized references should be consulted prior to final design.

9.2 REFERENCE STANDARDS FOR POLYETHYLENE MATERIALS AND PIPE

Because HDPE is a thermoplastic, both short- and long-term tests of its physical properties are necessary to determine how it will perform. It is therefore important to use reference standards when designing and specifying HDPE pipe for marine outfalls.

There are two major sources of standards for polyethylene resins and polyethylene pipe that are commonly used: (1) ISO (International Organization for Standardization) is a worldwide federation of national standards bodies and (2) ASTM International (American Society for Testing Materials). Their standards that are relevant to ocean outfalls are summarized in Tables 9.1 and 9.2.

Most polyethylene ocean outfalls are constructed of high density, high molecular weight polyethylene resins (HDPE), a thermoplastic material with many properties that are very different from other pipe materials commonly used for outfalls such as steel, ductile iron, and concrete. Understanding the unique characteristics of HDPE, its advantages and disadvantages, and how they affect the design and construction of outfalls is essential to ensure that they have a long useful life.

The classification system used by ISO for thermoplastics material (including polyethylene) is fundamentally different from that used by ASTM and direct conversion between the two is not possible. ISO uses a single classification number that is ten times the 97.5% lower confidence limit of the long term (50 year) strength determined by testing. A brief review of ISO terminology follows to facilitate a basic understanding of their PE pipe standards.

The *long-term strength* (σ_{LTHS}) is the basis for PE classification and for allowable design stress. It has dimensions of stress, expressed in megapascals (MPa). This property represents the 50% lower confidence limit of the long-term strength and is equal to the mean strength or predicted mean strength at 20°C for 50 years with internal water pressure.

The *lower confidence limit long-term strength* (σ_{LCL}) also has dimensions of stress, expressed in MPa. It is a material property that represents the 97.5% lower confidence limit of the long-term strength at 20°C for 50 years with internal water pressure.

Design of polyethylene outfalls 249

Table 9.1 ISO standards relevant to HDPE outfalls

161-1:1996	Thermoplastics pipes for the conveyance of fluids – Nominal outside diameters and nominal pressures Part 1: Metric series
1133:1997	Plastics – Determination of the melt mass-flow rate (MFR) and the melt volume-flow rate (MVR) of thermoplastics
1167:1996	Thermoplastics pipes for the conveyance of fluids – Resistance to internal pressure – Test method
2505-1:1994	Thermoplastics pipes – Longitudinal reversion – Part 1: Determination methods
2505-2:1994	Thermoplastics pipes – Longitudinal reversion – Part 2: Determination parameters
3126:2005	Plastics piping systems – Plastics components – Determination of dimensions
3607:1996	Polyethylene pipes – Tolerances on outside diameters and wall thicknesses
4065:1996	Thermoplastic pipes – Universal wall thickness table
4427:1996	Polyethylene (PE) pipes for water supply – Specifications
4607:1978	Plastics – Methods of exposure to natural weathering
6259-1:1997	Thermoplastics pipes – Determination of tensile properties – Part 1: General test method
6259-3:1997	Thermoplastics pipes – Determination of tensile properties – Part 3: Polyolefin pipes
6964:1986	Polyolefin pipes and fittings – Determination of carbon black content by calcinations and pyrolysis – Test method and basic specification
9080:2003	Plastics piping and ducting systems – Determination of the long-term hydrostatic strength of thermoplastics materials in pipe form by extrapolation
1:1922-1.1997	Thermoplastics pipes for the conveyance of fluids – Dimensions and tolerances – Part 1: Metric series
1:1922-2:1997	Thermoplastics pipes for the conveyance of fluids – Dimensions and tolerances – Part 2: Inch-based series
1:216-2:1995	Thermoplastics materials for pipes and fittings for pressure applications – Classification and designation – Overall service (design) coefficient
1376-1:1996	Plastics pipes and fittings – Pressure reduction factors for polyethylene pipeline systems for use at temperatures above 20°C
1855-3:2002	Method for the assessment of the degree of pigment or carbon black dispersion in polyolefin pipes, fittings and compounds

The *minimum required strength* (MRS) is the value of σ_{LCL}, rounded down to the next smaller value of ISO's R20 series of preferred numbers conforming to ISO 3 and ISO 497.

The *overall service design coefficient* is represented by the letter C (not to be confused with the Hazen-Williams C). It is a coefficient that accounts for service conditions and properties of piping system components other than those represented in the lower confidence limits. For most outfalls, $C = 1.6$.

Table 9.2 ASTM Standards Relevant to HDPE Outfalls

D 618	Practice for Conditioning Plastics for Testing
D 638	Test Method for Tensile Properties of Plastics
D 746	Test Method for Brittleness Temperature of Plastics and Elastomers by Impact
D 792	Test Methods for Density and Specific Gravity (Relative Density) of Plastics by Displacement
D 883	Terminology Relating to Plastics
D 1238	Specification for Polyethylene Plastics Extrusion Materials for Wire and Cable
D 1505	Test Method for Density of Plastics by the Density-Gradient Technique
D 1603	Test Method for Carbon Black in Olefin Plastics
D 1693	Test Method for Environmental Stress-Cracking of Ethylene Plastics
D 1898	Practice of Sampling for Plastics
D 2239	Polyethylene (PE) Plastic Pipe (SIDR-PR) Based on Controlled Inside Diameter
D 2683	Socket – Type Polyethylene Fittings for Outside Diameter-Controlled Polyethylene Pipe and Tubing
D 2837	Test Method for Obtaining Hydrostatic design Basis for Thermoplastic Pipe Materials
D 3035	Polyethylene (PE) Plastic Pipe (DR-PR) Based on Controlled Outside Diameter
D 3350	Polyethylene Plastic Pipe and Fittings Materials
D 3261	Butt Heat Fusion Polyethylene (PE) Plastic Fittings for Polyethylene (PE) Plastic Pipe and Tubing
D 4976	Specification for Polyethylene Plastics Molding and Extrusion Materials
D 5033	Guide for the Development of ASTM Standards Relating to Recycling and Use of Recycled Plastics
F 714	Standard Specification for Polyethylene (PE) Plastic Pipe (DR-PR) Based on Outside Diameter
F 1473	Test Method for notch Tensile Test to Measure the Resistance to Slow Growth of Polyethylene Pipes and Resins

The *allowable design stress* is defined as $\sigma_S = (MRS)/C$ rounded down to the next lower value of the R20 series of preferred numbers conforming to ISO 3 and ISO 497. The ISO classifies polyethylene resin by use of a number that is σ_{LCL} rounded down to the next smaller number of the series of preferred numbers (R10 or R20) and multiplying by 10. The two most common ISO classifications for ocean outfalls are PE 100 and PE 80. Unlike the ASTM classification number it does not give information regarding properties other than the lower confidence limit of the long-term strength. When using the ISO classification for purchasing HDPE pipe for ocean outfalls it is advisable to also include resin requirements for other important properties of polyethylene such a density, melt index, slow growth crack resistance, and impact strength.

Design of polyethylene outfalls

In the ASTM standards the properties of a specific resin are described by a six digit identification number followed by a letter as determined by use of Table 9.3. The first digit represents the resin's density, the second digit the melt index, the third the flexural modulus, the fourth the tensile strength, the fifth the slow crack growth resistance, and the sixth the hydrostatic strength. The code letter following the number designates color aspects.

It should be pointed out that there are two acceptable tests for slow crack growth resistance: the ESCR and the PENT tests. Likewise, hydrostatic strength classification can be based on either the *Hydrostatic Design Basis* (HDB) that is determined by ASTM D2837 or by the *Minimum Required Strength* (MRS) that is determined by the standard test ISO 12162. This is an important distinction because these designations are based on different standard temperatures and different testing procedures.

In addition ASTM 3350 has designated a code letter for color and UV stabilization:

A Natural
B Colored
C Black with 2% minimum carbon black
D Natural with UV stabilizer
E Colored with UV stabilizer

In both the ISO and ASTM standards, HDPE pipe is described by a specified exterior diameter and by a minimum wall thickness needed to obtain the pressure rating of the pipe. The pressure rating, P, of the pipe is:

$$P = \frac{2S}{\left(\frac{D}{t} - 1\right)} \tag{9.1}$$

S = hydrostatic design stress
D = outside diameter
t = minimum wall thickness
The dimension ratio (DR) is:

$$\mathrm{DR} = \frac{D}{t} \tag{9.2}$$

The ASTM hydrostatic pressure rating of pipes is based on a service design factor of 0.5 of the results of a sustained long-term (1,000 hour) pressure test (ASTM D-2827) that must result in at least 1,600 psi. In addition, a short-term pressure test must result in a 2,900 psi fiber stress before rupture. The ASTM design stress for HDPE is 800 psi at 23°C. In the ISO pressure ratings the allowable hydrostatic design stress (S) is 6.3 MPa for PE 100 resin and 5 MPa for PE 80 resin at 20°C and a design coefficient of 1.6 (that is commonly used for ocean outfalls).

Table 9.3 Cell classification limits of primary properties for polyethylene resins (ASTM D3350-04)

Property	Test method	Classification number					
		1	2	3	4	5	6
Density (g/cm³)	D1505	≤0.925	>0.925–0.940	>0.940–0.947	>0.947–0.955	>0.955	—
Melt index	D1238	>1.0	1.0–0.4	<0.4–0.15	<0.15	a	—
Flexural modulus (MPa)	D790	<138	138–<276	276–<552	552–<757	758–<1103	>1103
Tensile strength at yield (MPa)	D638	<15	15–<18	18–<21	21–<24	24–<28	>28
Slow growth crack resistance							
I. ESCR	D1693						
a. Test condition		A	B	C	C		
b. Test Duration (h)		48	24	192	600		
c. Failure, max (%)		50	50	20	20		
II. PENT (h)	F1473	0.1	1	3	10	30	100
Hydrostatic strength classification							
I. Hydrostatic design basis (HDB) at 23°C (MPa)	D2837	5.52	6.89	8.62	11.03		
II. Min. required strength (MRS) at 23°C (MPa)	ISO 12162					8	10

[a] Flow rate must be not greater than 4.0 g/10 min. when tested in accordance with D1238 condition 190/21.6.
1 MPa = 145 psi

Design of polyethylene outfalls 253

All pipes of the same class of material and the same DR will have the same pressure rating regardless of pipe diameter. Table 9.4 summarizes the nominal ASTM and ISO pressure ratings for the dimension ratios most commonly used for ocean outfalls.

Table 9.4 Nominal pressure ratings for HDPE pipe

Item	Dimension Ratio (DR)							
	33	32.5	26	21	17	15.5	13.6	11
ASTM Pressure rating for HDB = 1600 (psi)	–	50	64	80	100	110	–	160
ISO Nominal pressure rating for PE80, PN (bars)	3.2	–	4	–	6.3	–	8	10
ISO Nominal pressure rating for PE100, PN (bars)	–	–		6.3	8	–	10	12.5

1 bar = 14.5; psi = 0.1 MPa

9.3 SELECTING THE PIPE DIAMETER

The internal pipe diameter is based on many factors that include the outfall length, the present and future discharge, static head, available hydraulic head or acceptable head losses for pumping, and cleansing velocities in the pipe to prevent any significant deposition of suspended solids at the invert, or grease buildup on the pipe wall. When designing to accommodate the future peak flow, it is also important to check velocities at the present average and maximum flows to ensure that sufficient scour velocities occur on a daily basis during the first few years of operation. This is not always possible, especially with longer outfalls and high ratios of peak hourly design flow to present average hourly flow.

A useful method to "home in" on an outfall's optimal diameter that accommodates future peak flows with acceptable head loss and yet attains sufficient cleansing velocity during present average flows is to prepare a chart of a family of head loss versus discharge curves plus velocity curves intersecting the head loss curves. This is done for an appropriate range of internal pipe diameters. The present and design year average and peak hourly flows as well as the available gravity head or acceptable pumping head should also be indicated on the chart. The length of the outfall up to the diffuser plus the equivalent lengths of fittings and fusion weld beads should be included in the head loss calculations. The initial head for no flow is the sum of the density head between the seawater and the effluent at the discharge depth and the difference between median lower low tide and highest high tide, plus an estimate for the head loss in the diffuser. The procedures to calculate head loss in the outfall and diffuser are discussed in Section 9.7.5.

The graph can be used to determine the smallest internal pipe diameter for which the total head loss at peak flow is less than the available head for gravity

254   Marine Wastewater Outfalls and Treatment Systems

flow (or acceptable head for outfalls that rely on pumping) and to determine if cleansing velocities are achieved. Then the next largest manufactured pipe is chosen that also meets other requirements such as dimension ratio (DR), resin classification, etc.

In the example shown in Figure 9.1, the available head is 7.8 m, the present average flow is 0.18 m³/s and future peak flow is 0.41 m³/s. The five trial pipe diameters of 500, 525, 550, 575, and 600 mm were chosen based on experience. The range of flow varies from zero to 500 l/s, an amount somewhat greater than the peak hourly design flow. The initial zero flow head of 4 m is the sum of the density head (for the outfall depth), the vertical difference between high-high tide and mean lower low tide to which elevations on land are referenced, and a diffuser head loss estimate.

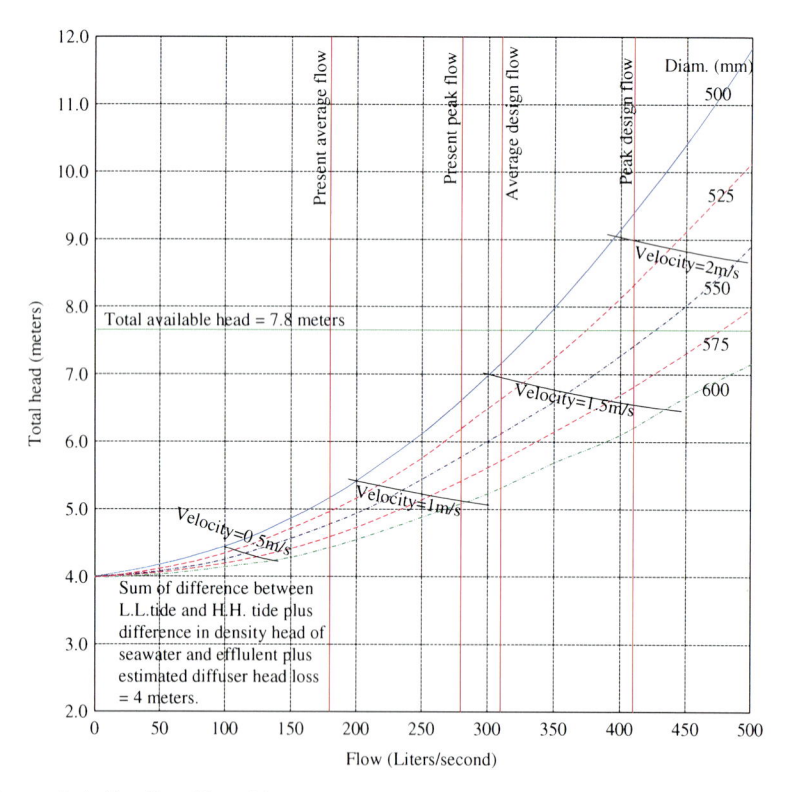

Figure 9.1 Family of head loss versus discharge and velocity curves for various pipe diameters at peak and average hourly flow

The head loss curves were obtained for each of the chosen diameters by application of the Darcy-Weisbach equation in conjunction with the Colebrook

Design of polyethylene outfalls   255

formula. These equations are discussed in Section 9.7.5. Losses due to fittings and weld beads were incorporated into the curves by adding these items as equivalent pipe length to the actual length of the pipe before calculating the friction head loss. Velocity curves for 0.5, 1.0, 1.5, and 2.0 m/s that intersect the head loss curves were also plotted on this chart to enable verification of self-cleaning velocities.

It can be seen that an internal diameter of 550 mm yields a total head at peak hourly flow that is slightly below the available head of 7.8 m, the pipe velocity is about 1.7 m/s at peak future flow, about 0.75 m/s at present average flow, and about 1.2 m/s at peak present flow.

Now it will be necessary to select a manufactured pipe with an internal diameter equal to or slightly larger than 550 mm. Tables 9.5 and 9.6 present typical internal diameters for pipes manufactured to ISO and ASTM standards. The available pipe with the closest match according to the tables would be DR 17, with a nominal outside diameter of 630 mm, and internal diameter of 551 mm. The DR is usually determined by other factors, however, such as the difference between external and internal pressures and allowable bending radius, so the final selection should be made accordingly.

Table 9.5 Typical internal diameters and weights for PE pipes (metric units)

Diameter (mm)		Diameter ratio (DR)									
		32.5		26.0		21.0		17.0		13.5	
Min	Max	ID (mm)	Wt (kg/m)	ID (mm)	Wt (kg/m)	ID (mm)	Wt (kg/m)	ID (mm)	Wt (kg/m)	ID (mm)	Wt (kg/m)
200	201.8	186	3.8	183	4.7	180	5.7	175	7.0	168	8.7
225	227.1	210	4.8	206	6.0	202	7.3	197	8.9	190	11.1
250	252.3	234	5.9	229	7.5	225	9	218	11	210	13.6
280	282.5	282	7.4	257	9.2	252	11	245	14	236	17.1
315	317.8	294	9.4	290	12	283	14	276	17	265	21.6
355	358.2	332	12	326	15	319	18	310	22	299	27.5
400	403.6	374	15	367	19	360	23	350	28	337	34.9
450	454.1	421	19	413	24	404	29	394	36	379	43.9
500	504.5	467	24	459	29	449	36	437	44	421	54.2
560	565.0	523	30	514	37	503	45	490	55	472	68.3
630	635.7	589	38	578	47	566	57	551	70	531	86.2
710	716.4	664	48	652	59	638	73	621	89	598	110.0
800	807.2	748	61	734	75	719	92	700	113	674	139
900	908.1	841	77	826	95	809	117	795	142	758	176
1000	1009.0	934	95	918	118	900	144	882	176	–	–
1200	1210.8	1122	136	1102	169	1079	207	1050	253	–	–
1400	1412.6	1310	–	1290	264	1260	283	–	–	–	–
1600	1614.4	1495	–	1472	328	1443	370	–	–	–	–
1800	–	1715	–	1658	368	–	–	–	–	–	–
2000	–	1871	–	1842	–	–	–	–	–	–	–

Table 9.6 Typical internal diameters and weights for PE pipe (English Units)

Diameter (inches)		\multicolumn{10}{c}{Dimension ratio (DR)}
		32.5		26.0		21.0		17.0		13.5	
Avg.	+/_	ID (in)	Wt (lb/ft)	ID (in)	Wt (lb/ft)	ID (in)	Wt (lb/ft)	ID (in)	Wt (lb/ft)	ID (in)	W. (lb/ft)
8.625	0.039	8.06	3.1	7.91	3.79	7.75	4.6	7.55	5.7	7.27	7.0
10.75	0.048	10.05	4.8	9.88	5.89	9.66	7.0	9.41	8.8	9.06	10.8
12.75	0.057	11.92	6.7	11.70	8.25	11.46	10.1	11.16	12.4	10.75	15.3
14	0.063	13.00	8.1	12.90	9.98	12.60	12.2	12.25	14.9	11.80	18.4
16	0.072	14.96	10.5	14.69	13.05	14.38	16.0	14.00	19.4	13.49	24.1
18	0.081	16.85	13.3	16.50	16.51	16.20	20.2	15.76	24.6	15.17	30.9
20	0.090	18.70	16.4	18.38	20.40	18.00	24.9	17.50	30.4	16.86	37.6
22	0.099	20.61	20.0	20.20	24.66	19.78	30.2	19.26	36.8	18.54	45.6
24	0.108	22.50	23.6	22.05	29.30	21.60	35.9	21.00	43.8	20.23	54.2
26	0.126	24.25	27.8	23.85	34.39	23.40	42.2	22.76	51.4	21.92	63.6
28	0.135	26.16	32.2	26.70	39.90	25.18	49.0	24.50	59.6	23.60	73.8
30	0.144	28.05	37.0	27.50	45.77	26.98	56.1	26.26	68.4	25.29	84.7
32	0.163	29.90	42.0	29.40	52.10	28.77	63.9	28.00	77.9	26.98	96.4
34	0.153	31.80	47.5	31.25	58.78	30.57	72.0	29.75	87.9	28.90	108.8
36	0.162	33.65	53.2	33.06	65.90	32.40	80.8	31.50	98.6	30.36	122.0
42	0.189	39.25	72.4	38.59	89.69	37.76	110.0	36.76	134.2		
48	0.216	44.89	94.6	44.10	117.5	43.15	143.6				
54	0.243	50.45	119.8	49.60	148.3	48.56	181.9				

9.4 STABILIZING AND PROTECTING THE OUTFALL

It is almost always necessary to stabilize marine outfalls against hydrodynamic oceanic forces to prevent movement and/or undermining beneath the pipe as this can also result in movement and/or induced stresses in the pipe. The main reasons to prevent pipe movement are to preclude loss of integrity of the pipe wall or joints and to avoid deformation that could restrict flow. HDPE pipe with thermally fused joints can withstand considerable movement without loss of integrity. In general the higher the molecular weight of the HDPE material, the better the pipe can withstand deformation. Because both the HDPE pipe and the wastewater are less dense than seawater, even in the absence of hydrodynamic forces it is still necessary to secure the pipe to the seabed to prevent flotation.

The pipe must be secured against the most severe hydrodynamic forces anticipated. There are four basic means to accomplish this:

- Bury the pipe in an excavated trench;
- Install the outfall pipe through directional drilling or micro-tunneling (see Chapter 15);

Design of polyethylene outfalls

- Attach sufficiently heavy ballast weights (usually concrete) to the pipe to resist movement due to oceanic forces;
- Attach the pipe to mechanical anchors or piling drilled or driven into the seabed.

Of these, concrete ballast weights are most commonly used for HDPE outfalls. Entrenchment, directional drilling, and microtunneling result in greater protection of the outfall, but are usually significantly more expensive than weights or anchors. They are mainly used when weights or anchors cannot provide adequate protection.

Selection of the stabilization method is based on many factors, including the magnitude of hydrodynamic forces, the seabed material, and the likelihood of damage by ship anchors or fishing equipment, environmental concerns, and onshore land use. The decision requires good engineering judgment as well as knowledge of the seabed geomorphology and a thorough understanding of the cause and magnitude of the hydrodynamic forces.

It is often necessary or convenient to use more than one stabilization method for an outfall. For example, it is relatively common to entrench the pipe from shore through the surf zone and to use concrete ballast weights beyond the surf zone.

9.4.1 Stabilization with concrete ballast weights

The basic strategy of ballast weights is to add sufficient weight to resist lateral and vertical lift forces due to currents and waves. The weights are usually made of reinforced concrete because it has sufficient density and is relatively resistance to seawater corrosion.

The basic shapes that are commonly used for concrete ballast weights are rectangular, circular, starred, trapezoidal, and combination as shown in Figure 9.2. There are numerous variations on these shapes that depend on concrete forming methodologies, the types of bolts or other connectors, and the specific class and condition of the seabed. Each shape has advantages and disadvantages.

Circular weights are used mainly when the outfall is buried in a trench from shore out beyond the surf zone or when it is intended for the outfall to settle into a soft bottom material. Circular ballasts are usually easier to place into a trench, especially trenches in rock, than the other shapes. This is because they do not have sharp corners that can hang up on irregularities on the trench wall. Circular ballasts are not appropriate for use on the open sea floor because they can roll and have little resistance to lateral forces.

Rectangular ballasts have greater resistance to movement from lateral forces than the circular, but less than the trapezoidal, combination, or starred shapes, and they are not as resistant to rotational forces as the trapezoidal or combination shapes. Their main advantages are that the formwork for the concrete is relatively easy to construct and the shape is relatively easy to handle and transport.

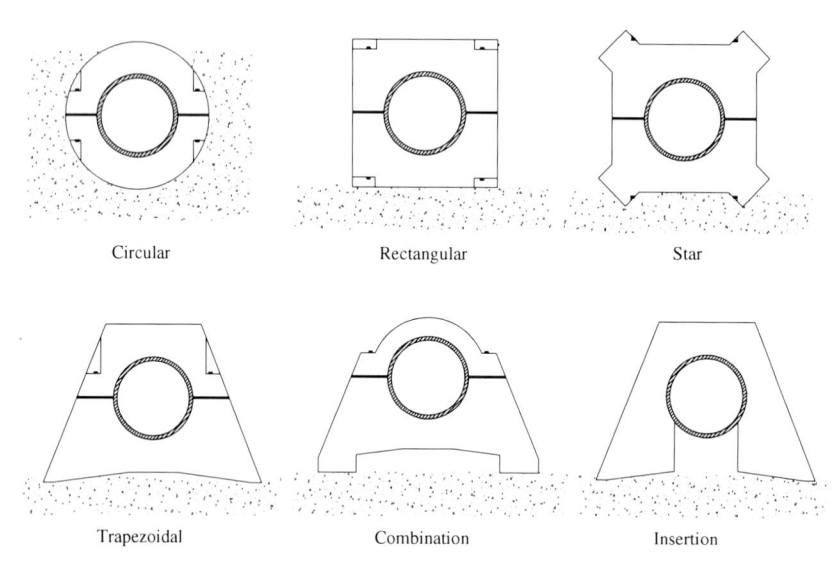

Figure 9.2 Basic shapes of concrete ballast weights for HDPE outfalls

The trapezoidal shape has the best resistance to lateral movement on sand, silt, clay, and gravel sea beds because the leading edge wedges down into the seabed. The trapezoidal and combination shapes are most resistant to rotation because their center of gravity is well below the center of buoyancy of the outfall, whereas for the circular, square, and star shapes they are coincident. The lower center of gravity also keeps the trapezoidal and combination ballast in a vertical position during flotation whereas the rectangular and starred tend to rotate to 45°. Its eccentricity also precludes slippage of the weight along the pipe during sinking. The trapezoidal and combination shapes also enable the use of shorter bolts than the square and star shapes. An exploded schematic of a typical trapezoidal ballast configuration is shown in Figure 9.3.

The insertion shape is useful for retrofitting an HDPE outfall with additional ballast weights. It is installed by compressing the barrel of the pipe with clamping devices, slipping the ballast over the pipe, and then removing the clamps allowing the plastic memory of HDPE to spring the pipe back to the circular shape.

Design of polyethylene outfalls

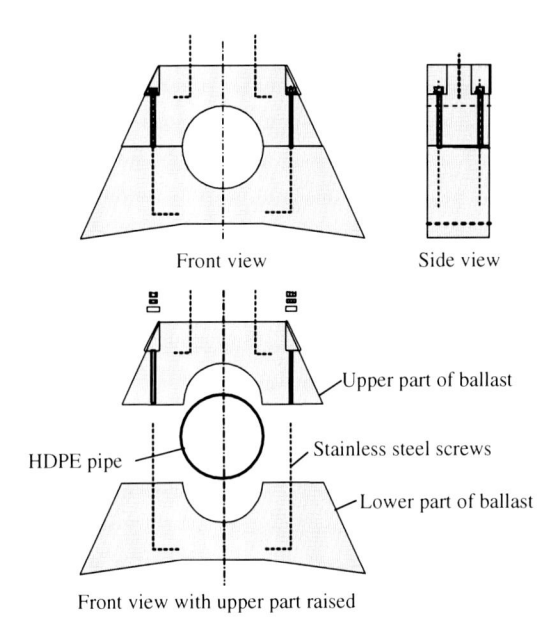

Figure 9.3 Schematic of trapezoidal ballast weight

The resistance of the weights and pipeline to horizontal movement due to horizontal forces is estimated by multiplying the total downward force on the ballast weight by a friction coefficient. These are not true friction coefficients because they take into account that the weight will settle into clay, silt, or sand, and a horizontal force will cause the leading edge to dig into the seabed thereby increasing resistance to lateral movement. The friction coefficient varies with the seabed material and shape of the ballast weight. Table 9.7 lists approximate ranges of the friction coefficients for various seabed materials and ballast weight shapes. Conversely, on a relatively smooth solid rock seabed, a thin accumulation of sand or other sediment will decrease resistance to sliding. If the rock surface is not smooth, but has protrusions or projections, the leading edge of the weight will catch on them, thereby increasing resistance to lateral movement.

The forces acting on an HDPE outfall stabilized by concrete ballast weights are shown in Figure 9.4. The resistance to horizontal movement is the sum of <u>all</u> vertical forces except the normal vertical forces of the seabed (under the most severe hydrodynamic conditions anticipated) multiplied by the appropriate friction coefficient. For stability, this resistance must exceed the horizontal forces imposed by currents and/or waves.

260   Marine Wastewater Outfalls and Treatment Systems

Table 9.7 Approximate friction coefficients for ballast weights

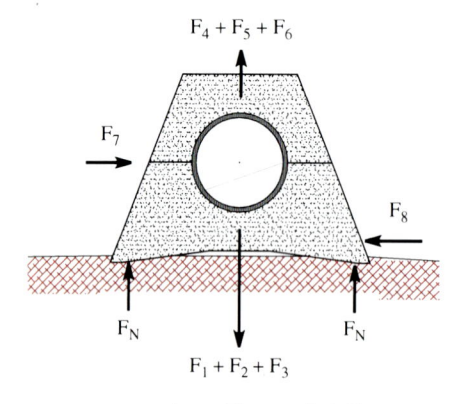

Shape of weight	Seabed material			
	Clay	Sand	Gravel	Rock
Circular	0.2–0.4	0.2–0.3	0.2	0.2
Rectangular	0.5–0.7	0.5–0.6	0.4–0.6	0.4–0.6
Starred	0.6–0.8	0.7–0.8	0.6–0.7	0.5–0.7
Trapezoidal	0.7–0.9	0.8–1.2	0.7–0.8	0.5–0.7

F_1 = Weight of concrete ballast
F_2 = Weight of HDPE pipe
F_3 = Weight of pipe's contents
F_4 = Weight of ambient water displaced by HDPE pipe
F_5 = Weight of ambient water displaced by concrete ballast
F_6 = Lift forces due to waves and/or current
F_7 = Horizontal forces due to waves and/or current
F_8 = Friction force
F_N = Normal forces from seabed

Figure 9.4 Forces on an HDPE outfall

To facilitate determination of the optimal size of the concrete ballast weights and their spacing it is convenient to carry out the calculations on a basis of unit (usually one meter) pipe length.

The stability relationship can be expressed as:

$$f(F_1 + F_2 + F_3 - F_4 - F_5 - F_6) \geq F_7$$

or

$$F_1 + F_2 + F_3 - F_4 - F_5 - F_6 \geq F_7/f \quad (9.3)$$

where:

F_1 = weight of concrete ballast/meter of outfall length (kN/m)
F_2 = weight of outfall pipe/meter of outfall length (kN/m)
F_3 = weight of pipe's contents/meter of outfall length (kN/m)
F_4 = weight of ambient water displaced by one meter of pipe (kN/m)
F_5 = weight of water displaced by the ballast/meter of outfall (kN/m)
F_6 = lift force per meter of outfall due to waves and/or currents (kN/m)
F_7 = horizontal force per meter due to currents and waves (kN/m)
f = friction coefficient (from Table 9.7)

Design of polyethylene outfalls 261

If we let:

γ_W = specific weight of the ambient water (kN/m³)
γ_C = specific weight of concrete (kN/m³)
γ_S = specific weight of the pipe's contents (kN/m³)
d = pipe internal diameter (m)
D = pipe external diameter (m)

Because the volume of the ballast and the volume of the water it displaces when totally submerged are the same, F_5 can be expressed in terms of F_1 by:

$$F_5 = F_1 \gamma_W / \gamma_C$$

and F_3 and F_4 by:

$$F_3 = \frac{\pi d_I^2}{4} \gamma_S \quad \text{and} \quad F_4 = \frac{\pi d_E^2}{4} \gamma_W$$

Substituting these relationships in Eq. (9.3) we obtain

$$F_1\left(1 - \frac{\gamma_W}{\gamma_C}\right) \geq \frac{F_7}{f} - F_2 - \frac{\pi d_I^2}{4} \gamma_S + \frac{\pi d_E^2}{4} \gamma_W + F_6$$

or
$$F_1 \geq \frac{\frac{F_7}{f} - F_2 - \frac{\pi d_I^2}{4} \gamma_S + \frac{\pi d_E^2}{4} \gamma_W + F_6}{1 - \frac{\gamma_W}{\gamma_C}} \quad (9.4)$$

The ballast weight per meter length must equal or exceed F_1 to ensure that the outfall is stable under all wave and current forces.

Ballast weights are rarely placed exactly at intervals of one meter. Their spacing depends on several factors that include the type and capacity of the equipment used for casting, handling, and attaching these weights to the outfall, the seabed topography, the outfall installation method, and the diameter and dimension ratio of the HDPE pipe. After the spacing is determined, the weight (F_1) that was calculated per meter of outfall length must be increased in proportion to the spacing. For example, if the spacing is four meters then the weight of the ballast (in air) should be $4F_1$.

It is also convenient to know the maximum weight per unit length that can be attached to an empty floating pipe before it begins to sink completely below the surface. This is the weight of water displaced by a unit length of submerged pipe filled with air minus the weight of a unit length of pipe.

$$W_{MAX} = (F_4 - F_2)\left(\frac{\gamma_C}{\gamma_C - \gamma_W}\right)$$

Substituting $F_4 = \pi d_E^2 \gamma_W/4$ into this equation converts it to a more useful form:

$$W_{MAX} = \left[\frac{\pi d_E^2}{4}\gamma_W - F_2\right]\left(\frac{\gamma_C}{\gamma_C - \gamma_W}\right) \qquad (9.5)$$

If the weight needed for stabilization (F_1) exceeds W_{MAX}, supplemental flotation will be required during transport, alignment, and submersion of the outfall. Supplemental flotation may also be required, however, for weights less than W_{MAX} to ensure that the minimum allowable radius of curvature is not violated during the submersion. It is always good practice to weigh the first concrete ballast fabricated and cured to find its actual (as opposed to theoretical) weight, and then adjust the spacing accordingly.

9.4.2 Stabilization with mechanical anchors

It is possible to stabilize an outfall by mechanical anchors set into the seabed. Different types are available for different classes of seabed materials; some of the more common are shown in Figure 9.5. They can be used alone or in conjunction with other stabilizing devices.

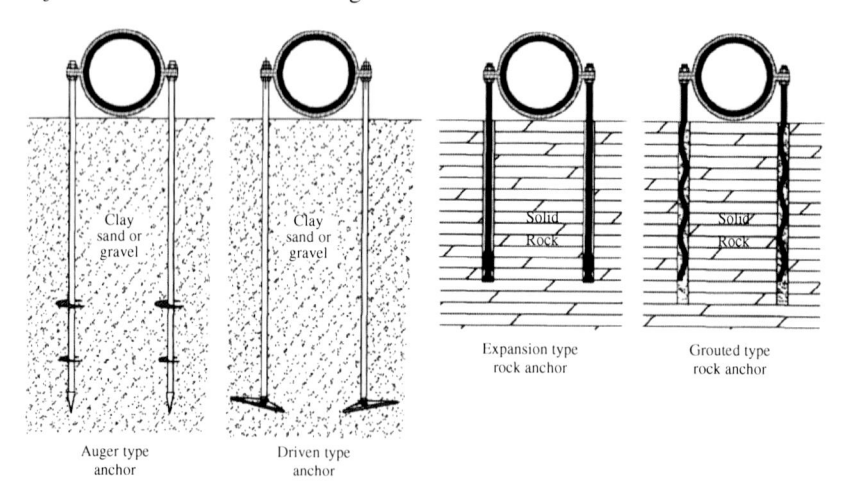

Figure 9.5 Mechanical anchors commonly used for marine outfalls

Mechanical anchors with HDPE are installed differently than with rigid pipe material such as ductile iron, concrete, and steel. Rigid pipes can be retained between the anchors with only an upper bridle, because the pipe is affixed firmly to the seabed by tightening down the nuts attaching the bridal to the anchor. This effectively offsets lateral forces. HDPE pipe, however, is not rigid and such a procedure will deform (flatten) the pipe. It is necessary to use an upper and lower bridle to fasten the pipe to the anchors and to install nuts and washers above and below each bridal connection to allow vertical adjustment.

The type of anchor selected depends on the seabed composition. Sand, clay, and gravel dictate use of either auger or driven-expansion anchors. The depth of penetration depends on the degree of material consolidation, the forces due to the design wave or current, and the depth of the active layer that will be affected by design currents and waves. Auger and expansion anchors have been tested in soft clay and sand, and manufacturers claim breakout (pullout) strength of up to 20,000 lbs (89 kN). This will vary considerably, however, depending on the depths of insertion and the seabed characteristics.

Hard, solid rock ocean floors require anchors placed in holes drilled into the rock. One type is retained in the hole by an expansion device that wedges against the walls of the hole. Another is designed for grouting the anchor into the hole and is shaped to resist pullout. The design load for expansion rock anchors varies with the diameter of the anchor as well as the strength of the rock. In strong solid monolithic rock, loads can be as high as 160 kN for a 25 mm diameter bolt in a 44 mm diameter hole and 326 kN for a 38 mm anchor bolt in a 76 mm hole. However, variation in the strength of the rock from one hole to another and inconsistencies in the rock such as voids, seams, and fissures can greatly affect the permissible loading. For this reason each anchor bolt should be individually tested after installation to verify the pullout strength.

Because mechanical anchors are fabricated of ferrous metal they require corrosion protection. The saddles used to attach the pipe to the anchors must also be protected and should preferably be of the same metal as the anchors. Hot dip galvanized coating is available and should be specified for the auger and driven type anchors. Additional galvanic corrosion protection in the form of anode packs or caps can extend the life of these anchors, especially where they attach to the outfall pipe. Expansion type anchor bolts in rock also need corrosion protection.

9.4.3 Protection by entrenchment

Installing an outfall in a trench can protect it from oceanic forces. Pipelines have been installed in trenches without backfilling if the currents and waves are not severe, but this is not recommended for outfalls. The type of backfill used depends on the severity of the oceanic conditions. Fine grained backfill such as silt or sand is less resistant to erosion than gravel, and gravel is less resistant to erosion than cobblestone. Figure 9.6 illustrates the deposition, erosion, and transport characteristics of various unconsolidated materials according to particle size and mean current velocity. Backfill material for trenches that is suitable for the anticipated highest current velocity should conform to the hatched portion of this graph designated deposition.

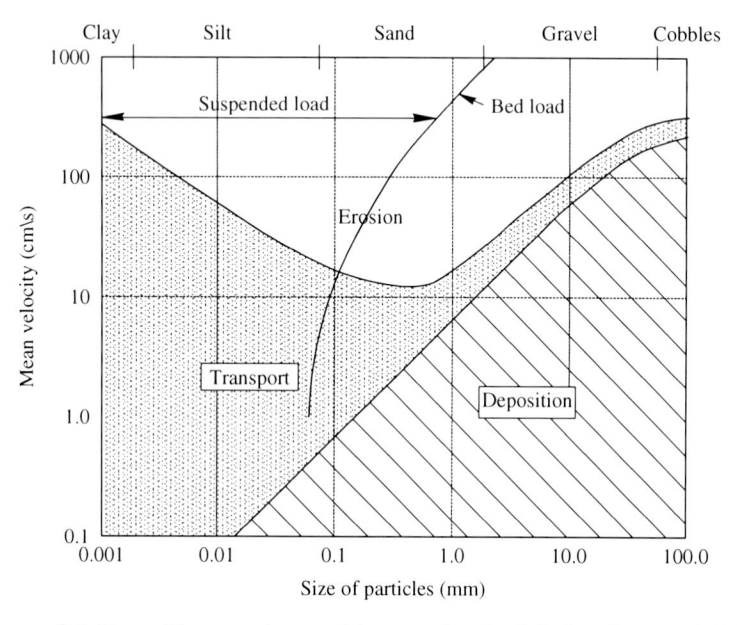

Figure 9.6 Deposition, erosion, and transport potential of various materials according to particle size and current velocity (after Trask, 1939 and Sternberg, 1972)

In a seabed of unconsolidated sedimentary deposits it is important to ensure that the depth of entrenchment is sufficiently below the active sediment layer that may be eroded and transported during storms, strong currents, or other causes. The depth of the active layer is greater for fine sand, somewhat less for hard clay and coarse sand, even less for gravel, and much less for cobbles. Accurate determination of the active layer may require considerable effort and expense in sampling and testing the seabed material and a solid understanding of the design wave and currents. When this information is not available it is a common practice for HDPE outfalls to have a trench depth 1.5 to 2.0 m greater than the outside pipe diameter, although even that might not be sufficient. Figure 9.7 illustrates an example of near shore seabed profile fluctuation due to wave action and ice scour over an interval of six years in Kotzebue, Alaska. The vertical perturbations of the profile were as much as 1.5 m.

Design of polyethylene outfalls

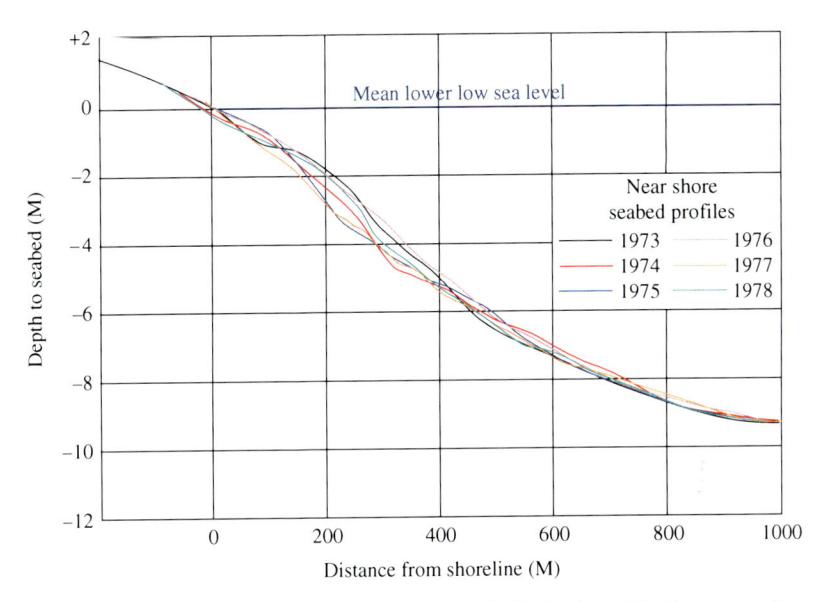

Figure 9.7 Nearshore seabed profile fluctuation in Kotzebue, Alaska over a 6 year period due to wave action and ice scour

When an outfall is partially buried in sand or sandy clay, erosion may actually be greater than if the outfall is supported slightly above the seabed on concrete ballasts. For this reason partial burial is not usually recommended. Figure 9.8 illustrates typical backfilling in an unconsolidated seabed.

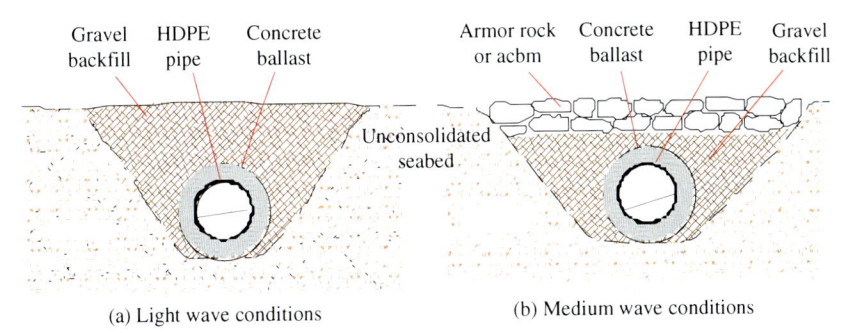

Figure 9.8 Typical trench backfilling in an unconsolidated seabed

If the backfill material could be subject to erosion the upper portion of the trench should be filled with armor rock or covered with an articulated concrete block mat (ACBM, see Section 9.4.4).

If the outfall crosses a solid rock seabed in shallow water that is exposed to large waves, it will be necessary to either anchor it to the seabed or excavate a trench and install the outfall in the trench. Entrenchment is more secure because it precludes impact by submerged objects driven by strong currents or large waves. Excavation of solid hard rock is usually done by explosives. This requires highly skilled and specialized personnel and equipment, is dangerous, quite costly (especially in deeper water), and usually requires special permits and an environmental impact analysis. Cutter head dredges (Chapter 10) can sometimes be used in shallow nearshore water to excavate softer rock, but this also requires special equipment and personnel, permits, and environmental considerations.

Figure 9.9 illustrates four typical backfilling methods for rock trenches under various wave conditions. Granular backfill must be carefully selected because the pressure exerted by large waves can penetrate it and cause the pipe to gradually work its way up to the surface of the fill material. HDPE pipe that has been ballasted too lightly is particularly vulnerable to this action. The backfill material may require protection by armor rock. Under severe wave conditions the upper part of the trench may need to be filled with concrete, and under extreme wave conditions, such as caused by powerful hurricanes, it may be necessary to completely encase the outfall in concrete. In this case, the trench depth can be shallower because the entire backfill then acts as ballast and wave penetration is precluded.

9.4.4 Articulated concrete block mats (ACBM)

An articulated concrete block mat (ACBM) consists of individual concrete blocks tied together by a grid of stainless steel cables or polypropylene rope. They have been used successfully for several decades to protect oil and gas pipelines resting on the ocean floor and since the early 1990s have been increasingly used to protect ocean outfalls.

ACBMs have been draped directly over outfall pipes resting on the seabed, as shown in Figure 9.10. When this is done with HDPE outfalls it is important to ensure that the wall thickness is sufficient to preclude pipe deformation from the weight of the ACBM. They have been used to provide supplemental protection of outfalls previously protected by armor rock, to protect backfill in an outfall trench, and to stabilize outfall diffusers.

A similar device is a strong double-layered synthetic fabric that is divided into cells (pockets) into which concrete or cement mortar is injected. This forms a mattress of concrete blocks tied together by the fabric that conforms to the surface on which it is placed. It has been used to stabilize ocean outfalls, as shown in Figure 9.11 on a diffuser.

Design of polyethylene outfalls 267

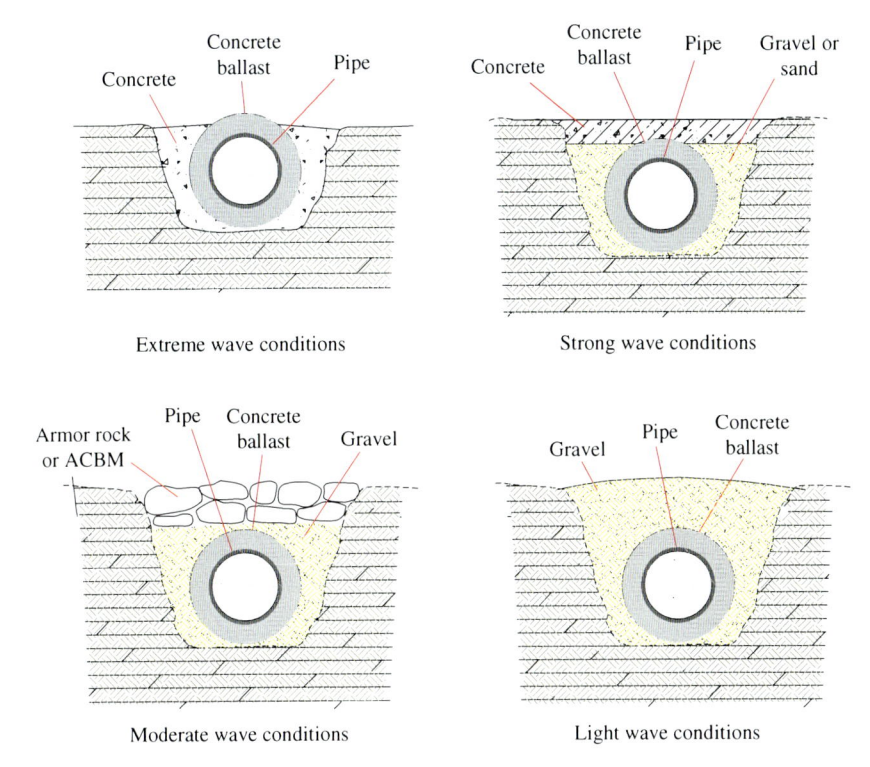

Figure 9.9 Typical trench backfilling in solid rock seabed

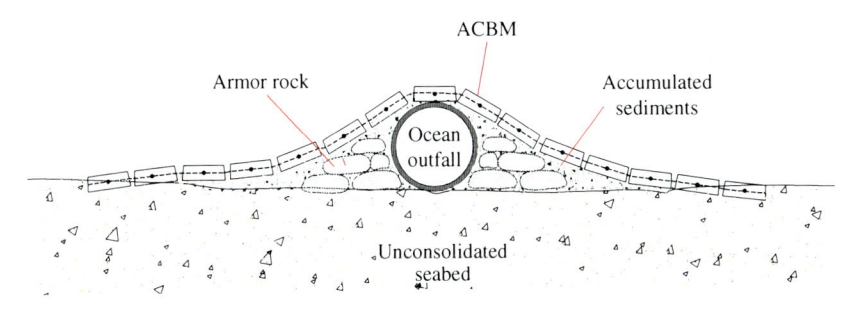

Figure 9.10 Supplemental protection by an articulated concrete block mat (ACBM) for an outfall previously protected by armor rock

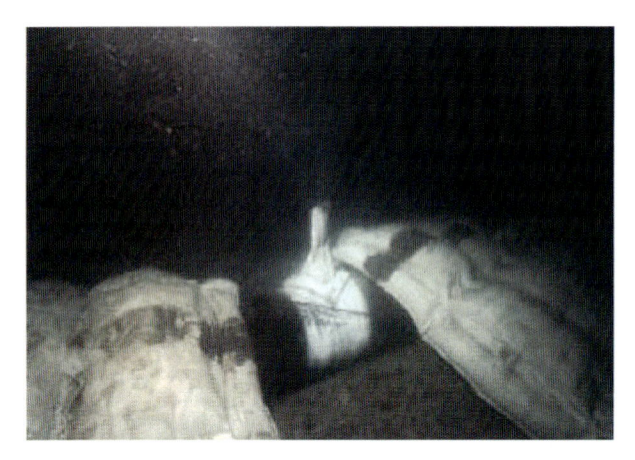

Figure 9.11 Concrete block mattress for stabilization of an outfall diffuser
(Courtesy Ivo Van Bastelaere, PEng)

One of the strongest testimonies of their effectiveness is the experience with a 914 mm diameter cast iron ocean outfall in Boca Raton, Florida that was retrofitted with ACBMs to protect a vulnerable unburied section located in water depths of 9 to 16 m. On August 24, 1992, shortly after installation, Andrew, a Category 5 hurricane, struck the Florida coast 60 miles to the south, subjecting this outfall to extreme wave action. The section protected by the ACBMs was undamaged, but in deeper water where the ACBM was not installed because of environmental concern for new coral growth on the ballast/armor rock, the rock was completely swept away. ACBMs were subsequently installed on this section (French et al., 1993).

ACBMs have a number of advantages. There is little uplift induced by waves, and unconsolidated seabed material will accumulate in the space between or within the individual concrete blocks thereby increasing stability. The concrete blocks provide a good substrate for growth of coral and other sessile marine organisms, a positive environmental factor. They conform to the surface on which they are applied and it is unnecessary to pin them to the seabed. Finally, ACBMs can be lifted off the pipe to enable pipeline repairs and then reinstalled.

9.5 STRESSES ON AN HDPE OUTFALL

An HDPE outfall is subjected to stress in a number of different ways. After it is installed, current and wave forces cause beam stress in the pipe between the ballasts or mechanical anchors. Installation causes stress in two ways. During flotation and transportation, the buoyant upward force of the empty pipe and the downward force of the ballast weights cause a bending stress. The pipe will also be bent during sinking and may be bent to avoid obstacles, resulting in strain and stress that depends on the radius of curvature attained.

9.5.1 Pipe stress due to currents and waves

An outfall resting on the seabed is exposed to forces due to waves and currents as previously described in Chapter 8. A pipe stabilized by either concrete ballasts or mechanical anchors will act as a fixed-end beam subject to loading of drag and inertia forces perpendicular to its axis, as shown in Figure 9.12.

Uniform load W = horizontal force per unit length of pipe due to cuurent and/or waves with fixed ends at concrete ballast weights (Plan view)

$+WL^2/24$ $+WL/2$

$-WL^2/12$ Moment diagram $-WL^2/12$ $-WL/2$ Shear diagram

Moment of inertia

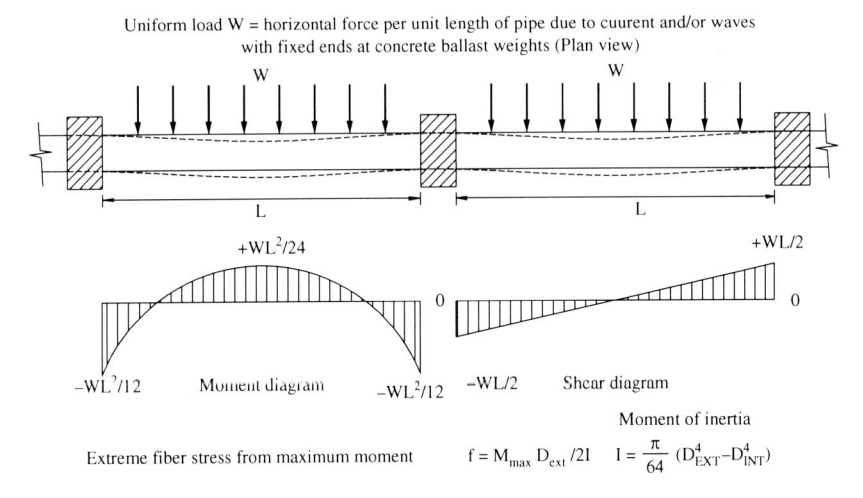

Extreme fiber stress from maximum moment $f = M_{max} D_{ext}/2I$ $I = \dfrac{\pi}{64}(D_{EXT}^4 - D_{INT}^4)$

Figure 9.12 Loading on outfall pipe anchored to the seabed due to horizontal drag and inertia forces

The loading is the force per unit length of pipe calculated for the worst-case scenario for current and/or waves. The maximum shear stress and the maximum tensile and compressive stresses from bending occur next to the concrete ballasts. This is evident from the loading, shear, and moment diagrams in Figure 9.12. Because the cross section of a round pipe is such an unfavorable geometry with a relatively small moment of inertia, the critical stress will be the extreme fiber stress due to the bending moment rather than shear. After the extreme fiber stress is calculated it must be compared to the allowable stress for the PE resin of which the pipe is made. Problems with beam stress become more likely as the distance between the mechanical anchors or ballasts increases because the bending moment varies with the square of the distance between supports. If the allowable stress is exceeded, it will be necessary to either decrease the spacing between the weights or use a pipe with a greater wall thickness (smaller DR).

Because polyethylene is a plastic, the elastic modulus used in calculating the stress varies considerably with the duration of the loading and temperature. Table 9.8 lists examples of elastic modulus for different loading durations and temperatures for a typical resin used in HDPE pipe and can be used for preliminary analysis. The designer of an HDPE outfall should obtain such a table from the pipe manufacturer, however.

Table 9.8 Typical elastic modulus (MPa) for a PE resin for various durations and temperatures

Load duration	Temperature			
	4°C	16°C	23°C	38°C
Short-term	1,170	900	761	695
10 hours	550	425	398	325
100 hours	490	375	356	290
1000 hours	415	324	308	247
1 year	368	280	265	215
10 years	310	236	220	180
50 years	275	210	198	161

For an outfall that is subject to a steady current throughout its life, the allowable stress should be based on a 50-year lifetime. For a loading that occurs for several days during severe storm conditions the allowable stress should probably be based on a load duration of 100 hours.

9.5.2 Stress during flotation

For outfalls stabilized with concrete ballasts, the stress in the pipe wall during flotation and transportation can be greater than that due to waves and currents after installation, especially with heavy ballasts at large intervals. The stress during flotation must therefore be investigated.

The stress during flotation is shown in Figure 9.13. It is very similar to the stress caused by drag and inertia forces. In relatively calm seas the forces during flotation are transient, and the elastic modulus from Table 9.8 should correspond to load durations comparable to the time that the pipe will be floating. For an outfall launched from land at the job site that is immediately aligned and submerged this would probably be the 100-hour load duration. If the outfall with weights attached is to be assembled and stored while floating and sections added, then the elastic modulus should be based on a commensurately longer duration.

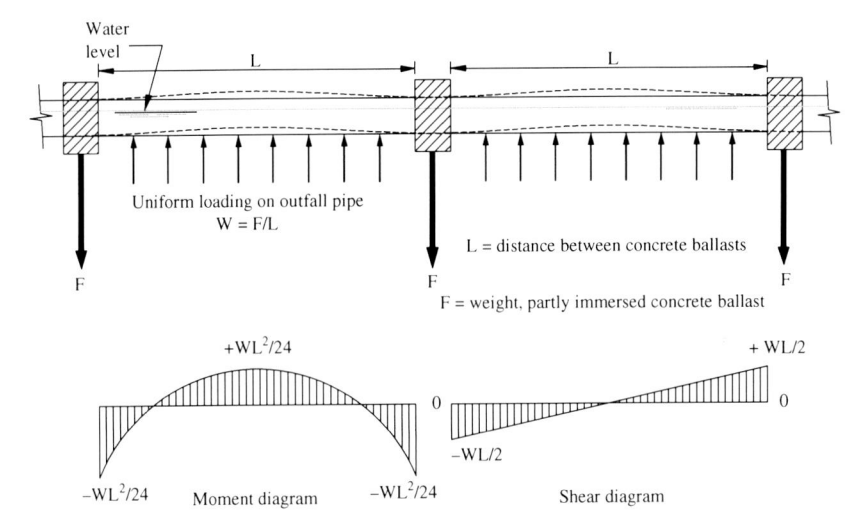

Figure 9.13 Loading on a floating outfall with weights attached, relative to the immersed weight of the concrete ballasts and the weight of the pipe

9.5.3 Bending stresses

A very important characteristic of HDPE for marine outfalls is its flexibility. It enables the pipe to curve around obstacles, sharp escarpments, or other problem areas. It also allows considerable movement after installation without loss of pipe integrity. When HDPE pipe is bent, however, there will be strain with commensurate stress in the pipe wall that increase as the radius of curvature decreases. It is important that the stress and strain due to bending be kept below the level that could cause failure of the pipe wall.

HDPE pipe manufacturers usually provide guidelines for the minimum allowable bend radius expressed as a multiplier of the outside pipe diameter. Their recommendations range widely, from 25 to 70 times the pipe diameter. The differences are due to the magnitude of the safety factor assumed, the internal pressure the pipe will be subjected to, and the long and short term strength characteristics of the resins used.

The two major concerns associated with cold-bending HDPE pipe are the minimum allowable bend radius to avoid pipe wall buckling and the stress due to bending. These must be considered along with stresses due to other causes such as internal pressure and hydrodynamic forces.

9.5.3.1 Minimum bending radius

To understand the bending limitations of HDPE pipe it is helpful to review the mechanism of stress and strain caused by bending. When a pipe is bent, the pipe wall on the inside of the bend undergoes compression strain and the wall on the outside, tensile strain. Assuming the strains from compression and tension are equal (not strictly valid, but sufficiently close for present purposes), from simple geometric proportion the maximum strain occurs in the outer wall and can be expressed by:

$$\varepsilon_s = \frac{D}{2R} \tag{9.6}$$

where ε_s is the maximum strain, R the radius of curvature of the bend, and D the outside diameter.

It is essential to avoid bending the pipe to a radius less than that which results in buckling. Outfalls are especially vulnerable to buckling during launching and sinking onto the seabed and precautions should be taken to preclude this. Two types of buckling can occur during bending. Radial buckling occurs in the hoop direction and usually when the internal pressure is equal to or less than the external pressure. Axial buckling can occur when the internal pressure is significantly greater than the external pressure.

The ratio of the bending radius (measured to the central axis of the pipe) to the external pipe diameter, a, is:

$$a = \frac{R}{D} \tag{9.7}$$

and substituting Eq. (9.7) into (9.6) gives:

$$\varepsilon_s = \frac{1}{2a} \tag{9.8}$$

The tendency for buckling in a pipe subjected to bending is inversely proportional to the rigidity of the pipe wall, which is a function of the dimension ratio. The strain at which radial buckling occurs in a bent pipe can be estimated by:

$$\varepsilon_{rcrit} = 0.28\left(\frac{t}{D_m}\right) \tag{9.9}$$

where t is the wall thickness and D_m = mean pipe diameter.

Since the strain due to the bending moment is an axial strain and the ratio of radial to axial strain is Poisson's ratio, u, then:

$$\varepsilon_{rcrit} = \varepsilon_{acrit} u \qquad (9.10)$$

The critical strain in the axial direction is obtained by combining Eqs. (9.9) and (9.10):

$$\varepsilon_{acrit} = \frac{0.28}{u}\left(\frac{t}{D_m}\right) \qquad (9.11)$$

Because the dimension ratio DR = D/s, and the mean diameter $D_m = D-t$, and for HDPE pipe, Poisson's ratio is 0.5, Eq. (9.11) can be written as:

$$\varepsilon_{acrit} = \frac{0.56}{DR - 1} \qquad (9.12)$$

This relationship can also be expressed as the minimum bend radius that can be tolerated without radial buckling. Substituting Eq. (9.12) into (9.8) and solving for a gives the relationship between the critical ratio of radius of curvature to the external pipe diameter and the dimension ratio:

$$a_{crit} = \frac{DR - 1}{1.12} \qquad (9.13)$$

Because good engineering always introduces a margin of safety before failure occurs, a safety factor (SF) is inserted into Eq. (9.13) to obtain.

$$a_{allowable} = SF\frac{DR - 1}{1.12} \qquad (9.14)$$

SF = 1.5 is commonly used but this may be increased to compensate for other uncertainties inherent to the specific application.

Table 9.9 presents the allowable ratio of bend radius to pipe diameter for various common dimension ratios and safety factors of 1.5 and 2. This table can be used for preliminary design, but for final design it is important to use the minimum allowable radius provided by the pipe manufacturer.

9.5.3.2 Stress due to bending

When HDPE pipe is bent to make a horizontal curve to avoid an obstacle, or a vertical curve for transition from one slope to another, the bend will be permanent. The bend radius should be more than the minimum to accommodate the additional stress due to the internal pressure. To analyze the combined effect it is first necessary to determine the stress that is induced by the actual bend radius and then to add the stress due to internal pressure.

Table 9.9 Allowable ratio of bend radius to pipe diameter

Dimension ratio (DR)	Minimum allowable ratio	
	SF = 1.5	SF = 2.0
33	43	59
32.5	42	56
26	34	45
22	28	37
21	27	36
17	21	28
15.5	19	26
13.5	15	21
11	13	17

The modulus of elasticity, E (MPa) is defined as stress divided by strain:

$$\sigma_b = E\varepsilon_s \tag{9.15}$$

where σ_b is the stress due to bending (MPa). Substituting Eq. (9.6) into (9.15) gives:

$$\sigma_b = E\frac{D}{2R} \tag{9.16}$$

The value for E should be selected from the resin or pipe manufacturer's equivalent of Table 9.8 in accordance with the duration that the pipe will be bent and its temperature. For a permanent bend in a pressurized outfall the bending stress should be calculated using a value for 50 years duration. This should be added to the stress due to the internal pressure at design flow and compared to the minimum required strength (MRS) defined by ISO or the hydrostatic design basis (HDB) defined by ASTM.

9.6 HYDRAULIC CONSIDERATIONS

9.6.1 Criteria

There are a number of hydraulic criteria to be met for an outfall and diffuser. An important one is to preclude seawater intrusion into the ports once the outfall is flowing. This can be achieved by maintaining a total port area less than the outfall pipe cross-sectional area and the densimetric Froude numbers of each individual port F_j greater than one, where:

$$F_j = \frac{V_j}{\sqrt{g \frac{\Delta\rho}{\rho} d_j}} \qquad (9.17)$$

where V_j is the jet exit velocity, g the acceleration due to gravity, $\Delta\rho$ the density difference between sewage and receiving water, ρ the receiving water density at the port level, and d_j the port diameter. The ratio of total port area to cross sectional area of the pipe is typically between about 1/3 and 2/3, depending on bottom slope. The velocity in the pipe should be sufficient at peak flow to scour and prevent deposition of particles. The total head loss in the outfall should be kept small to minimize head loss and therefore pumping costs. There should be a uniform division of flow between the ports at all flow rates. The calculation procedures to accomplish these objectives are summarized in Section 9.7.5.

9.6.1.1 Self-cleaning velocities

It is important that the velocity in an outfall be sufficient to prevent excessive deposits of sediment and grease and/or microbial growth in the pipe. The velocity varies diurnally as the wastewater flow rate varies, so a self-cleansing velocity should be achieved long enough during the higher flow to flush deposits that might have settled during lower flow. If this does not occur it will eventually become necessary to send a pig (cleaning swab) through the outfall or to increase the velocity (through pumping, dosing siphons, or other measures) at regular intervals to prevent pipe constriction or closure (as shown dramatically in Figure 9.14). Two other cases that exemplify the importance of cleansing velocities are the San Francisco, California, outfall which is about half-filled with sediments, and the Chimbote, Peru outfall that was completely clogged with grease discharged by a fish meal plant.

The accumulation and rate of deposition in an outfall depends on four main factors: (1) the flow velocity, (2) the amount, grain size, and specific gravity of suspended inorganic particles, (3) the amount and characteristics of grease, oil, and other hydrophobic substances in the effluent, and (4) the temperature difference between the effluent and the seawater. Wastewater characteristics vary widely depending on the effluent source, the type and condition of the sewerage system, and the level of treatment.

Considerable research and many scientific articles are available on the transport and deposition of suspended matter and self-cleansing velocities in sanitary sewers. They show that the flushing velocity may depend on pipe diameter. For smaller pipes the design may better be governed by shear stress (erosion criteria for cohesive particles) whereas for larger pipe sizes it may best be governed by either the suspended sediment transport criterion or the bedload and deposited bed criterion (Butler *et al.*, 1996). Unfortunately, there is little practical research regarding self-cleaning velocities to prevent grease buildup.

Figure 9.14 Extreme case of diffuser failure due to clogging (Courtesy Pedro Campos)

Recommended minimum velocities to ensure self cleansing as a function of internal pipe diameter are shown in Figure 9.15 for effluent that has received treatment by milliscreen and grease trap or better. It is based on the research carried out under controlled conditions and the author's practical experience.

Cleansing velocities should be achieved for several hours a day. If this is not feasible during the initial years of operation, then facilities for insertion and removal of a cleaning pig should be included in the outfall design and a maintenance schedule determined for its use until flows increase to a level that consistently result in adequate cleaning velocities. Treatment to remove sand, sediment, and grease from the effluent will help minimize problems due to deposition. Removal of grease and other buoyant material also helps to maintain acceptable aesthetic conditions in the receiving waters.

9.6.1.2 *Elimination of entrained or trapped air*

In addition to removing sediment and grease, it is important to attain velocities high enough to transport entrained or entrapped air (or gases) completely through the outfall. Air usually enters an outfall by entrainment at the entrance, and gas may form due to biological action within the outfall The greatest concern is for air or gas being trapped at high points because this can partially or even totally block the flow of water. It can also cause vertical pipe displacement that will exacerbate the problem. It can even cause re-flotation if a sufficiently long section of the outfall is involved. There are several strategies to preclude these problems.

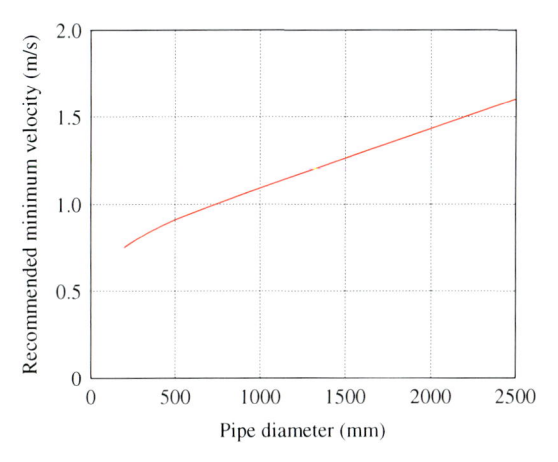

Figure 9.15 Recommended minimum velocities at peak hourly flow for self-cleaning HDPE outfalls

Where possible the outfall should have a constant downward slope thereby avoiding high points. Then during low flow any entrained air or gas bubbles can migrate back to the entrance or toward the diffuser and escape. The outfall entrance (headworks) should be designed to avoid air entrainment and provide for its release before the effluent enters the ocean outfall pipe, particularly if a hydraulic jump occurs near the entrance. This can be accomplished by including a surge and air separation chamber, as shown in Figure 9.16. Such chambers are also effective for gravity flow where the effluent contains significant amounts of entrained air due to turbulence or a hydraulic jump in open channel flow.

Surge chambers are always recommended when the effluent is pumped and there is possibility of sudden pump shutoff, or any other situation that may result in sudden cessation of flow. The momentum of the flowing water causes flow to continue after the pump or driving force is terminated. This can cause sudden low pressure resulting in air sucked into the pipe, or high pressures (water hammer) that can cause pipe failure. The volume of the chamber and the outlet depth below sea level must be sufficient to always prevent water in the chamber from dropping below the crown of the outlet.

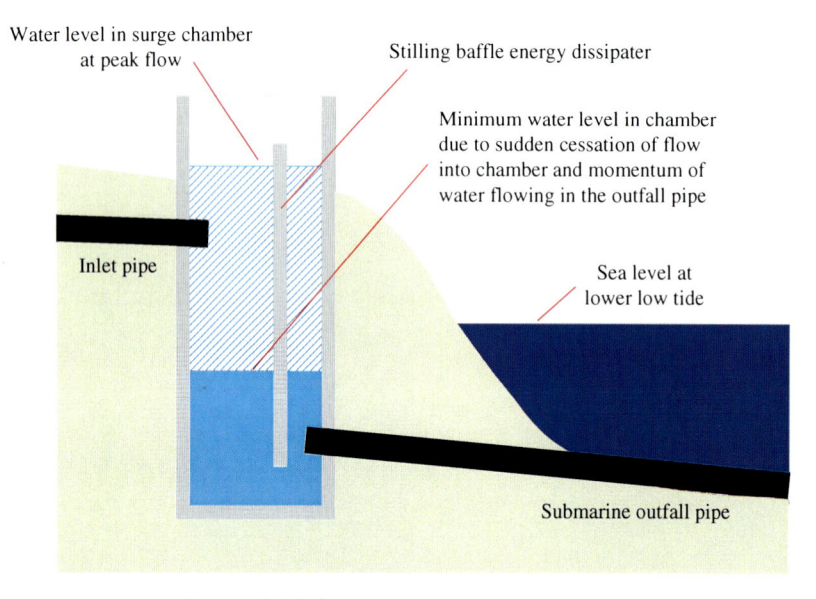

Figure 9.16 Schematic of surge chamber

A surge chamber may not be required if sudden flow termination cannot occur, such as gravity flow from a treatment plant with flow equalization or where a long sloped gravity sewer leads into the outfall and serves as the surge chamber.

It should be noted that surge-air release chambers will not prevent trapped gas at high points if the gases are generated within the outfall. This is not common, but it can occur in small diameter outfalls serving small communities or industries that have relatively warm sewage, where the flow ceases for long periods (usually during the night), and where the receiving waters are also warm.

Entrained air and gas bubbles can be transported through the outfall if the velocity is high enough. The critical velocity to achieve this is a function of the internal diameter and the outfall slope. It can be estimated by:

$$V_C = K\sqrt{gd} \tag{9.18}$$

where V_C is the critical velocity for transporting air bubbles, and K is a factor that depends on the pipe slope as shown in Figure 9.17.

The critical velocity for transporting entrained air should be calculated for the greatest angle of slope (the steepest downward gradient) in the outfall. This should then be compared to the velocities calculated for the present peak and average flows.

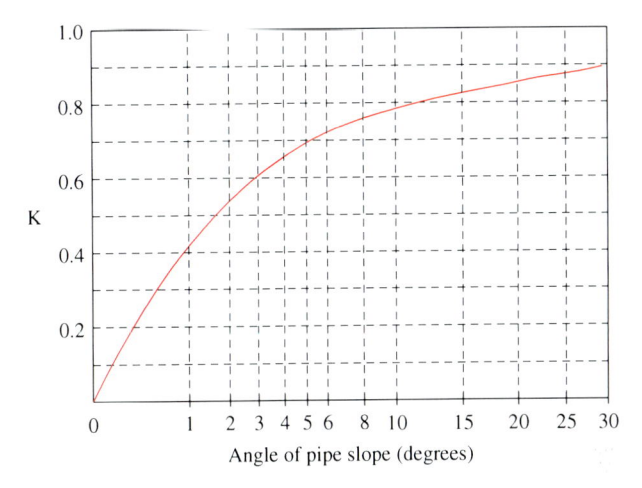

Figure 9.17 The coefficient *K* in the entrained air transport equation for various slopes of outfall pipes

9.7 DESIGNING THE DIFFUSER

9.7.1 Introduction

The diffuser design includes designing to meet dilution requirements, port and/or riser configurations, diffuser orientation, hydraulic considerations, and structural integrity.

Designing for dilution was discussed in Chapters 3 and 4. It is usually an iterative process that is carried out by means of computer-aided dilution models such as NRFIELD (RSB), UM3, DKHW, FRFIELD and CORMIX. The inputs include diffuser and ambient variables. The ambient variables are determined by local oceanographic conditions, measured as in Chapter 5. A range of diffuser variables, including port diameter, number and spacing (which determines the diffuser length) are selected based on the guidelines in Chapter 3, and the computer program run to determine near and far field dilutions. After a number of trials are run, the design that yields the best results is chosen.

This section discusses design of the diffuser details such as of the geometrical diffuser configuration, the type of ports used, and internal hydraulics.

9.7.2 Diffuser configuration

Numerous diffuser configurations have been used to achieve optimal dilution under the prevailing oceanographic and geomorphologic conditions. Some common ones are illustrated in Figure 9.18. As discussed in Chapter 3, the near field dilution depends primarily on diffuser length; the diffusers illustrated all have the same total length. The effect of diffuser orientation relative to the currents was discussed in Chapter 3. Orientation perpendicular to the current gives highest dilution and parallel gives lowest, and the difference between them increases as the current speed increases.

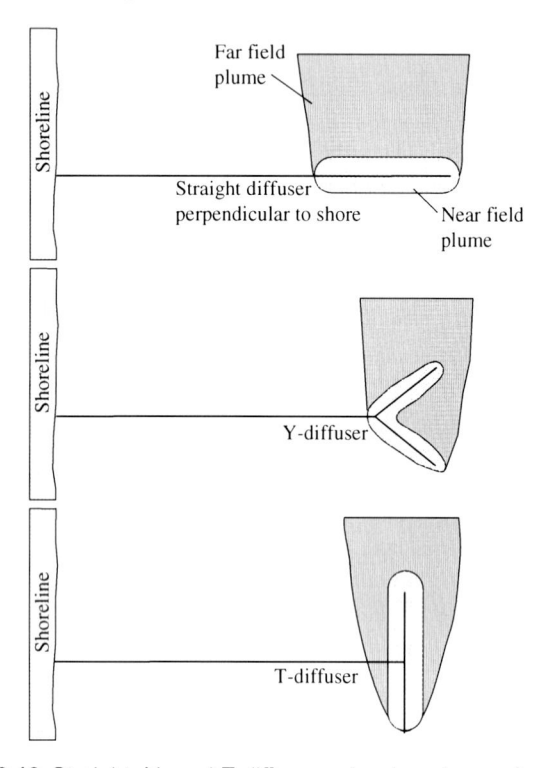

Figure 9.18 Straight, Y, and T-diffusers showing plumes for a current parallel to shore

The simplest configuration is a linear extension of the outfall pipe that has had ports added to it. As currents usually run approximately parallel to the shore, this may be advantageous (top schematic of Figure 9.18). If the current is onshore, this configuration gives lower dilution, but onshore currents are usually not sustained. Wye diffusers (middle schematic) are sometimes used to intersect a wide swath of current regardless of the current direction. Another configuration that has occasionally been used to accomplish this is a single curved diffuser that gradually changes from a direction perpendicular to shore to parallel. Wye diffusers have an advantage that a gate for removal of a pig can be added to the junction structure and gates can be installed to isolate one diffuser arm for maintenance. In the unlikely event that currents mostly run directly onshore, it may be advantageous to orient the diffuser parallel to shore as shown in the bottom schematic. In reality, coastal currents vary considerably in speed and direction (Chapters 4 and 5). The best diffuser orientation is obtained by running a time-series model such as NRFIELD and varying the diffuser orientation to find the best near field dilution. In a tidal environment the diffuser is best oriented perpendicular to the first principal (the most energetic) current component. An example is the San Francisco ocean outfall diffuser (see Figure 5.4).

Elevation changes along the diffuser affect its internal hydraulics, especially the distribution of flow between ports. A diffuser that is perpendicular to the coastline is usually perpendicular to the bathymetric contour lines (isobaths) so that the ports will be at different depths. The shallower ports will discharge more than the deeper ones with the difference increasing with increasing slope. To obtain approximately uniform flow it is necessary to vary the port diameters from smaller inshore to larger offshore, as discussed in Section 9.7.5. A diffuser parallel to shore often runs along an isobath with a more uniform discharge among the ports. This configuration shortens the length of the diffuser from the outfall pipe to the final port(s) which often helps to achieve more uniform flow among the ports. It also enables easy installation of a gate for pig removal as well as isolation gates installed at the junction.

9.7.3 Diffuser design objectives

There are a number of diffuser design objectives:

- Reasonably uniform flow from each port at all flow rates;
- Prevention of seawater intrusion to preclude the establishment and growth of marine organisms within the diffuser;
- Preclude entrance of oceanic sediments such as silt or sand into the diffuser;
- Preclude accumulation of wastewater sediments in the diffuser;
- Minimize probability of physical damage from hydrodynamic ocean forces, fishing gear, and boat anchors;
- Placement and orientation of the diffuser to maximize dilution and minimize transport of wastewater to beaches, shellfish harvesting, or environmentally sensitive areas.

The first objective, uniform discharge from the ports, can be achieved if the following conditions are met: (1) All the diffuser ports are at the same depth, meaning the diffuser lies on an isobath, (2) the head loss in the diffuser pipe is small (in comparison to the head loss of the ports) and approximately constant between the ports, and (3) the effluent flow rate is fairly constant with time. If the diffuser ports cannot be at the same depth, it is still possible to approximate uniform flow by varying the port diameters. It is not practical to give each port a different diameter so ports are grouped with each group having a different diameter from the other group. The smaller diameters are usually at the shallower locations.

The second condition of maintaining small and nearly constant hydraulic friction losses in the pipe between the ports can be approximated in several ways. One is to use a Wye-diffuser to reduce the distance that effluent must travel along the diffuser to the terminal port as illustrated in Figure 9.18. To attain small friction loss in the diffuser it is necessary to have low flow velocities but this can result in sediment deposition in the diffuser. A strategy to reduce deposition in the diffuser is to use a tapered diffuser whose diameter gradually decreases towards the terminal end. A constant taper is not feasible so it is usually approximated by diffuser sections whose diameters decrease in steps, as illustrated in Figure 9.19.

The second objective, preventing the entrance of seawater into the diffuser can be accomplished in several ways. The easiest is ensuring that the port densimetric Froude numbers (Eq. 9.17) always exceed unity. If the flow in the early years is too low to accomplish this, some ports can be initially closed off. Another method is to install check valves on each port. This may be important for a diffuser with a very large depth variation or very long risers or a flow that varies widely or is intermittent, because then effluent can flow from the higher ports and seawater flow into the lower ports. The most practical type of check valve for ocean outfalls is the "duckbill" type because it is made of material not affected by marine corrosion and requires lesser maintenance. Two typical duckbill valve configurations are shown in Figure 9.20 (see also Figure 9.25). The hydraulics of duckbill valves are discussed in Section 9.7.6.

Sediment buildup in the diffuser is prevented by ensuring sufficient velocity to transport sediment along the diffuser and out the ports (see Section 9.6.1.1). A tapered diffuser, such as shown in Figure 9.19, can accomplish this. A terminal port larger than the preceding ones located at the pipe invert can aid in flushing the terminal end of the diffuser.

Seabed material such as silt, sand, or gravel can be prevented from entering the diffuser in several ways. If the diffuser is laid on the seabed, the ports should be high enough to preclude entrance of this material. If the diffuser is buried,

Design of polyethylene outfalls

or if sediments are actively moving near the diffuser, the ports should be equipped with risers that reach above the anticipated deposits. Riser designs are discussed below.

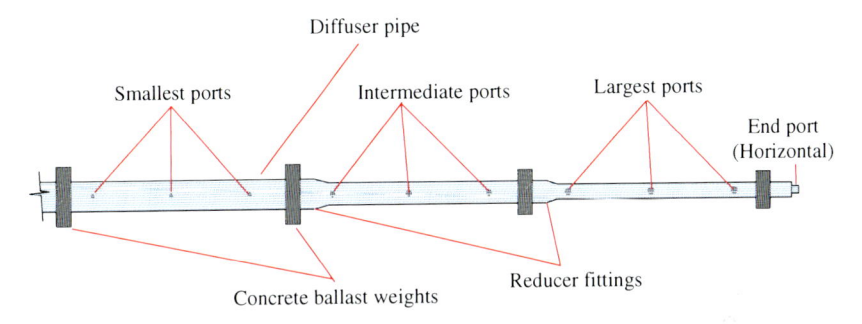

Figure 9.19 A tapered HDPE diffuser

Figure 9.20 "Duckbill" check valves (courtesy of Tideflex Technologies)

Reducing the probability of damage to the diffuser by oceanic forces should be viewed in light of the anticipated current and wave dynamics. As previously discussed, if the diffuser is located in water deeper than about 30 m the diffuser can usually be stabilized against oceanic forces in the same manner as the main pipeline. In very severe hydrodynamic conditions such as imposed by hurricanes or when the outfall is located in relatively shallow water, it may be necessary to bury the diffuser. When the diffuser is located on a solid rock seabed it may be necessary to use mechanical anchors (Figure 9.5) or articulated concrete block mats (Figures 9.10 and 9.11) for stabilization.

If the pipe diameter is small, and the ports discharge horizontally, it may be necessary to elevate the diffuser or risers somewhat to ensure the jets do not impact or erode the seabed. Alternatively, the jets can be pointed slightly upwards, say at 5° to the horizontal.

For all marine outfalls it is recommended that warning buoys be permanently anchored over the diffuser. The buoy should have a sign, authorized by the proper governmental agency, prohibiting the anchoring of boats, water contact sports, and fishing within a prescribed distance from the sign. In addition, it is recommended that the outfall be clearly indicated on all nautical charts of the area and that fishermen and water sport businesses be notified to not use this area.

9.7.4 Port configurations

Figure 9.21 shows eight port configurations commonly used in HDPE diffusers. Types (a), (b), (c), and (d) are used in diffusers that are supported on concrete weights resting on the seabed, and (e), (f), and (g) are used for buried diffusers or diffusers where significant sediment movement or buildup occurs. Type (h) is a terminal port.

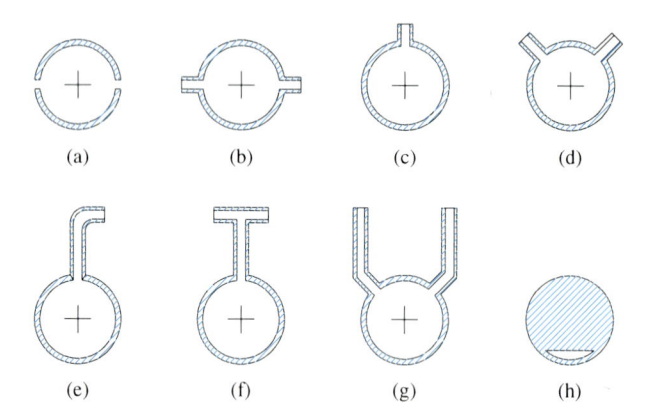

Figure 9.21 Typical diffuser port configurations

Type (a), which is simply an orifice located at the springline of the pipe, has several advantages. For depths less than about 40 m the diffuser (without the ports cut into the pipe wall) can be installed by flotation and submersion as an integral part of the outfall. After it is resting on the seabed, divers can drill the orifices in the sidewall. Weak areas caused by the orifices are then avoided and are not subject to bending stresses during pipe submergence. Another advantage is that the port does not project from the pipe wall and is thereby less likely to be snagged by fishing nets. A disadvantage is that the discharge directly from the pipe wall can foster growth of marine organisms adjacent to the port.

If the flow is low during the initial years of operation, not all of the ports may be necessary. Some ports can be closed off (or not opened) until later when the flow has increased and divers can remove the caps or plugs. The port exit velocity can then be maintained high enough to prevent the entrance of seawater and marine organisms into the diffuser.

Port (h) is a special port located at the terminal end of the diffuser. It should have approximately twice the cross-sectional area of the preceding port to promote self-cleaning velocities at the end of the diffuser and should be at the pipe invert to aid in sediment removal. For a constant diameter diffuser, this port can be formed by cutting a slice off the bottom of a blind flange. The advantage of using a blind flange is that it can be removed to facilitate access for cleaning sediment deposits from the interior of the diffuser. For a tapered diffuser the desired end port area is frequently obtained by installing a prefabricated reducer on the diffuser end. An opening end gate can also be added to the diffuser to aid in flushing and maintenance.

It may be advantageous for buried diffusers with riser type ports (e), (f), and (g) to have the ports constructed of a flexible material that bends instead of breaking if impacted by a dragging anchor or fishing net.

An adaption of type (f) port is the cluster port. This configuration utilizes a riser that contains four to eight ports; examples are shown in Figure 9.22. Clustered ports can be used to reduce the number of risers, and therefore their total cost. They are spaced so that the average port spacing is the same as if uniformly distributed along the diffuser. For example, if the riser contains four ports, the risers are spaced twice as far apart as if the risers were Tee-shaped each with two ports. Clustered ports are used if risers are used in order to reduce their number and cost. They are particularly useful with tunneled outfalls with long, expensive, risers such as the Boston outfall (see Figure 4.24). Experiments (Roberts and Snyder, 1993) show that ports can be clustered up to eight per riser with no significant effect on dilution.

The port spacing and diameter are determined by the analyses described in Chapter 3. Three common port arrangements are shown in Figure 9.23 for which the average port spacing is the same. The ports can discharge vertically or, preferably, horizontally. A vertical discharge may increase the plume rise somewhat when the receiving water is stratified. Horizontal discharge results in the highest initial dilution. The difference can be significant for diffusers in shallow water, but it decreases as the water depth increases. There is little difference in dilution between alternating ports and opposing ports if the number of ports, the diameter of the ports, and the length of the diffuser are the same. Pipes with alternating ports may have greater resistance to radial buckling, however, which may be important during diffuser installation.

(a) Two ports (Tideflex technologies) (b) Boston outfall riser cap

Figure 9.22 Risers with multiple ports

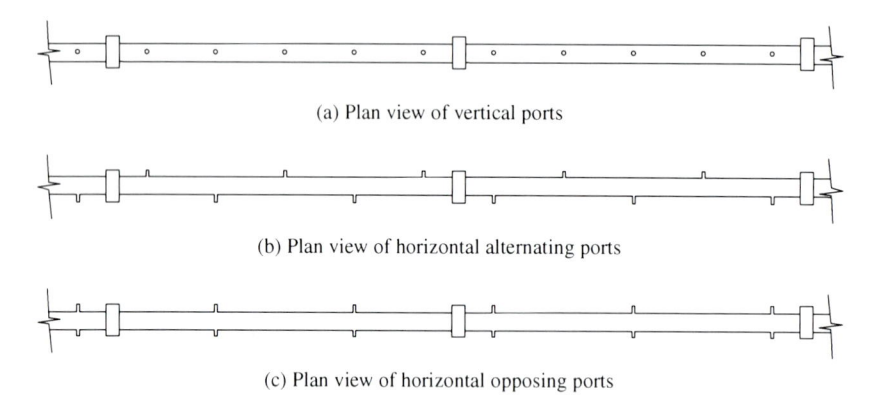

(a) Plan view of vertical ports

(b) Plan view of horizontal alternating ports

(c) Plan view of horizontal opposing ports

Figure 9.23 Three typical port arrangements for HDPE outfalls

9.7.5 Hydraulic calculations

The details of the hydraulic design of diffusers have been discussed thoroughly elsewhere (Fischer *et al.*, 1979), and are only summarized here. The diffuser manifold hydraulics are first calculated as follows.

Design of polyethylene outfalls

The discharge from a port, Q_j depends on the port design, the velocity and pressure in the diffuser, and the port elevation relative to the previous one. It can be computed from:

$$Q_j = C_D A_j \sqrt{2gE} \qquad (9.19)$$

where C_D is the port discharge coefficient, A_j the port area, and E the difference in total head across the port.

The discharge coefficient depends on the port design and the ratio of the velocity head in the diffuser to the total head E. According to Brooks (1970), for sharp-edged ports:

$$C_D = 0.63 - 0.58 \frac{V_j^2}{2gE} \qquad (9.20)$$

and for bell-mouth ports:

$$C_D = 0.975 \left(1 - \frac{V_j^2}{2gE}\right)^{3/8} \qquad (9.21)$$

where V_j is the velocity in the diffuser pipe. Different equations apply if check valves are used (see Section 9.7.6).

The head loss due to friction in the pipe can be calculated from the Darcy-Weisbach equation:

$$h_f = f \frac{l}{d} \frac{V^2}{2g} \qquad (9.22)$$

where h_f is the head loss over length l of pipe, d the internal diameter, and f the friction factor.

The friction factor f depends on the pipe Reynolds number $\mathrm{Re} = Vd/v$ where v is the kinematic viscosity, and k the wall roughness height. Various equations are used to estimate f, such as the Colebrook equation that is valid for the entire range of turbulent flow:

$$\frac{1}{f^{1/2}} = -2 \log\left(\frac{k/d}{3.7} + \frac{2.51}{\mathrm{Re}\, f^{1/2}}\right)$$

Solving the Colebrook equation requires iteration or a graphical solution by use of the Moody Chart, however. For conditions typical of outfalls a more convenient equation that does not require iteration is:

$$f = \frac{0.25}{\left[\log\left(\frac{k/d}{3.7} + \frac{5.74}{\mathrm{Re}^{0.9}}\right)\right]^2} \qquad (9.23)$$

The density head is:

$$h_d = \frac{\Delta\rho}{\rho}\Delta z \qquad (9.24)$$

where $\Delta\rho/\rho$ is the relative density difference between wastewater and seawater (approximately equal to 0.026) and Δz is the height difference between two points along the diffuser.

The analysis is a trial-and-error procedure that begins at the offshore end of the diffuser and proceeds step-by-step back up the diffuser to the first port. The solution procedure is to first guess the value of the head E at the diffuser end. The discharge from the end port is then computed from Eqs. (9.19) and (9.20) or (9.21) and the velocity in the pipe upstream of this port is calculated. The head at the next upstream port is found by adding the heads computed from Eqs. (9.22), (9.23), (9.24) to the assumed head. The discharge from the upstream port is then calculated, the cumulative flow computed and the procedure is repeated. This is continued until the most inshore port is reached. Finally, the friction head loss and density head in the main outfall pipe is calculated from Eqs. (9.22), (9.23), (9.24) and added to the diffuser head. The first guess of the head is then varied by trial and error until the desired flow rate is achieved. Next, the outfall and port diameters are varied to keep the flow per port reasonably uniform and maintain an adequate flushing velocity in the diffuser. The diffuser pipe and outfall diameters are not usually varied continually. Typically, the diffuser might consist of three sections each of constant pipe and port diameters.

These calculations can be done readily in a spreadsheet program or by the many computer programs that are available. For example, the PLUMEHYD computer program included in the DOS version of U.S. EPA "Third Edition of Dilution Models for Effluent Discharges," or CorHyd (Bleninger, 2006). These programs can analyze multiport diffusers, sharp-edged or bell mouth ports, and multi-segmented diffusers with varying internal diameters. The procedure is repeated for the various design flow rates.

Examples are shown in Figure 9.24. For this example, the main outfall pipe is 1.82 m internal diameter HDPE with an assumed roughness height of 0.5 mm. The total diffuser length is 244 m, consisting of 674 ports each 180 mm diameter spaced 7.6 m apart (on each side of the diffuser). The diffuser is tapered in three sections. The first section (from inshore) is 1820 mm diameter, 84 m long, containing 22 ports, the second is 1600 mm diameter, 84 m long, containing 22 ports, and the third 1100 mm diameter, 76 m long, containing 20 ports. The minimum, average, and maximum design flows were respectively: 2.3, 4.7, and 6.8 m^3/s. The diffuser is laid on a horizontal bottom.

Design of polyethylene outfalls 289

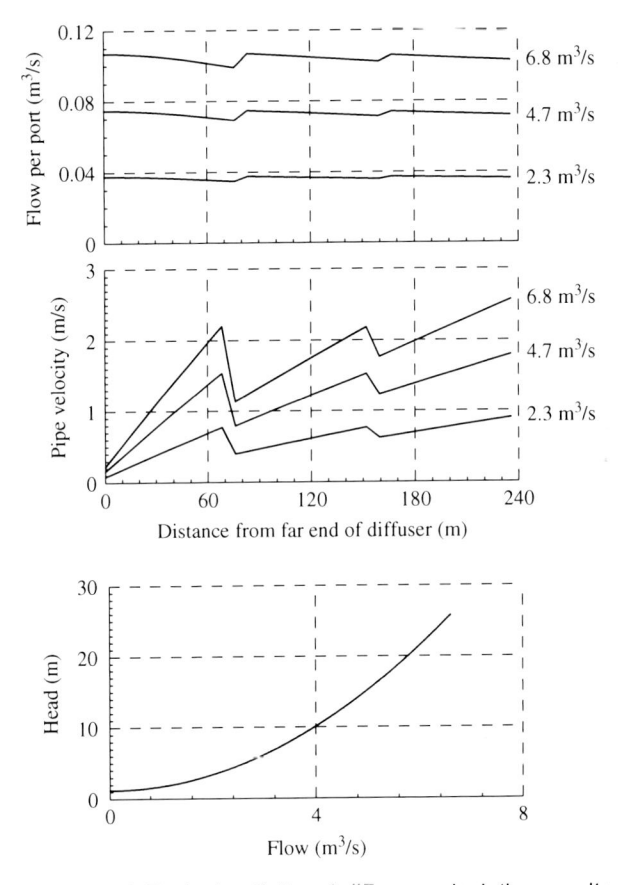

Figure 9.24 Typical outfall and diffuser calculation results

It can be seen that the flow distribution among the ports varied by less than about 10%. The velocity in the pipe remains above 0.6 m/s (the chosen criteria for flushing for this case) until the last 50 m or so at the lowest flow rate. (It could be argued that tapering of this last section is not necessary, as deposition there will cause the diffuser to "self taper"). The head loss increases rapidly as the flow increases (in proportion to the square of flow) from 1.2 m at zero flow (the density head). The densimetric Froude numbers of the ports (not shown) are always significantly greater than one so will always flow full and not allow seawater intrusion.

Equations other than Darcy-Weisbach (Eq. 9.22) are sometimes used to compute friction head loss in pipes. A common one is the Hazen-Williams equation:

$$V = 0.85 C R^{0.63} S^{0.54} \qquad (9.25)$$

where S is head loss per meter of pipe length (m/m), C is the Hazen-Williams coefficient, R is the hydraulic radius equal to $d/4$ (m), and V is the velocity (m/s).

New HDPE pipe is quite smooth, but its flow resistance increases with age due to accumulation of grease, biofilm, or other substances on the pipe wall. The roughness depends on the resin and the manufacturing equipment and process. The absolute roughness height k of new HDPE pipe ranges from about 0.05 to 0.10 mm but for pipe that has been conveying sewage effluent it is usually assumed to be about 0.50 mm. For new HDPE pipe a Hazen-Williams coefficient $C = 155$ is assumed but it progressively decreases to about 140 after transporting sewage for some time.

9.7.6 Check valves

Typical "duckbill" check valves are shown in Figure 9.25. As discussed previously, they are frequently employed on marine outfalls. They are closed at zero flow and gradually open as the flow increases. This allows a higher jet velocity at low flow and can maintain the densimetric Froude number (Eq. 9.17) greater than unity at all flows, preventing seawater intrusion. They also improve saltwater purging from tunneled outfalls. Their flow characteristics must be provided by the manufacturer. For example, Figure 9.26 shows typical variations of open area, jet velocity, and head loss across the valve as functions of flow rate, and a comparison with a fixed diameter orifice. Duckbill valves are available in a wide range of sizes from 25 to 2400 mm and up to 30 different hydraulic variations of geometry and relative stiffness (Duer, 2000). The valves are customized to produce the required jet velocity, headloss, and effective open area characteristics. The consulting engineer should coordinate the sizing with the manufacturer and verify that the manufacturer has conducted independent hydraulic testing on valves of the sizes under consideration and on the hydraulic variations within each size.

Design of polyethylene outfalls 291

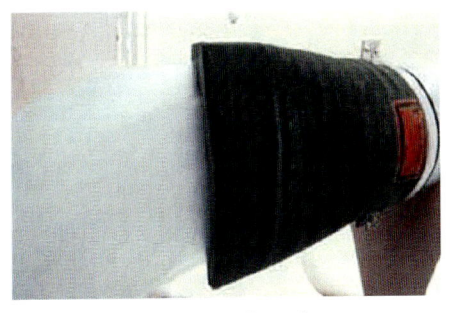

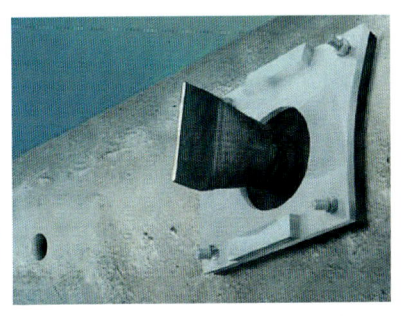

(a) Valve configuration

(b) Bolted to sidewall with flange plate

(c) Various sizes

Figure 9.25 "Duckbill" check valves (Courtesy Tideflex Technologies)

Special procedures are needed to compute internal hydraulics with check valves. For a 100 mm diameter valve similar to that shown in Figure 9.26, the variation of the flow per port with head was approximated by the following equation:

$$E = 0.117 Q_j + 0.000532 Q_j^2$$

or
$$Q_j = 8.06E - 0.145 E^2 \tag{9.26}$$

where E is the head loss across the port in meters, and Q_j the flow rate in liters per second.

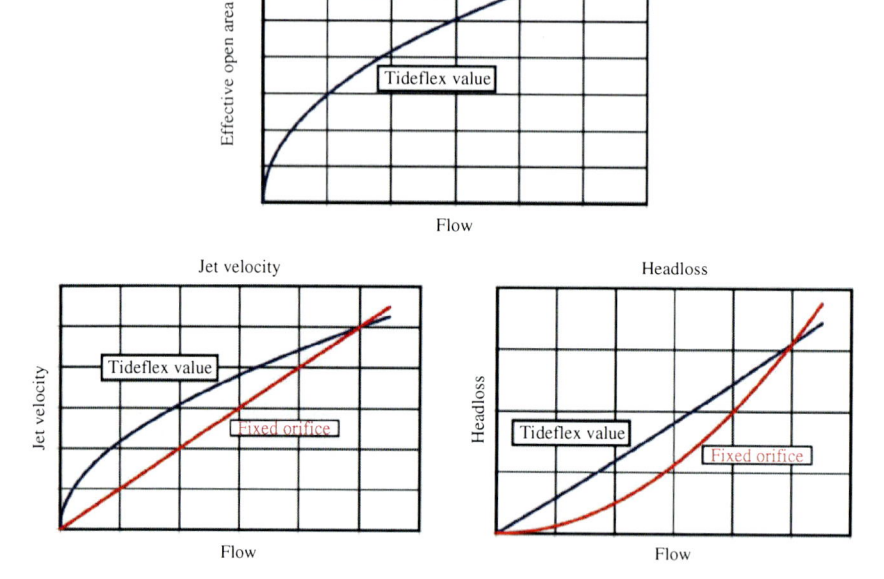

Figure 9.26 Flow characteristics of a typical duckbill check valve (Courtesy Tideflex Technologies)

To calculate the internal diffuser hydraulics, the same procedure as in Section 9.7.5 is followed, except that Eq. (9.19) is replaced by Eq. (9.26). After computing the port flows, the jet velocities, V_j are calculated from:

$$V_j = \sqrt{\frac{2gE}{K}} \qquad (9.27)$$

where K is the valve loss coefficient, which testing has shown to be equal to be approximately one. The port area, A_j is then given by:

$$A_j = \frac{Q_j}{V_j} \qquad (9.28)$$

and the diameter, d_j, of an equivalent round port is given by:

$$d_j = \sqrt{\frac{4A_j}{\pi}} \qquad (9.29)$$

The calculations then proceed as previously carried out for simple orifices. For more discussion of the hydraulics and calculation procedures with check valves, see Lee *et al.* (1998, 2001), and Bleninger (2006).

9.8 EXTREMELY DEEP OUTFALLS

Coastal communities on archipelagos or volcanic islands are often located where the seabed drops off rapidly with slopes often greater than 1:4. The outfall will be in very deep water to achieve adequate distance from shore to avoid beach pollution, contamination of water sport areas, pollution of shellfish harvesting waters, or threats to environmentally sensitive areas. The term extremely deep is defined here as depths greater than 60 m or even exceeding 150 meters. Outfalls at extreme depths and steep seabed slopes require special considerations during planning, design, and construction.

The "normal" range of diffuser depths for many communities is 20 to 40 m. For extremely steep seabeds this depth may occur less than 200 m offshore, a distance that is insufficient to avoid beach contamination. It is therefore necessary to go to greater depths to gain farther distance. This introduces the following considerations:

- Extremely steep slopes may require more secure stabilization of HDPE outfalls;
- A steep seabed does not dissipate the energy of large waves until they are very close to shore. This results in strong erosive forces nearshore;
- When accurate placement of the outfall is necessary to avoid obstacles, escarpments, etc., expensive specialized equipment and remote sensing and installation techniques are needed;
- Construction becomes more difficult because much of the outfall is deeper than the maximum working depth of scuba divers;
- If deep water diving is required it must be done using mixed gas and decompression chambers which is difficult and expensive. This is also true for future inspection and repairs;
- If inspections must be made below the working depths for divers, they must be made by a remote operated vehicle (ROV) or a manned submersible;
- The head due to the density difference between seawater and effluent (Eq. 9.24) can become a significant component of the head required;
- The internal pressure during submersion is much greater than for conventional depths. This usually necessitates a thicker pipe wall;
- Accuracy of bathymetric maps is critical because placement of the outfall on the seabed at depths over 60 m must be done with remote sensing methods.

There are also a number of advantages associated with very deep outfalls on steep slopes:

- The length is usually considerably shorter than an outfall on a gentle slope;
- It is usually possible to trap the plume well below the water surface, preventing encroachment onto beaches, shellfish harvesting waters, or other shallow environmentally sensitive areas;
- Because the wave-induced forces decrease with increasing ocean depth the stabilization requirements such as trenching, ballast weights, mechanical anchors, and armor rock are significantly less;
- For small flows, it may be possible to obtain sufficiently high initial dilution with a single port, i.e. an open-ended pipe, thereby precluding the need for a multiport diffuser.

Examples of plume calculations for an extremely deep outfall are shown in Figures 9.27 and 9.28. These are from an analysis using the UM3 model in Visual Plumes (Frick *et al.*, 2003) using actual oceanographic data consisting of 15 density profiles and respective current speeds and directions. The wastewater flow rate is 0.49 m^3/s and it has been treated for removal of grease and floatables. The discharge is from a single 0.55 m diameter port (the end of the outfall pipe) at a depth of 140 m located 500 m offshore. The current speed at this depth is between 2 and 7 cm/s in a direction perpendicular to the direction of discharge. The discharge is horizontal but the seabed slopes down at 40° at this location.

Figure 9.27 shows the plume trajectories. The centerlines of the 15 plume trajectories are shown as solid lines and the upper and lower plume boundaries of the trajectory envelopes as dotted lines. The lower plume boundaries are 25 m above the seabed. Figure 9.28 shows dilutions versus distance. Average dilutions are indicated by the solid line and centerline dilution by dotted lines. Average dilutions range from about 740 to 1,500 at a distance of only 160 m from the end of the outfall; centerline dilutions are about 380 at this same distance.

Because the upper boundary of the plume is deeply trapped below a depth of 54 m, the effluent plume will not reach the shore or beach areas, and is well below the range of scuba diving and water sports activities. The high dilution and distance from environmentally sensitive coral areas ensure their protection. The lower boundary of the plume is about 25 m above the seabed so it will not directly impact the seabed or benthic organisms.

Design of polyethylene outfalls 295

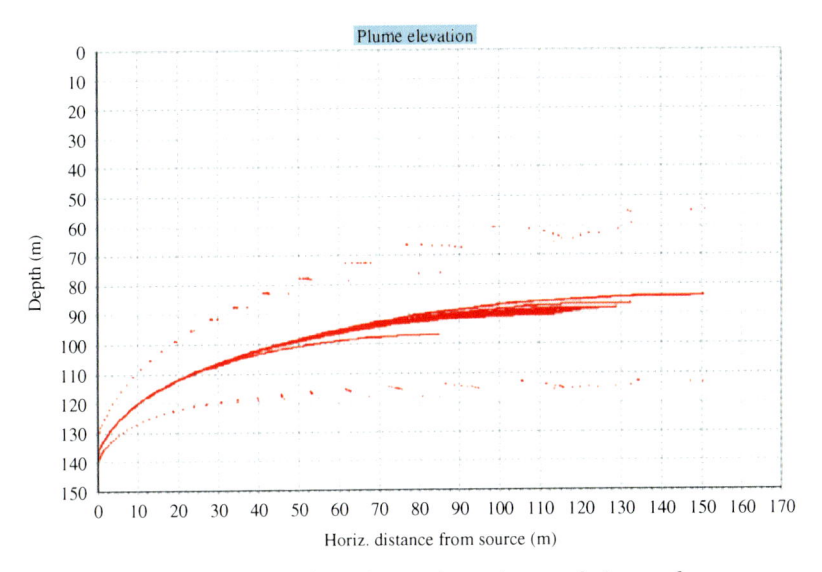

Figure 9.27 Centerline trajectories and envelopes of plumes from a very deep outfall

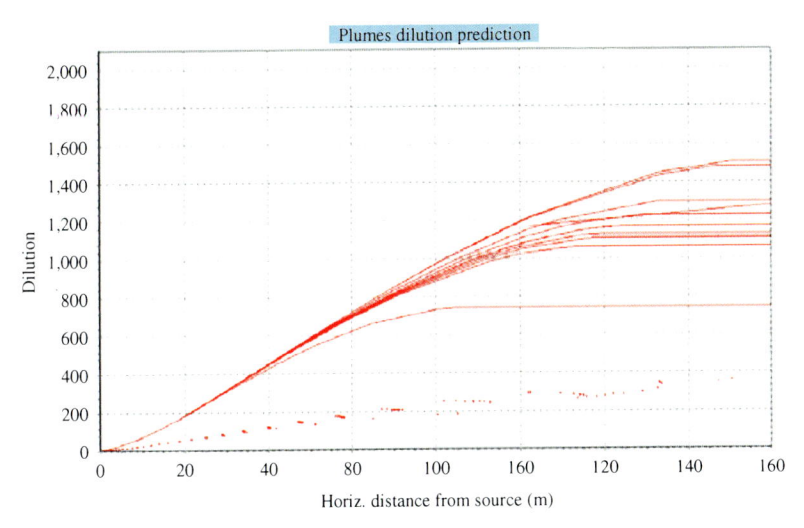

Figure 9.28 Near field dilution for a very deep outfall

The first step to evaluate the feasibility of an extremely deep outfall is to obtain an accurate bathymetric chart that includes any areas to be protected. Then a rapid analysis is made to determine if the outfall is structurally feasible. If it is, salinity-temperature-density profiles along with current velocity and direction measurements should be obtained throughout the water column at times of the year that includes seasonal extremes. Experience indicates that density profiles of nearshore water columns more than 100 m deep can fluctuate widely over two hours or less and that current velocity and direction commonly fluctuate both with depth and with the hour of the day. Therefore it is important that data be collected at sufficiently short intervals of time to be representative.

Times series analysis by Visual Plumes (or an equivalent computer program) can then be used to determine the near and far field plume characteristics for various discharge depths for the density profiles and current data to determine if the dilution is adequate.

It should be pointed out that all the density profiles in the example above were slightly stratified. Even a very weak stratification will result in a trapped plume for a very deep outfall. If the water is not stratified (the density is constant throughout the water column), then the plume would surface. This would be quite unusual for ocean depths greater than 100 m, however, as even small temperature differences will be sufficient to submerge a plume. It is therefore very important to obtain accurate density profiles and current velocities that are representative of conditions throughout the year.

For extreme depth multiport diffusers on a steep seabed check valves may be required because of the large elevation differences between the ports. When discharging into slightly stratified water, the individual plumes from each port often become trapped at different depths resulting in high vertical dispersion and very high initial dilutions. Outfalls discharging into depths greater than 100 m will almost always result in a trapped plume sufficiently deep and distant from shore and with high dilution that will protect beaches and water sports areas.

9.9 VERY SHALLOW OUTFALLS

The opposite case from the previous one is a very flat seabed which prevents discharge at "normal" depths (20 to 60 m) at a reasonable distance from shore. They may require an outfall four km or more long to obtain such depths.

For most small coastal communities (defined as requiring an outfall diameter less than 360 mm) very long outfalls are generally neither economically nor operationally feasible. Not only is the construction cost usually prohibitive but the head loss may be excessive, resulting in operational complications and prohibitive pumping costs. Increasing the diameter to reduce head loss is usually not viable because this increases the construction cost and may decrease the flow velocity below that required for self cleaning. It can also aggravate operational problems faced by communities with a small permanent population that has designed an outfall to serve relatively large populations during tourist season.

For these small communities it is usually possible to decrease the outfall length by using a diffuser designed for discharge into relatively shallow ocean depths (between 10 and 17 m) that still attains sufficient near field dilution to protect beaches and environmentally sensitive areas. This is accomplished by increasing the diffuser length.

Plumes from shallow diffusers will usually surface and near field dilution may be limited. Because far field dilutions can be small (see Section 3.3.3.2) it is important that initial dilution be maximized to ensure that contaminant levels at sensitive areas are lower than those established by health and/or environmental standards. Near field dilution can be maximized in several ways:

- Making the diffuser as long as possible;
- Orienting the axis of the diffuser as close as possible at 90° to the predominant current;
- Using orifices that discharge horizontally.

A frequently cited rule of thumb for port diameters is that the minimum diameter should be 100 mm. This is almost never appropriate for small diameter shallow diffusers because there would be too few ports. Spacing a few large ports far apart will not increase dilution as it is then governed by point-plume characteristics that do not merge. To maximize dilution, the plumes should merge to a line plume (see Section 3.3.2) and this is accomplished by using the smallest feasible port diameters. If the effluent has been treated by milliscreening and grease removal, port diameters as small as 40 mm can be used for shallow small diameter diffusers. The orifices should discharge horizontally because this gives a higher near field dilution.

The port spacing should be based on having the plumes merge just before impacting the water surface. According to Section 3.3.2, for a surfacing plume this occurs when $s/H > 0.3$, and if stratified, $s/l_b > 0.3$. Visual Plumes can be used to determine the near and far field dilutions.

It is possible to use a single straight line diffuser, but, from a standpoint of internal friction losses and maintaining a uniform flow distribution, it may be advantageous to use a Wye or otherwise branched diffuser.

Another important consideration for very shallow diffusers is that the bottom pressure under a wave crest can be considerably greater than that under the wave trough and this differential can cause seawater to enter the diffuser. To avoid this it is recommended that check valves (such as the duckbill valve) be installed on the ports. It should also be noted that wave action which passes sequentially over the ports will cause non-uniform discharge among the ports, although the average discharge between the ports will remain constant.

Wave forces can be very important for shallow diffusers and can cause instability. The greater the wave height and the shallower the diffuser the greater the concern. The methodologies for stabilization are basically the same as outlined in Sections 9.4.3, 9.4.4, 9.4.5, and 9.5. The main difference is to ensure the stabilization methods do not interfere with the port discharges. When entrenchment of the diffuser is necessary, risers like those shown in Figures 9.20 and 9.21 (e), (f), and (g) are necessary to raise the ports above the seabed.

The diffuser is designed for the maximum design flow but there may be a large difference between initial and future flows. During the early years of a long shallow diffuser, the flow may not reach the end, increasing the likelihood of sedimentation in the pipe and intrusion of marine organisms. This is easily remedied, as mentioned in previous sections, by blocking some inshore ports during the early years. Because the depth is shallow, a diver can remove some of the caps or plugs as the flow increases.

An advantage of shallow diffusers is that diver bottom time is usually not a major concern during construction, inspection, or repair of the diffuser. A disadvantage is that the shallower depth increases the likelihood of damage from waves, small boat anchors, and fishing equipment.

10
Ocean outfall construction

10.1 INTRODUCTION

The purpose of this chapter is to review the major methods of marine outfall pipeline construction, the principal variables that will be encountered and how they affect construction, and the most important steps that should be taken to help ensure an orderly, efficient, and effective construction enterprise. Planning, design, and construction of outfalls are closely linked and must be considered together to ensure a trouble free facility with a long useful life that is economical and efficient. The construction phase must comply with the planning and design requirements. Conversely, the planning and design process should consider construction practicality and feasibility.

There are many options for planning, designing, and constructing an ocean outfall. Three basic ones are summarized below to illustrate how construction requirements and responsibilities differ for each of them.

© 2010 IWA Publishing. *Marine Wastewater Outfalls and Treatment Systems.* By Philip JW Roberts, Henry J Salas, Fred M Reiff, Menahem Libhaber, Alejandro Labbe, and James C Thomson. ISBN: 9781843391890. Published by IWA Publishing, London, UK.

One option is for the entity that will be the owner and operator of the outfall (usually a water and sewer authority) to carry out the planning, design, and construction utilizing their own personnel. Water and sewer authorities may elect this option particularly if the outfall is small diameter, they have qualified personnel with previous successful experience in the planning, design, and construction of marine outfalls, sea conditions are normally favorable for outfall installation, and the geomorphology of the seabed is well documented and suitable for outfalls. In this option the personnel responsible for planning and design are also frequently responsible for construction. When this option is well orchestrated by experienced and qualified personnel, the time from the initial planning stages to initiation of operation and maintenance can be greatly compressed, and, because the risks involved are assumed by the owner, the overall cost can be very low. A distinct advantage of this option is that if unexpected conditions are encountered during construction that make it necessary to modify the design, it is usually possible to do this quickly. Another is that with each succeeding outfall, the process usually becomes more efficient and cost effective. A variation of this option that is sometimes employed by a relatively inexperienced entity is to retain an overall expert in planning, design, and construction to guide, oversee, teach, and perhaps supervise, the entire project.

The water and sewer authority (or other responsible entity) will often retain a qualified engineering consulting firm to carry out or assist with the planning and design phases whereas construction will be accomplished by a construction contract, based on the plans and specifications prepared by the consulting firm. The construction contract can be awarded through a competitive bidding or a negotiation process. This option becomes more attractive with increasing outfall diameter and length, more difficult sea conditions, insufficient oceanographic information, important environmental considerations, and limited local experience with ocean outfalls. In this arrangement, the authority may either choose to inspect the work being carried out by the contractor with their own personnel, or it might retain the consulting firm to carry out the inspection. The major advantage to this option is the increased likelihood of the different phases of the project being carried out by experts with both broad and in-depth experience of all aspects of outfalls.

Another option is for the authority or agency to negotiate a "turnkey contract" in which the contractor has full responsibility for all aspects of outfall design, construction, and commissioning. A turnkey contract may even include responsibility for planning of the outfall. For this option, the authority may carry out the necessary inspection to ensure that the work is being carried out in compliance with the contract requirements, or retain a third party to provide this service.

The planning and design phases determine the onshore outfall location, the diffuser location, the outfall diameter and length, the general if not the specific outfall alignment, the diffuser details, the material and class of the pipe to be used, the measures to protect the outfall from the hydrodynamic and corrosive forces of the marine environment, and how water quality standards and environmental impact requirements will be met during the construction and the operational phases. The outputs from these two phases culminate in the project's construction plans and specifications, and the construction phase achieves the requirements established in the plans and specifications. Construction plans and specifications rarely dictate the exact methodology to construct the outfall, but the diameter, weight, and material of the pipe, the sea conditions, the seabed geomorphology, environmental requirements, and competing uses of the ocean strongly influence or determine the most feasible construction methodology.

10.2 IMPORTANT PRELIMINARY CONSIDERATIONS FOR CONSTRUCTION

There are a number of factors that should be considered as early as possible during construction of an ocean outfall. Failure to do so has resulted in project delays, cost overruns, accidents, litigation, and even project failure.

10.2.1 Permits and compliance with regulations

Construction of outfalls is affected by government regulations and permitting processes at local and national levels. Examples are permits for transportation and use of explosives, environmental protection, use of ports and harbors, connection to or disruption of utilities, worker health and safety requirements, and traffic control at ingress and egress to the work site. These should be reviewed, considered, and addressed as early as possible because some of them require a long lead time.

10.2.2 Seabed conditions

It is important that the construction contractor have detailed information of the submarine conditions not only on the selected outfall route but also in the general area. Bathymetric charts, seabed geomorphology charts, navigational charts, subsea videos, sonar imaging, and diver reports can provide crucial information pertinent to construction. Any such information that was obtained during the planning and design phases of the project should be made available to the entity responsible for construction.

It is a good practice for the construction contractor to dive the site as soon as possible to verify or supplement information already available. The divers should permanently mark any anomaly, problem area, or potential obstruction and accurately establish their location. The contractor should also take a pre-construction video of the submarine route to help identify potential problem areas as well as document the conditions before construction begins. For complex seabeds, if an accurate bathymetric chart of the seabed is not available, the contractor should have one prepared using multibeam sonar and GPS.

If the construction involves submarine excavation, the contractor should consider taking bottom samples and drilling boreholes to enable selection of the best method of excavation. If this involves the use of explosives, test blasts to determine the most suitable explosive should be carried out as early as possible. Likewise, if mechanical anchors are to be used to stabilize the outfall, the suitability of the seabed for their installation should be verified as soon as possible.

10.2.3 Sea conditions, wind, waves, tides, and currents

Although waves, tides, currents, and wind should have been considered during design, the focus would usually have been on the extremes that might be encountered over the useful life of the outfall. Construction activities can be adversely affected, however, or even prevented altogether, by much lower levels of winds, waves, and currents. The contractor is more interested in seasonal, daily, or hourly probabilities of significant perturbations to enable scheduling construction activities to take advantage of the most favorable times.

Wind speed and direction are extremely important, particularly for float and sink installation. It is important for the contractor to be able to make short-term projections of wind speed and direction, and changes that are likely to occur during the day. Historical meteorological data can be useful for this purpose.

Where underwater work by divers is required, visibility is very important for work efficiency, so the likelihood and effect of stormwater runoff and resuspension of seabed sediments by currents and waves must be considered.

Tide tables are not only valuable for scheduling activities but they are crucially important for work in shallow, nearshore areas and the intertidal zone. Tidally-driven currents can also adversely affect construction in offshore waters.

Wind, waves, currents, and tides must all be considered because they can adversely affect construction in many ways and can strongly influence the choice of construction techniques and methodology. They determine to a great extent whether or not it is necessary to construct a temporary trestle in the surf zone. The height of anticipated waves determines the height and strength of such a trestle to ensure its adequacy and safety. They determine the construction method of the temporary trestle itself. Even moderate waves can adversely affect the lay barge method of outfall installation and make it difficult, if not impossible, due to instability of the barge. Even moderate waves can cause considerable water movement near the seabed when the ocean depth is less than about 15 m. The drag on a diver from the current, the eddies that are induced around the outfall pipe, and the reversal of current direction with each passing wave can preclude useful work.

10.2.4 Surf zone conditions

Two terms used somewhat interchangeably are "surf zone" and "inshore zone". The surf zone is defined as extending from the shoreline to a depth where approaching waves begin to break. This is not an exact definition since waves of different heights break at different depths. The inshore zone is defined as water less than about 6 m deep, or shoreward of the depth in which a barge or ship can work. This is also not a precise definition.

Because water conditions in the surf zone influence construction decisions in many ways, it is important to know how they vary with the seasons of the year, tides, winds, time of day, and with local and remote sea conditions. It is virtually impossible for divers to work under breaking waves. A trench excavated in a sandy seabed will rapidly fill in the surf zone so it becomes necessary to install sheet piling to keep it open. Rip currents and cross shore littoral currents can make accurate outfall placement difficult. Surf zone conditions frequently make it necessary to construct a trestle to excavate a trench for outfall pipe installation. The trestle length, water depth, and height above sea level are determined by the surf zone conditions.

10.2.5 Construction offices

An office for the management, coordination, and supervision of construction activities should be made available early in the construction phase. This can be a trailer, a temporary building, or rental of an existing building. The construction office(s) should be as close as possible to the major construction activities. They should have adequate space, appropriate furniture, computer equipment, files, and office equipment to carry out the work. It should have communication systems such as telephone and internet, and heating and air conditioning appropriate for the local climate.

10.2.6 Supply of pipe and fittings

Pipe quality is critically important for outfalls, so the purchase order or contract should contain appropriate reference standards. Inspection of the manufacturing process is sometimes warranted. In addition, the contractor (or purchasing entity) should request quality certificates regarding the manufacturing process as well as the quality of the raw materials. These should be accompanied by properly certified results of the manufacturer's laboratory tests carried out on the pipe. For HDPE pipe it is also recommended that, as soon as the pipe arrives at the work site, samples should be taken and destructive and non-destructive tests carried out to certify the pipe quality.

Regardless of the type, class, and pipe diameter, the arrangements for purchase and transport of the pipe should be made as early as possible. Pipe for ocean outfalls is usually not available from stock and is usually made specifically. Therefore, the purchase order or contract should contain a clause that requires the manufacturer (or supplier) to provide a firm date for pipe fabrication so that arrangements for inspection and testing and acceptance can be carried out in a timely manner. Likewise the dates of shipment and arrival should be clearly established. When international shipment is involved, clearance through customs must be considered when scheduling the project.

The purchase order or contract should also contain requirements for shipping that include loading, handling, and packaging or containerization so as to preclude damage to the pipe in transit. Most pipes are manufactured in standard lengths that are suitable for ocean and land transport. The standard 12 m length of HDPE pipe fits nicely into shipping containers as shown in Figure 10.1, and this provides adequate protection during shipment, unloading, and local transportation by truck or rail.

Loading, offloading, and handling of the pipe should be carried out in a manner that prevents damage. This is especially important for pipe made of softer materials such as HDPE, PP, and GRP as well as steel pipes with protective coatings or pipe with integral joints that are vulnerable to damage. Figure 10.2 illustrates the use of nylon straps to minimize damage to HDPE pipe during offloading from a container.

There are a number of manufacturers that produce polyethylene pipe in lengths up 500 m. Because the specific gravity of polyethylene is less than one, it is possible to tow a number of these very long pipe lengths behind a seagoing vessel over the open ocean directly to the project site (Figure 10.3). This has the advantage of eliminating or greatly reducing the number of butt fusion joints to be made at the site but has the disadvantage of exposing the pipe to potential abuse and damage during transportation.

Ocean outfall construction

Figure 10.1 HDPE pipe loaded into container for ocean transport

Figure 10.2 Unloading HDPE pipe from a container using nylon straps to protect the pipe from abrasion

306   Marine Wastewater Outfalls and Treatment Systems

(a) Pipe extrusion

(b) Transport by sea

Figure 10.3 Production and transport of long lengths of HDPE pipe (courtesy Pipelife Norway)

10.2.7 Storage of materials and equipment

Regardless of the materials used for the outfall or the construction methodology employed there will be a need for a storage yard for materials, supplies, and equipment. The construction methods will influence the location(s) for site storage that can be used to best advantage. If the outfall is assembled on land before installation in the ocean, the storage yard should preferably be located where the assembly (construction) takes place. If the outfall is constructed offshore the storage yard should be near a port or harbor from which pipe, supplies, and equipment can be loaded onto barges to transport them to the offshore installation site.

The storage yard should be arranged to facilitate transport of materials or equipment into or out of it. It must have aisles that are wide enough to accommodate the maneuvering of forklifts, cranes, and trucks as they handle and transport the pipe and other materials. The area should be fenced and guarded against theft and vandalism during non-working hours. It should have appropriate shelter for supplies that can be damaged by rain, wind, sun, and dust. It is usually advantageous to have the construction office located on or adjacent to the storage yard.

10.2.8 Public relations

Construction of an ocean outfall can have a strong impact on the local community, especially if it is in a densely populated area. The magnitude of the works and the fact that the outfall signifies an important advancement for the municipality generates high expectations among the local authorities, the media, and the inhabitants.

Local authorities should be notified well in advance of any construction activities that might interfere with the daily routine of the local inhabitants and advised of measures that will be taken to minimize the impact. Where feasible the local media can be utilized to inform the public to steer clear of the area when interference with their daily lives is likely. If disruption of local utilities is likely, the public should be forewarned and the affected utility should be restored to at pre-outfall conditions as soon as possible.

Demobilization is the final stage of outfall construction. This includes dismantling temporary structures and removal of all equipment, machinery, tools, supplies, and refuse. Even if the outfall is perfectly constructed, the general public will only be able to observe the conditions on land. The demobilization should be thoroughly carried out because abandoned equipment and materials or trash left behind will be associated with slipshod work and disrespect for the community.

10.3 OUTFALL INSTALLATION METHODS

Various methods have been successfully used for outfall installation. Most have been adapted from those originally developed for installation of submarine pipelines for the offshore petroleum industry. The choice depends mostly on the type of pipe used, its diameter and length, the water depth, sea conditions, and seabed characteristics.

The following sections describe the methods most commonly used and review their salient features and the situations where they would be favorable or unfavorable. Installation by flotation and submersion method is covered in the most detail because it is most commonly used for HDPE submarine outfalls, and is particularly appropriate for the numerous small and medium sized coastal communities that could benefit from an ocean outfall.

10.3.1 Flotation and submersion

Installation by flotation and submersion is used primarily for outfalls of high density polyethylene (HDPE) and polypropylene (PP) pipe, but it has occasionally been used for steel and ductile iron, usually with supplemental flotation. This method is based on the principal that a pipe filled with air will float when the weight of the pipe is less than the weight of water displaced by the pipe. Only HDPE will be covered here because of its predominance for small to medium diameter outfalls.

The buoyancy of empty HDPE pipe is usually sufficient to support the weight of the concrete ballasts attached to the pipe to stabilize it against the hydrodynamic forces of the ocean as it rests on the seabed (see Chapter 9). If the weight of the pipe and concrete ballasts is too heavy to be supported by the buoyancy of the empty pipe it is necessary to provide temporary supplemental flotation during transportation and submersion. An alternative is to not attach all of the ballast weights so the outfall will float and then have divers attach them underwater after submersion. A quicker and cheaper option is to design concrete ballasts that allow easy addition of supplemental concrete weights to them after the pipe is submerged.

For the flotation and submersion method the outfall is usually assembled with ballasts and fittings on land that is suitable for launching it directly into the water. Two basic options are (1) to assemble the entire outfall as a single unit for subsequent launching and transport to the submersion site, and (2) to assemble the outfall in two or more sections that are launched and transported separately to the outfall site where they are connected together during the submersion process. Flotation and submersion of HDPE outfalls consists of the following components:

- Preparation of temporary working platform and launching facility;
- Butt fusion welding to join individual pipe sections;

- Fabricating concrete ballasts and attaching them to the pipe;
- Temporary storage and testing of the assembled outfall;
- Launching the completed outfall into the ocean;
- Towing the floating outfall to the desired location and aligning it to its predetermined route;
- Submerging the outfall onto the designated seabed alignment;
- Post submerging activities.

10.3.1.1 Temporary working platform and launch facility

The distance between the outfall assembly site and the final outfall location on the seabed is an important consideration. Ideally, the assembly site would be located where the outfall is to be installed, but frequently such a site is not available. If a remote site must be used, it should be as close as possible to the submersion site so that navigation time over water is minimized and the pipe is exposed to less risk, particularly over deep waters. The land distance between these points should also be minimal to facilitate movement of personnel and machinery. Other aspects to be considered include distance from the urban center so that any needed resources can be obtained easily and quickly. It should also have easy access to communications facilities, electrical power, and other needs. Proximity to port facilities and the availability of tug boats and deep sea vessels for use in launching the pipe and the other maritime works are also important. The work platform must be guarded at all times to protect the general public against accidental injury as well as to protect the outfall from damage or vandalism.

An HDPE outfall must be assembled on a flat, relatively level surface (hereafter called a working platform) to ensure correct axial alignment of the pipe lengths as they are joined together. The working platform should be designed to facilitate movement of either the connected string of pipe or the butt fusion welding equipment as new pipe sections are being joined to the pipe string. Such a platform may also be used to "store" the completed segment(s) in a manner that enables testing and facilitates launching the outfall into the sea. There are various options that are commonly selected for a working platform.

One that is widely used in Chile and other countries is a temporary rail system that terminates in a trestle and ramp at the shoreline. The entire outfall is assembled using the rail system as the working platform and on completion it is pulled over the trestle and ramp and launched into the ocean. Figure 10.4 portrays the rail system used in 2004 for the installation of an outfall in Quintero, Chile. Ideally it should be perpendicular to the shoreline at the launching site. Its length is determined by the length of the outfall. If it is not feasible to construct a platform as long as the entire outfall, it is usually constructed as several parallel rail systems that are ½, ⅓, or ¼ the outfall length and wide enough to allow movement of machinery, equipment, and pipe for assembling each outfall section.

Figure 10.4 Temporary rail system used in constructing the outfall for Quintero, Chile (courtesy of Belfi-Montec)

With this option the butt fusion welding operation can either be stationary with the assembled outfall moved from it over the rails, or it can move on the rails after each joint is completed with the completed sections of the outfall remaining stationary until launching. The rail system is usually constructed of lightweight rails fastened to railroad ties or sleepers resting on a prepared flat and level, or nearly level, surface. The separation between the rails and spacing of the ties or sleepers should be appropriate for the pipe diameter and concrete ballasts as well as the butt welding equipment. Carts are placed on the rails to hold and move the pipe while it is being welded and the concrete ballasts are being attached.

Another option is a short working platform, usually no longer than three or four pipe lengths, set up close to a calm protected water body (such as a canal, estuary, bay, or harbor) in which the completed outfall will be temporarily stored before transport to the submersion site. The platform is designed for stationary butt fusion operation, attachment of concrete ballasts, and launching each completed section into the water after the concrete ballasts have been attached. It is frequently set up on the end of an existing dock or pier that has been fitted with a temporary ramp or trestle designed to carefully feed the outfall into the water as illustrated conceptually in Figure 10.5.

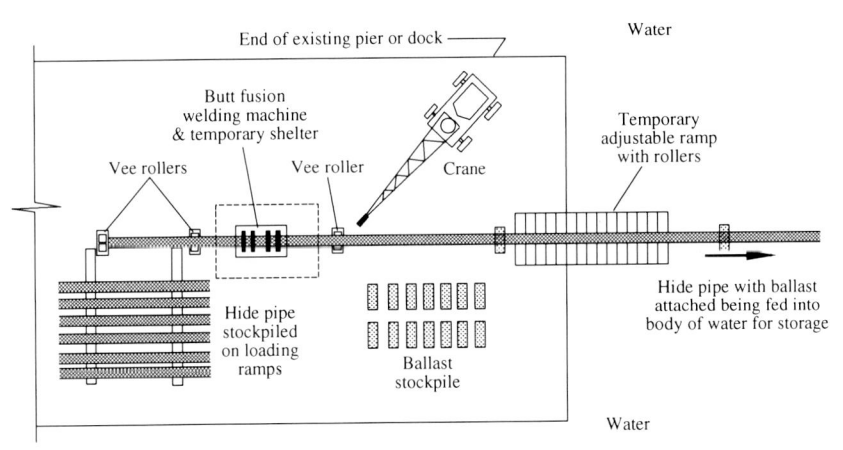

Figure 10.5 Temporary outfall assembly platform on an existing pier or dock

This option usually utilizes a crane or forklift along with rollers, dollies, and/or rails to facilitate handling the pipe and materials during the connection of the pipe sections, attachment of ballast weights, and feeding the assembled outfall into the water.

Adequate anchorage must be provided for the floating outfall to secure it against movement by wind, waves, or currents. The water body used for storage must connect to the sea where the outfall is to be installed. This option is also sometimes used for launching the outfall directly to its final submersion site if the sea conditions are dependably calm and the diameter and length of the outfall are small enough that assembly takes less than one or two days.

An existing paved roadway or a shoulder of a roadway can sometimes be used for a working platform. An example is shown in Figure 10.6, in Viña del Mar, Chile, in 1996. Ideally the roadway should be straight and level and terminate in a paved ramp leading to the sea (such as a launching ramp for boats carried on trailers) otherwise a temporary launching facility that connects the roadway to the sea must be constructed. This option may benefit the contractor, as it reduces the temporary works needed. But it requires the support and collaboration of local authorities and considerable effort on the part of the contractor to ensure public safety, minimize interference with traffic, and avoid accidents or hazardous situations, especially during outfall launching.

Figure 10.6 Outfall assembled alongside a major thoroughfare in Viña del Mar, Chile (1,500 m long by 1,200 mm OD) (courtesy Belfi-Montec)

10.3.1.2 Butt fusion welding HDPE pipe

HDPE pipe outfalls are connected together by a process known as butt fusion welding using equipment manufactured specifically for this purpose. When this is properly done the joint has a quick burst strength that is over 85% of the long term burst strength of the pipe itself.

Quality control of the butt fusion process is essential because a poorly made joint can have less than 50% of the strength of the pipe. The welding process requires careful alignment of the axes of the pipe lengths to be welded, milling (planing) of the pipe ends so they will abut squarely and precisely, heating the two ends to the melting point of the HDPE resin, joining the melted ends together under pressure, and slowly cooling the joint. The heating temperature, time, and pressure must be carefully controlled so that the physical properties of the HDPE are retained after the joint cools. Figure 10.7 graphically illustrates the relationship of the applied pressure (compressive stress on the butt end pipe walls) to the elapsed time during the principal stages of the welding process.

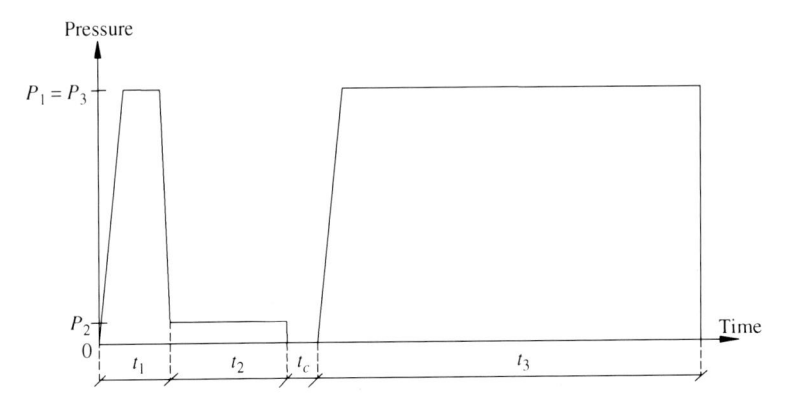

Figure 10.7 Graphical representation of butt fusion welding process

Welding commences after the butt ends of the pipes have been squarely milled to a near perfect match and the heating element (also referred to as mirror) has reached the temperature required by the pipe manufacturer. This is commonly 220°C for HDPE pipe. The heating element is placed between the butt ends of the pipe which are moved against the heating element with a pressure P_1 for a time t_1. As the HDPE begins to melt, beads of molten material form around the pipe circumference where they contact the heating element. When the desired bead width (e) is reached, the pressure is lowered to a level P_2 for a period of time t_2 known as the heat soak. This time is dependent on the pipe wall thickness. The pipe butts are than separated from the heating element, the heating element is removed, and the molten pipe ends are brought together. This all occurs as quickly as possible (t_c) to minimize heat loss. The butt ends are brought together with a pressure P_3 (that is not applied too abruptly) and this pressure is maintained for a time t_3 while the joint cools. It is important that the operator of the welding machine not cut this time short to increase productivity because the duration of the cooling time strongly affects the joint strength. The time required varies with both the pipe wall thickness and the ambient temperature.

Different HDPE pipe manufacturers use a number of different resins which have a range of physical-chemical properties. It is recommended that the specific pipe manufacturer's directions for butt fusion welding be faithfully followed.

Widely accepted butt fusion welding parameters that have been derived from laboratory testing and practical experience are summarized in Table 10.1 and Figure 10.8

Table 10.1 Typical butt fusion welding parameters for HDPE pipe (refer to Figure 10.7)

Parameter	Value
Heating element temperature	220°C
Initial pressure, p_1	0.18 MPa
Initial heating time, t_1	Until a fusion bead width e is reached
Bead width, e	$e = 0.5 + 0.1\,s$ where s is pipe wall thickness in mm
Heat soak time, t_2	15 s (seconds) where s is pipe wall thickness in mm
Change over time, t_c	Less than 20 seconds
Welding pressure, p_3	0.18 MPa
Cooling time, t_3	$t_3 = 10 + k_c\,0.5\,s$ (minutes) where s is pipe wall thickness in mm and k_c is derived from Figure 10.8 (Janson, 2005)

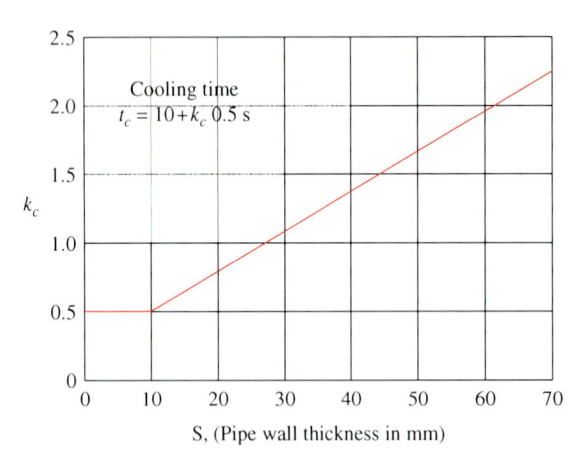

Figure 10.8 Coefficient k_c in cooling time equation (Janson, 2005)

The welding process should be carried out in an orderly manner. The pipe ends must be thoroughly cleaned before milling to help prevent contamination of the pipe ends after they are milled. It is essential that the ends of the pipes be protected from contamination from dust and especially water after milling before heating, and particularly during the time t_c when they are most vulnerable. To protect against adverse climatic conditions, welding should be carried out in a small, enclosed building that allows cooling to take place at an adequate temperature without being affected by the wind.

Butt fusion welding produces a double melt bead on the inside and outside of the pipe. When it is necessary or desirable to remove this bead the pipe and bead should first be completely cooled to ambient temperature to prevent a notch forming at the fusion joint due to polyethylene shrinkage during cooling. The bead should be removed down to (or just above) and never below the surface of the pipe. Special tools are commercially available for this purpose.

10.3.1.3 Concrete ballasts

Concrete ballasts are mounted onto HDPE outfalls to provide stability against the hydrodynamic forces of the ocean and the buoyancy of polyethylene. Determination of the weight and spacing of the ballasts is covered in Section 9.4.1.

The ballasts can be fabricated at the outfall assembly site or at a remotely located concrete plant and hauled to the site. The cement must be specifically intended for a marine environment, and the concrete mix, particularly the cement content and the water-cement ratio, must be carefully controlled to ensure the hardened concrete's suitability for a marine environment. High early strength cement is commonly used to shorten the time between pouring and form removal. The manufacture of concrete ballast can be divided into five steps:

- Assembling the forms;
- Placing reinforcing steel in the forms;
- Pouring the concrete into the forms;
- Curing the concrete;
- Removing and cleaning the formwork.

The forms should be designed to facilitate rapid assembly and removal from the hardened concrete. They should be built to preclude bleeding of cement mortar and potential honeycombing effects. Since the forms will be used many times, they are usually made of metal. The number of forms should be compatible with the scheduled time between production of the ballasts and their attachment to the outfall so as to not delay other project activities.

The reinforcing steel should be bent and tied as a cage before it is placed in the forms. It should have an epoxy coating to help prevent salt water corrosion.

It is important that the reinforcing be carefully placed in the forms to ensure sufficient concrete cover.

It is critically important that the concrete be properly cured with adequate moisture at all times to obtain the designated strength and durability. Concrete ballasts should not be hauled or attached to the outfall until sufficiently cured to ensure adequate strength to withstand the associated stresses imposed.

The concrete ballasts should be individually numbered and the number should be correlated with the concrete sampling and testing. The numbers should be stenciled on the ballast upon removal from the form. They should be large enough to be easily read when submerged, so epoxy paint suitable for the marine environment should be used. The ballast numbers should also be correlated with their locations on the pipe.

Concrete ballasts are usually designed in two pieces that clamp directly to the pipe by means of bolts or post tensioned cables running through them. There are numerous variations on ballast shape, but the four basic ones are round, rectangular, starred, and trapezoidal, as shown in Figure 9.2. The round, rectangular, or starred shaped pieces can be split vertically or horizontally, but the trapezoidal shape is usually split horizontally to facilitate assembly (Figure 9.3).

The bolts must be either hot dipped galvanized steel or stainless steel. Both should be capped with a sacrificial zinc anode for corrosion protection. The mass of the anode should be sufficient for a useful life of at least 20 years.

There are two schools of thought regarding the annular space between the bolts and the holes in the concrete ballasts to accommodate them. One is to leave the annular space open so the bolts can easily be removed and replaced if necessary. The other is to fill the annular space with cement grout when the ballasts are being clamped to the pipe to provide better corrosion protection, but this must be done right the first time because the bolt cannot be readily removed once the grout is hardened. When the ballast pieces are fastened with post-tensioned cables, the annular space is always filled with cement grout and sacrificial anodes are attached to the fastening-tensioning system at each end of the cable.

A neoprene gasket is usually inserted between the concrete ballasts and the pipe when they are being clamped to the pipe. This is to prevent the ballasts slipping downward during pipe submersion and to protect the HDPE pipe material from abrasion by the rough concrete. To prevent slipping, the gasket is usually either glued to the ballast, or the ballast is designed with a recess into which the gasket is fitted before the ballast is fastened to the pipe.

The methods used to mount the ballast onto the outfall depend mainly on the pipe diameter and the ballast weights. Large ballasts usually require a crane, as illustrated in Figure 10.9, or a specially equipped backhoe to lift and place the

Ocean outfall construction

ballast on the pipe. For outfalls with diameters 300 mm or less it is usually possible to mount the ballasts by hand using a simple locally constructed gantry as shown in Figure 10.10 that was used for launching a 3,800 m HDPE pipeline in Placencia, Belize.

Figure 10.9 Mounting concrete ballasts on the Loma Larga outfall, Valparaiso, Chile (courtesy Belfi-Montec)

Figure 10.10 Locally constructed gantry used for launching a small diameter submarine pipeline 3,800 m long for Placencia, Belize

318 Marine Wastewater Outfalls and Treatment Systems

The concrete ballasts can be attached to the pipe at different locations. For outfalls assembled onshore, they are normally attached immediately to each length of pipe after it has been joined to the previous string. For long pipe lengths that are towed directly from the manufacturing factory to the job site, the ballasts can be attached offshore as shown in Figure 10.11. This is done by means of a special barge that lifts the pipe out of the water over the bow, feeds it across the deck where the ballasts are attached, and then feeds the assembled outfall over the stern and back into the water. This has been done for pipe diameters up to 2 m. Another barge is needed to transport the ballasts from shore to the attachment barge. Supplemental ballasts are sometimes attached to an outfall after submersion to enable underwater mounting.

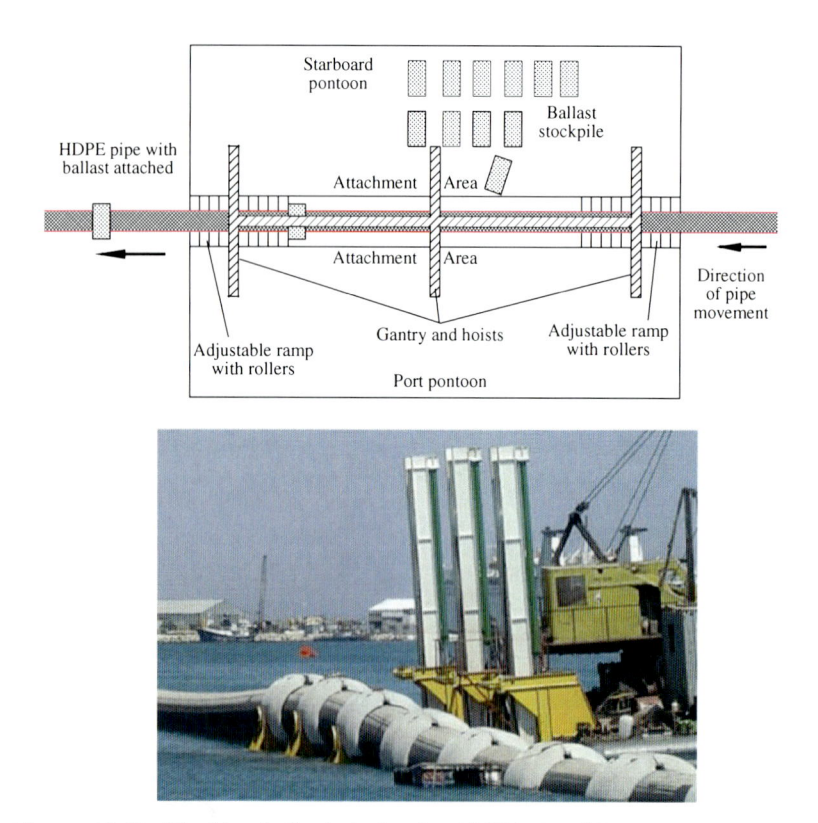

Figure 10.11 Attaching ballasts to floating HDPE pipe (Photo courtesy EDT)

10.3.1.4 Testing the outfall

The quality of butt fused joints of an HDPE outfall must be tested before it is towed into place for submersion. This is usually done by an ultrasound scan or hydraulic pressure testing.

Ultrasound testing requires specialized equipment and a qualified technician that can operate it and interpret the results. It should be carried out around the outside circumference of the pipe immediately after completion of each joint. If the inspector has any doubts about the weld quality the test should also be conducted from the inside of the pipe. The major advantage of ultrasound testing over hydrostatic testing is that the joints are tested immediately after the newly formed joint is cooled. Then if a defect is found the joint can be cut out and immediately redone. This requires much less set up time for correcting a defective joint than that which follows hydrostatic testing.

The hydraulic pressure test is usually considered more reliable, particularly if it is conducted at a pressure that is 50% greater than the pressure rating of the pipe being tested. In this test the outfall is first tightly sealed at both ends and then filled with water, being careful to evacuate all air from the pipe. A pump (usually a positive displacement pump) is then used to pressurize the pipe to the designated level and hold that value for a specified duration. The test is usually conducted on land, where all leaks are readily visible. If the test is to be conducted while the outfall is on water it is a good idea to add a visible dye, such as Rhodamine, to the test water to enable rapid identification of any leaks. The major disadvantage of this test is that if a defective joint is found, the removal and re-welding process can be difficult and time consuming because the entire outfall must be handled.

10.3.1.5 Launching the outfall

After passing the pressure test, the water is evacuated from the pipe and its ends are tightly sealed to entrap the air. A special pulling head, such as shown in Figure 10.12, is then attached to the end of the pipe for towing by a tugboat. It must be designed and fabricated for the particular pipe being used to enable it to safely transfer the pulling force from the tugboat to prevent pipe buckling during submersion as well as provide the force necessary to tow the outfall to its final location.

The end seals must be suitable for the maximum internal pressure that will be encountered during submersion. They also must permit controlled entrance or release of air or water as necessary during submersion. This is usually accomplished by blind flanges fitted with ball valves for controlling the flow of air and/or water, and pressure gages for monitoring the internal air pressure during submersion.

Figure 10.12 Pulling head for pipe launching

The height and length of the structure from which the outfall is launched depend on the height of breaking waves and tidal variations. It is better to construct the launching structure a little higher and longer than seems necessary than to risk the possibility of a sudden unexpected deterioration of sea conditions that could suspend the launching operation while it is in progress.

Long, large diameter outfalls with heavy concrete ballasts usually require more force to launch them than can be applied by a tugboat. The additional force is commonly applied by bulldozers or a system of winches, pulleys, and cables.

To execute the launching as planned, prior coordination meetings with all personnel in charge of the boats as well as those in charge of the onshore activities are essential.

The window of opportunity for launching, towing, and sinking an HDPE outfall by flotation and submersion is frequently narrow so it is very important to avoid delays or problems that might jeopardize the operation. Preparation of a detailed checklist of all critical items, and verification that each is taken care of before launching commences is essential for success. The following checklist includes most items that are commonly included:

- Check meteorological forecasts and projected ocean conditions for the potential launch dates and hours, particularly wind, waves, tides, and currents at the launch site, the towing route, and the submersion site to ensure that conditions are likely to be favorable;
- Confirm that all the ballasts are in their correct position and tightly affixed to the pipe to preclude slipping, and verify that a numbering system for future reference has been stenciled on them;
- Confirm that sacrificial anodes for corrosion control have been fastened to each ballast bolt;
- Check and confirm the secure fastenings of the pulling head to the outfall and the towline connected to the tugboat;
- Confirm that equipment and machinery used for pulling and braking the outfall as it is being launched are in their proper positions and are fully operational;
- Confirm that the air compressor to be used in controlling the rate of submersion has sufficient capacity to supply air at the necessary rate and pressure, that the compressor is in good working order, and that it is in or will be in its designated location;
- Confirm that the air and water valves used for controlling the submersion are the appropriate size and that they are properly installed and operational;
- If a pump is to be used to inject water during submersion confirm that it is in place and operational;
- Confirm that marker buoys are in place to guide the transport and positioning of the outfall;
- Confirm that the boat that is responsible for directing the outfall positioning has the necessary GPS, multibeam sonar, and equipment to communicate with the divers;
- Confirm that the communication equipment between the divers and the boat controlling the air release-injection is operative.

10.3.1.6 Towing and positioning the outfall

After the outfall is completely assembled and it has passed the testing protocol, has been launched, and the sea conditions are suitable, it is towed from its launch or storage site to its final submersion location. In preparation for this, it is important to determine and explore the route for towing the outfall well in advance. When planning the route, nautical and bathymetric charts should be consulted to disclose any problem areas such as submerged reefs and narrow passages. Areas suitable for emergency submersion and reflotation of the outfall, in case of unforeseen rough sea conditions or other unanticipated problems, should be identified in advance. It is helpful to mark the route with temporary buoys prior to towing particularly near potential problem areas. Towing should be planned so that the outfall arrives at its submersion site at a time of day when the sea conditions are anticipated to be most favorable.

The towing must be coordinated with the coast guard or naval authorities, harbor masters, or other pertinent maritime authorities. Measures should be taken to minimize adverse effects that transport of the outfall may have on other users of the sea. This becomes increasingly important with very long outfalls.

Towing and positioning an outfall for submergence requires a number of vessels as illustrated in Figure 10.13. Most critical is the tugboat that is attached to the pulling head to tow the outfall. This vessel should be equipped with GPS navigation to ensure that the chosen route is followed. It must have sufficient power to pull the outfall against currents, waves, and wind and also provide the necessary axial force to the outfall as it is being submerged. Smaller attendant vessels are also necessary at appropriate intervals along the outfall to keep it from drifting too far off course because of currents, waves, or wind. These vessels are critically important if the outfall has to pass through narrow curved channels, or around reefs, islets, or other obstacles. Their number and size should be appropriate for the outfall length and diameter, anticipated sea conditions, and the complexity of the chosen route.

There is a considerable difference in the towing operation for an outfall that is being launched from shore near its final location and one that is towed a long distance from a remote site. A long route from a distant location often requires a secondary tugboat at the trailing outfall end, especially for large diameters. Towing an outfall after launching it from a nearby site rarely requires more than one tugboat.

10.3.1.7 Submerging the outfall

After the outfall is floating on the ocean surface directly over its designated alignment, water is introduced into one end of the pipe and air is released from the other end causing the section with the introduced water to sink to the seabed. Figure 10.14 illustrates the sequence of submersion from shore to deeper water which is the usual direction. It is possible to sink an outfall from deep water towards shore but this requires a high rate of air injection at the opposite end to offset rapid sinking due to the sudden large water inflow caused by the rapidly increasing pressure as the outfall sinks.

Figure 10.13 Floating and positioning an outfall for submersion in Quintero, Chile

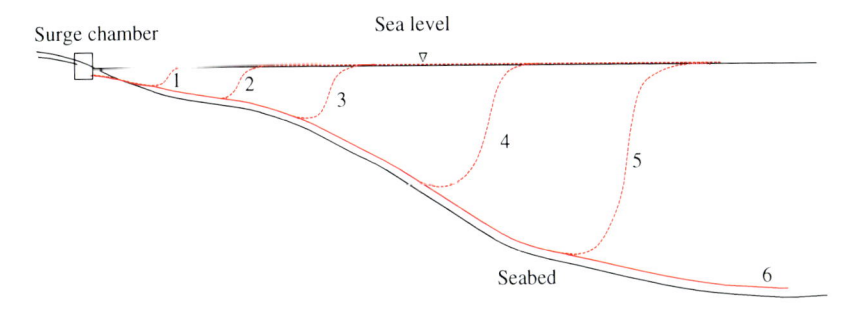

Figure 10.14 Submersion sequence for an HDPE outfall

The rate of submergence is controlled by adjusting the air release rate and/or the water inflow rate. As the outfall is submerged, it is positioned by divers that observe its descent and instruct the boats to move it.

Submersion must be done in a carefully prescribed manner to prevent damage to the outfall. The air pressure must be carefully controlled and adjusted according to the load imposed by the ballast weights, the water depth and the pulling force applied to the outfall so as to avoid a bending radius that becomes smaller than its allowable minimum. This is necessary to prevent the pipe wall buckling.

During submersion the water entering the pipe at one end will cause that end to sink and as the water advances in the pipe it will compress the air trapped in the pipe. This results in an S shape, illustrated in Figure 10.15, between the portion of the outfall resting on the seabed and that floating on the surface.

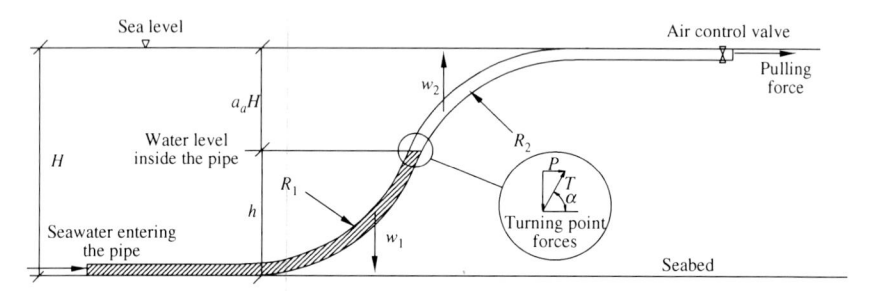

Figure 10.15 Relationship between air fill ratio and forces acting on an HDPE submarine outfall during submersion

The sinking rate should be maintained slow and nearly constant to prevent undue accelerations that result in forces that overly stress the pipe. A rule of thumb, based on practical experience and commonly recommended, is a submergence of 0.3 m per second, which is slightly more than 1,000 m per hour. Various formulae are provided by HDPE pipe manufacturers that may yield higher maximum sinking speeds, but because they have no theoretical basis and involve several assumptions and approximations, the authors recommend a sinking rate of 0.3 m per second even though it is somewhat conservative. Furthermore, other considerations such as waves, currents, visibility, and ensuring proper outfall alignment make this a reasonable sinking velocity.

It is important that the radius of curvature of both the upper and lower parts of the outfall become less than the allowable minimum. Rearranging Eq. (9.14), and including a safety factor of 1.5, the minimum allowable bending radius can be expressed as:

$$R_{min} = 1.34 d_e (DR - 1) \qquad (10.1)$$

where

R_{min} is the minimum allowable bending radius (m)
d_e is the external diameter of the HDPE pipe (m)
DR is the dimension ratio of the pipe (d_e/s)
s is the wall thickness (m).

Maintaining an adequately large bending radius during submersion is usually accomplished by applying a pulling force on the floating end of the pipe. The required force depends on the water depth and the ballast weight, also known as the degree of loading, needed to withstand the hydrodynamic forces imposed by the ocean. The determination of that ballast weight is covered in Chapter 9.

The air fill ratio is a parameter used to determine the relationship between the radius of the upper and lower portions of the S curve and the pulling force on the pipe. It is the ratio of the sum of the submerged weights of the concrete ballasts plus the submerged weight of the pipe per unit length to the buoyancy of the air volume per unit length in an air filled pipe. This is conceptually illustrated in Figure 10.15.

The air fill ratio, a_a, is defined by:

$$a_a = \frac{w_{cs} + w_{ps}}{\pi \frac{d_i^2}{4} \gamma_w} \qquad (10.2)$$

where

a_a is the air fill ratio
w_{cs} is the weight of concrete ballast submerged in seawater per unit length
w_{ps} is the submerged weight of pipe in seawater per unit length
d_i is the internal pipe diameter
γ_w is the specific gravity of the seawater

The air fill ratio tells us what fraction of the internal pipe volume has to be filled with air to maintain neutral buoyancy of the outfall with the ballast weights attached to the pipe.

The pulling force to maintain a radius larger than the allowable minimum can be estimated by:

$$P_1 = w_1 R_{\min} \qquad (10.3)$$

$$P_2 = w_2 R_{\min} \qquad (10.4)$$

Equation (10.3) (P_1) is used when the air fill ratio is equal or greater than 0.5 and Eq. (10.3) (P_2) is used when the air fill ratio is less than 0.5. In these equations:

P_1 is the necessary pulling force (N) for $a_a > 0.5$
P_2 is the necessary pulling force (N) for $a_a < 0.5$
w_1 is the net weight of the water filled section (N/m)
w_2 is the net buoyancy of the air filled section (N/m)

and w_1 is determined from:

$$w_1 = a_a \pi \frac{d_i^2}{4} \gamma_w \tag{10.5}$$

where

a_a is the air fill ratio
d_i is the internal diameter of the pipe (m)
g_w is the specific gravity of the ambient water (N/m^3)
The relationship between w_1 and w_2 is given by:

$$w_2 = \frac{1-a_a}{a_a} w_1 \tag{10.6}$$

Keeping the sinking velocity constant usually entails controlling either the rate that air is released from the pipe or the rate that pressurized air is injected into the pipe. Whether release or injection is necessary is determined mainly by the weight of the ballasting and the seabed topography. For example, in the case of submersion away from shore onto a seabed with a uniform gentle slope with relatively light ballast weights, the sinking velocity can usually be controlled simply by adjusting the rate of air release. If the ocean depth increases rapidly nearshore, due to a steep bottom slope or an escarpment, but thereafter continues at a gentle slope for a long distance and where heavy ballasting near the shore is required for stability against the hydrodynamic forces of large waves, it will be necessary to inject air into the outfall to control the initial phase of sinking. This is because the heavy ballasting requires only a small depth (or volume) of internal water to give negative net buoyancy, and the volume of air being compressed is very large compared to the volume of water entering. There is then relatively little increase in air pressure (in accordance with Boyles Law) so the pressure differential will rapidly increase and the submersion will accelerate unless measures are taken to prevent it. Controlling the water inflow could slow the rate of descent, but this by itself is not satisfactory because the ambient water pressure will soon exceed the air pressure inside the pipe, possibly causing it to buckle.

This situation can be mitigated in one of two ways. One is to inject air under pressure to balance the internal and external pressures. Another is to add supplemental flotation to counterbalance the heavy ballasting, so that a large height of water inside the pipe is required to submerge it.

If the outfall does not settle properly onto its designated alignment, it will be necessary to refloat it by injecting air and releasing water from its other end so that it can be repositioned.

10.3.1.8 *Post submergence activities*

An underwater inspection by divers should be done immediately after the outfall is installed to confirm that it is resting in its correct position. Special attention should be given to placement in an excavated trench to confirm that it rests on the bottom center of the trench. The inspection should ensure that the concrete ballasts are resting on the seabed and none are suspended above it. It should also ensure that the bottom of the pipe is not pressing against rocks, coral formations, or other objects that could abrade, puncture, or otherwise damage the pipe wall. It should also verify that no ballasts have been displaced by sliding down the pipe during submersion. It should verify that all sacrificial anodes for the ballast fastening bolts are in place. It is recommended that the inspection include videos taken from both sides of the outfall, with close-ups if necessary to show details. The videos should display the ballast numbers to accurately locate any problem areas.

If the outfall is installed in sections, inspections should be carried out immediately after each is installed with special emphasis on the connections between sections. It should ensure that all the sacrificial anodes are correctly in place on the bolts used for these connections.

Adjustments or corrections of any discrepancies should be carried out as needed. This may entail attachment of air-lift bags to facilitate lateral adjustment of outfall location. In extreme cases the injection of air to re-float a section or even the entire outfall and then submerge it correctly may be needed.

If more ballasting is required for stability than can be attached to a floating outfall, the supplemental ballasts should be mounted underwater at their designated locations at this time.

Trenches should be backfilled after a notice to proceed is given, based on the inspections.

10.3.2 Bottom pull method

The bottom pull method for installing submarine outfalls is conceptually depicted in Figure 10.16. It is applicable for steel pipe with welded joints, generally less than 2 m diameter, with or without protective coatings such as mortar or concrete. It is particularly appropriate for outfalls less than 500 m long insituations where the outfall can be assembled onshore, perpendicular to the shoreline at the outfall site, on a uniform gentle gradient and the entire outfall subsequently dragged into the ocean, directly into place on the chosen submarine route.

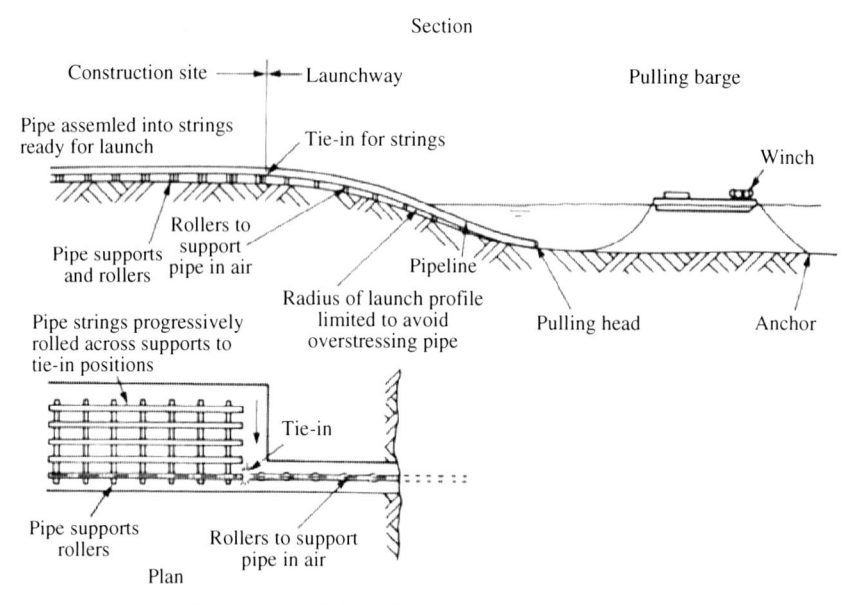

Figure 10.16 Bottom pull installation (Neville-Jones and Chitty, 1996)

Bottom pull has also been used successfully for outfalls as long as 4,800 m. This is done by fabricating several sections of the outfall parallel to each other on land, then pulling one section into the ocean, pausing, welding a second section onto the first section, testing the weld, preparing and applying the pipe coating to the welded area and then pulling the combined section. This is repeated until the whole outfall is in place. This entails pulling the entire outfall in the final stage which can require considerable force. An example is shown in Figure 10.17.

A variation of the bottom pull method used for longer outfalls is to fabricate the pipe sections with specially designed weld-neck anchor flanges on each end. Then the first section (usually the diffuser) is pulled into place at the far end of the outfall route. A subsequent section is then pulled up so that its leading end abuts the trailing end of the first section. The adjacent ends are then closely aligned for connection by underwater welding or mechanical joints by means of a specially designed alignment frame that is lowered into place from a support barge. Divers carry out the alignment, make the connection, test the joint, and complete the joint by applying a protective coating. This process is repeated with succeeding pipe strings until the entire outfall is installed. This method requires highly trained, experienced divers to make and test the underwater connections.

Figure 10.17 Steel outfall 1.34 m diameter, 4,800 m long to be installed by bottom pull in Salvador, Brazil

All bottom pull methods require construction of a temporary launch way onshore to support the pipe and prevent damage to the pipe or coatings as it is being dragged over land. Various types of launch ways have been used such as rails and dollies, rollers, and skids. When the seabed is rough and sharp and there is possibility that it might damage the protective coating, skids are fastened to the outfall as it is being dragged into place.

Bottom pull requires an easily detachable and re-attachable pulling head specifically designed to securely attach the pulling cable(s) and preclude damage to the end of the pipe string(s). The pulling operation is usually accomplished in one of two ways. One is by a powerful winch located onshore with steel cables that run through a pulley that is securely anchored beyond the outfall end. Another is by a securely anchored offshore barge or tugboat that either pulls the pipe toward the barge with a winch and cable or by attaching to the pulling head with cables and pulling itself towards previously set anchors. Bottom pull methods are limited to straight outfall routes; routes with curves are not possible.

The length of pipe that can be pulled by these methods depends on the submerged weight of the pipe, friction between the pipe and seabed, the power of the winch, the strength of the cable, and/or the allowable tension on the outfall pipe itself. Temporary supplemental flotation is frequently attached to the outfall pipe during pulling to increase its buoyancy. This both lessens friction, thereby reducing the force needed to pull the pipe string, and reduces the likelihood of damage to the underside coating on the pipe. Experience has shown that if the submerged weight of the outfall is more than about 25 kg/m^2 of contact area with the seabed, damage to a coating is likely.

When bottom pull is used in conjunction with subsea trenches, it is important to use marker buoys and/or GPS systems as well as divers to monitor the installation to properly align the outfall in the trench. It is also advisable to install the outfall as soon as possible after the trench has been excavated to avoid significant movement of material into the trench by wave and current action. This is especially important in the surf zone of sandy shorelines because lengthy delays can result in considerable accumulation of sediment in the trench, resulting in the outfall being laid too high and not being fully protected by burial.

10.3.3 Mobile jack-up platforms

Mobile jack-up platforms (Figure 10.18) are primarily used for oil field drilling operations but have also been used successfully for placing outfalls on the seabed in waters up to 60 m deep. Such platforms are quite large, often more than 50 m long, by 30 m wide, by 5 m deep. Because they are quite costly, their use for submarine outfalls has generally been limited to the installation of concrete pipe with diameters larger than 2.5 m.

The platforms are equipped with four long legs that can be extended downward or raised upward by a jacking system, a crane for handling and connecting the pipe sections pipe (and excavating the seabed if necessary), and a bin and tremie for placement of bedding or backfill material. The platform is floated into place with the legs raised and large fore and aft anchors are attached to maintain it in position. Then the legs are lowered to the seabed and the platform is jacked up and leveled a sufficient height above the ocean surface so that it is above the highest anticipated wave.

Installation of outfalls by bottom assembly is used in conjunction with jack-up platforms. This requires highly skilled divers to control and guide the underwater activities. While the seabed is being prepared by excavation and/or placement of bedding rock, several lengths of pipe are connected together on the deck of the platform and the string is then attached to a strongback, a heavy beam. The assembly is lifted by crane and lowered into place on the seabed. It is then joined to previously placed pipe by pulling it with a cable attached to the aft anchor(s). Alternatively, pipe lengths may be placed one at a time. After a number of pipe sections are installed, ballast rock is placed by use of a tremie, guided by the divers.

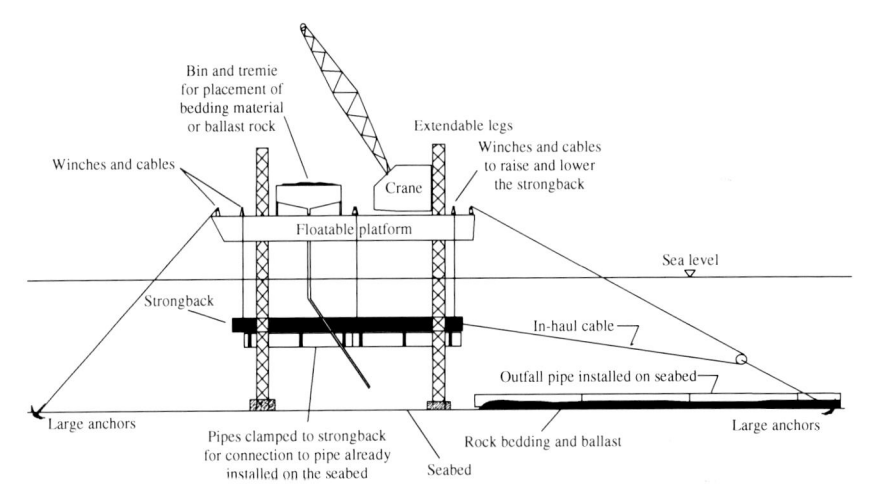

Figure 10.18 Mobile jack-up platform concept

To continue the laying operation it is necessary to move the platform forward. This is accomplished by lowering the floating platform into the water, jacking the legs up until they are well above the seabed and then pulling the platform forward with the line to the fore anchor and releasing line on the aft anchor. This procedure is repeated until the entire outfall is installed.

The main advantage of this method is that the platform on which the crane is operating is very stable and this facilitates assembly of the pipe sections and placement of bedding material and armor rock. However, alignment and insertion of each pipe section requires considerable care and expertise, and the difficulty of doing this without damaging the pipe ends increases greatly if there are bottom currents or wave surges.

10.3.4 Outfall installation from a lay barge

Lay barges (Figure 10.19), which are commonly used by the offshore oil and gas industry to install pipelines in water as shallow as 2 m and as deep as 200 m, have also been used for ocean outfalls. They are used primarily for pipe that has some flexibility in the joints between pipe sections or has some flexibility itself. It has been used to successfully installation of steel pipe; concrete coated steel pipe, and ductile iron pipe. It can also be used for HDPE or polypropylene pipe but would probably only be advantageous over the float and sink method for outfalls with concrete weights that exceed the maximum flotation capacity of the empty pipe. This method of installation is limited to weather and sea conditions that are not too severe.

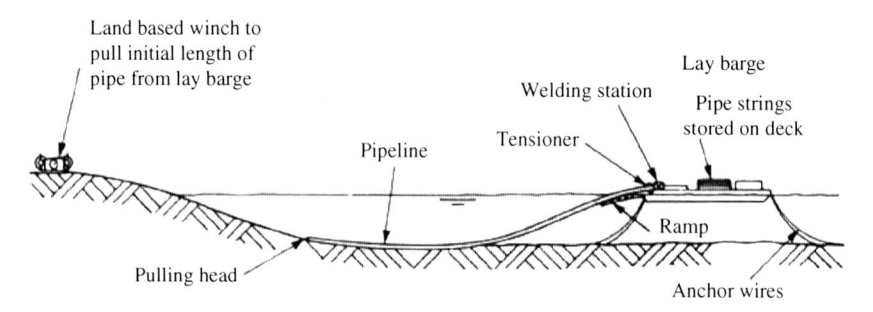

Figure 10.19 Continual placement from a lay barge
(Neville-Jones and Chitty, 1996)

At least two vessels are involved, the lay barge and an attendant barge. The attendant barge ferries personal and pipe sections, fittings, and other materials from shore to the lay barge. The lay barge is positioned and held in place by fore and aft lines attached to large anchors that were previously set (by the attendant barge or a tugboat) on the designated route. The pipe sections are joined together on the deck or working platform in assembly line fashion. After a joint is satisfactorily completed the connected pipe is fed over the aft end and the barge is progressively moved forward over the designated route by taking up line on the fore anchors and feeding it out on the aft anchors. For deep installation, the lay barge is equipped with an axial tensioner located in the assembly line before the end of the barge. This maintains an axial force in the pipe to enable it to descend to the seabed without violating the minimum allowable curvature so as to avoid buckling of the pipe, failure of a pipe joint, or spalling off of the pipe's protective coating.

Ocean outfall construction

The aft end of the lay barge is equipped with a ramp (commonly called a stinger) that can be raised or lowered into the sea. The purpose of the stinger is to preclude excessive flexure and stress in the pipe or pipe joint as the pipe is being fed out. The type of pipe being installed, the allowable pipe curvature, the water depth, and the anticipated sea conditions determine the length, angle, and design of the stinger. There are three basic types of stingers being used: straight, curved, and articulating that can change shape. In general, the nearer the end of the stinger to the seabed the less the stress in the pipe.

Successful placement of the outfall pipe on the seabed depends on secure anchoring because excessive barge movement can not only cause the pipeline to drift off course but also excessive stress in the pipeline as it is being lowered. Extreme uncontrolled movement can cause failure of the stinger. It is common practice to set 8 to 12 anchors in water deeper than 50 m and 6 to 8 anchors in water shallower than 50 m. The anchors must be heavy enough to prevent slippage and are attached to barge winches with heavy steel cables. Accurate barge positioning is essential; GPS are now commonly used for this purpose.

A typical lay barge can install pipes up to 1.2 m diameter but some have been modified for even larger diameters. Because of the high costs associated with lay barges their use has primarily been for longer outfalls. The rate of laying pipe with lay barges is mostly controlled by the time required to join pipe sections. It is thus disadvantageous to install pipe that requires a long time to make and test a pipe joint.

10.3.5 Installation from a floating crane barge

The floating crane method is a bottom assembly technique that has been used for pipes with a large range of diameters made of various materials including reinforced concrete, prestressed concrete, cement coated steel, cast iron, and GRP. Compared to other methods it is slow and expensive. A crane is mounted on a barge that is located over the installation site and anchored in place in a similar manner to that previously described for the lay barge. An attendant barge ferries personnel, pipe sections, fittings, and other necessary materials from shore to the crane barge. The assembly and connection of the outfall pipe is carried out in the same manner as for the jack-up platform. A separate barge is also used for bottom preparation such as excavation placing of bedding material, and placement of armor rock.

The main disadvantage of this method is that even slightly rough sea conditions can move the barge enough to make if difficult if not impossible to align the pipe section being lowered to that already installed, and it greatly increases the likelihood of damage to the pipe ends. Overcoming this necessitates the use of a special handling frame to which a section of pipe is attached before being placed in the water. The pipe is suspended in the handling frame with hydraulically activated travelers that can move the pipe vertically and horizontally at either end and also move it back and forth. The frame with the pipe attached is lowered by the crane onto the seabed into position near the end of the assembled pipe. The hydraulic rams then move the pipe into position and align it to the end of the previously installed pipe. After the joint is completed, the frame is detached from the pipe, lifted back onto the barge and the process is repeated.

The alignment and joint completion require visual observation. Some handling frames require the use of divers to communicate instructions to an operator located on the barge. A disadvantage of this is the limited time that divers can remain on the bottom and the uncertainties associated with indirect observation. Some frames are equipped with a water tight observation-control chamber at atmospheric pressure so the operator can directly observe the joint being made and not be limited by diver bottom time.

10.4 TEMPORARY TRESTLES

Local geomorphology as well as the difficulty of contending with breaking waves, tides, currents, varying winds, and movement of sand or bottom sediment often necessitates construction of a temporary structure. It is also often necessary when the outfall requires entrenchment from the shoreline to the ocean and/or nearshore waters that are too shallow for pipe laying barges. A temporary structure is also usually required for launching an outfall that has been assembled on land into the ocean.

A temporary trestle-type pier can serve all of these purposes for any type or diameter of outfall pipe. It can be used as a stable platform to support heavy equipment for excavation of a subsea trench, to shore up the walls of excavated trenches prone to collapse, to block wave or current-transported sand and sediment from filling the trench before the outfall pipe is installed in it, and as a launching ramp for an outfall assembled on land. It can also serve as a stable platform to support bottom installation of pipe one length at a time, and from which to place bedding material or armor rock. The trestle must of course be designed to support the heaviest loads imposed by the equipment to be used on it as well as withstand the hydrodynamic forces due to waves and currents.

The trestle can be used for launching an HDPE pipe for flotation and submergence as shown in Figure 10.20. In that case, the terminal end must be fitted with a ramp or rail system that provides a gradual transition into the water to ensure that the minimum allowable bending radius of the pipe is not violated. The ramp should extend sufficiently below low tide to ensure that the pipe and concrete ballast are floating before they reach the end of the ramp.

Figure 10.20 Temporary launching trestle for HDPE pipe

Trestle construction usually involves driving steel H piles into the seabed at prescribed longitudinal and lateral intervals. Horizontal steel beams are bolted between the piles to form pile bents, and longitudinal steel beams are bolted to the bents to form the support for the working platform and any rails on which the cranes, backhoes, trucks, etc will operate. The trestle continues seaward until the designated water depth or distance from shore is reached. It is imperative that the platform be well above the maximum anticipated wave height to preclude damage or destruction during heavy seas.

If trenching beneath the trestle is required for the outfall pipe, sheet piling can be driven into the seabed to the prescribed depth by a second crane, usually with a vibratory hammer, while the trestle is being advanced. The seabed material can be excavated from between the parallel walls formed by the sheet piling using the crane with a clamshell, a long boom backhoe, or a hydraulic dredge while the pier is being advanced seaward.

10.5 SUBSEA EXCAVATION

The transition of an outfall from land to the ocean almost always requires subsea excavation. The reasons include, but are not limited to:

- Entrenchment of the outfall pipe to protect it from oceanic forces, damage from ship anchors, and impact from submerged objects driven by waves and currents;
- Removal of obstacles such as reefs and rocks from the selected alignment;
- Maintaining a constant downward slope and elimination of high points that could cause air to be trapped in the outfall pipe;
- Elimination of abrupt changes in the slope at escarpments or other anomalies that would cause the outfall to violate curvature limitations;
- Removal of seabed material that is too weak or otherwise unsuitable to support the outfall.

Subsea excavation can be a difficult, dangerous, time-consuming, and expensive endeavor and can account for a significant portion of the total cost of outfall construction, particularly where sea conditions and the geomorphology of the seabed are unfavorable for excavation. If it is not thoroughly planned, carefully scheduled, and properly executed it can result in major project delays, difficulties, and cost overruns. Submarine excavation should ideally be scheduled for the time of year that historically has the most favorable sea conditions. Qualified personnel with subsea excavation experience are required to carry out this work in an efficient and safe manner.

The equipment and excavation methodology must be appropriate for the specific situation. The following factors must be considered:

- Type of pipe used for the outfall;
- Class of material to be excavated;
- Water depth;
- Depth of the excavation;
- Volume of material to be excavated;
- Distance from the shoreline;
- Anticipated wind and wave conditions;
- Environmental restriction and limitations;
- Uses of the ocean nearby.

10.5.1 Excavation in an unconsolidated seabed

The term unconsolidated seabed refers to seabed material composed of sand, silt, clay, gravel, and/or cobbles. All of these materials can be excavated with draglines, clamshell buckets, and backhoes. Sand, silt, and gravel can also be excavated with cutterhead dredges, hydraulic dredges, and specially constructed sub-sea trenching machines. Excavation with hydraulic jets and jet sleds is limited to sand and silt.

Dragline buckets (Figure 10.21) can be operated by a crane or can be drawn by a system comprising an onshore winch, cables, and pulleys anchored offshore. An onshore crane's usefulness is limited to nearshore excavation according to the length of its boom. The winch-cable-pulley system, however, can excavate from shore out to the site where the pulley is anchored. Its use is primarily for a seabed of sand, hard clay, gravel, or cobbles.

Figure 10.21 Dragline bucket

For ocean outfalls, dragline buckets are limited to straight trenches in calm waters where there is little likelihood of bottom material moving back into the trench due to waves or currents. In this system the empty bucket is first drawn backwards offshore by a cable fed through the offshore pulley and then pulled shoreward, filling it as it is dragged over the seabed. The bucket contents are then dumped on land.

Dragline bucket excavation has two major disadvantages:

- The long distance that the bucket must travel before it is dumped makes it relatively inefficient;
- It must excavate the entire trench before the outfall can be installed.

338   Marine Wastewater Outfalls and Treatment Systems

Clamshell buckets (Figure 10.22) are suspended from the boom of a crane with steel cables that can open or close the jaws and raise and lower the bucket. The crane drops the bucket, with the jaws open, onto the seabed and the impact digs the jaws in. When the bucket is lifted, the jaws close and take a bite out of the seabed that fills it. The bucket is then lifted and moved by the crane's boom to discharge to a haul barge or sidecast to the seabed at a sufficient distance from the excavation.

Figure 10.22 An open clamshell bucket

Because the crane must be located close to the excavation it is mounted on a barge, a jack-up platform, or a trestle. This method is commonly utilized for excavating the material between parallel rows of sheet piling in conjunction with a trestle. It can be utilized from the shore up to a depth of about 50 m.

Hydraulic suction type dredges with jet nozzle cutter heads (Figures 10.23, 10.24, and 10.25) are used extensively for excavating trenches in seabeds consisting of sand, silt, or clay. They use powerful water pumps that discharge high velocity water streams through multiple jet nozzles in a cutter head to erode the seabed. The head includes a bucket which is attached to a pipe connected to the suction side of pumps that "vacuum" up the slurry and discharge it away from the trench area. As the dredge moves forward the cutter head position and the excavation depth determined from a computerized control room on the ship maintain them at their prescribed values.

Ocean outfall construction 339

Figure 10.23 A modern jet nozzle suction dredge (courtesy of Bob Dunbar)

Mechanical cutter head suction dredges are similar to jet nozzle cutterhead dredges but utilize a rotating cutterhead at the end of the suction pipe to mechanically excavate the seabed. Figure 10.26 shows a small dredge ready to be launched from a dock into the water. The cutterhead is powered by a shaft attached to a motor located above water. The cutterhead greatly increases the rate of excavation of sand, silt, and gravel and can also excavate clay, hardpan, and even soft rock. Large cutterhead dredges with 10,000 hp pumps and 2,500 hp motors can excavate as much as 3,000 m^3 of material per hour and discharge it as far away as 5 km. The depth in which they can operate depends upon the dredge design. Some are capable of working in ocean depths up to 60 m, but smaller units have successfully excavated trenches in depths as little as 1 m.

Jet sleds are mainly used for trenches beneath a pipeline *after* it has been placed on the seabed. They are limited to seabeds of sand, silt, small gravel, and soft marine clay. This equipment has multiple jets and eductors mounted on both sides of a frame that is designed to straddle a pipe resting on the seabed. High-pressure pumps located on a surface-support barge supply water to the jets and eductors. The high velocity water jets are directed inward and downward beneath the pipeline to loosen and erode the seabed material beneath both the pipeline and the frame of the sled.

The support barge tows the sled along the pipeline, and eductors suck up the slurry created by the jets and both the jet sled and the pipeline settle into the excavation. The eductors can either be aimed to cast the slurry sideways so that the suspended material settles back onto the seabed on either side of the trench, or, if the excavated material is suitable for bedding the pipe, it can be directed to backfill around the pipe after it is resting in the bottom of the newly excavated trench. Figure 10.27 is a bottom view of the jet nozzles on the runners (skids) of a large jet sled and Figure 10.28 shows deployment of a small jet sled from the stern of the control barge.

Figure 10.24 Jet nozzle cutter head and suction pipe (courtesy of Bob Dunbar)

Jet sleds are equipped with guides or sensors that signal the operators on the support barge of the sled's alignment with the pipe so it can be constantly monitored and adjusted while the sled is moving. Larger jet sleds can excavate up to two m of depth on the first pass in sand or silt seabeds; deeper excavations usually require several passes. Jet sleds are only appropriate for pipe that can withstand some deformation during settlement into the trench without damage to the pipe wall, joints, or coating. The efficiency of jet sleds decreases rapidly with increasing water depth. They are mainly used in depths less than 100 m but have been used in the petroleum industry in depths up to 300 m.

Figure 10.25 Testing the jet nozzle cutter head on a hydraulic suction dredge (courtesy of Bob Dunbar)

Figure 10.26 Small mechanical cutter head dredge (courtesy of Mud Cat)

Jet sleds are occasionally used for excavating a trench before the pipeline is installed but this loses the advantage of excavating and backfilling on a single pass. This use is also limited to calm protected water where there is little possibility of wave and currents washing the excavated material back into the trench.

Figure 10.27 Jet nozzles on the bottom of a jet sled (courtesy of Bob Dunbar)

Figure 10.28 Deployment of a small jet sled from the stern of the control barge (courtesy of Bob Dunbar)

10.5.2 Subsea excavation in rock

The term rock is used in this section to refer not only to igneous, sedimentary, and metamorphic rock, but also to hard pan, caliches, hard coral, and large cobbles and boulders. These materials cannot be excavated with the methods covered under the previous section on unconsolidated seabeds and require the use of explosives (blasting), hydraulic percussion drills, pneumatic percussion drills, or chippers (also called drop hammers).

There are two main categories of blasting that are used for subsea excavation: drilled holes charged with explosives, and shaped charges. Both have been used successfully in excavations for ocean outfalls.

The composition of rock seabeds can vary widely. This can obviously occur between different outfall sites but also frequently occurs within an outfall site in both the horizontal and vertical direction. Layers of sand and/or clay can be covered with layers of hard limestone or sandstone. Volcanic, igneous, and metamorphic rock formations are frequently covered with a layer of coral or limestone in tropical seas. Some rock seabeds contain numerous voids and cavities. These and numerous other factors greatly complicate the choice of class of explosive, amount of explosive, spacing and depth of charges, and the sequence and timing of blasting. Subsea excavation with explosives is a highly specialized profession that requires considerable training and experience; it should never be attempted by anyone lacking them.

Sometimes, by use of different velocity explosives and detonation timing delays, it is possible to blast and shatter rock and lift most of it out of the excavation in a single sequence. More often, however, it is necessary to clean some of the shattered rock out of the excavation. The size and quantity of the shattered rock to be removed dictate the method of removal. This can be by jet eductors, clamshell buckets, grappling claws on cables connected to cranes, or even by divers.

After each blast the site should be inspected by qualified specialists to determine the next sequence of events. This includes whether or not additional blasting is needed to achieve the desired depth and width of excavation, the method used to remove loosened material from the trench, and whether or not to authorize immediate continuation of the blasting excavation process.

An alternative to blasting that is sometimes used for smaller diameter outfalls traversing rock in shallow waters is use of hydraulic percussion drills (similar to pavement breakers). These are attached to a backhoe boom with the backhoe mounted on a barge or jack-up platform. After the rock is sufficiently fractured by the percussion drill, a bucket is attached to the boom and the loose rock is removed. The barge is then moved over the next site to be excavated and the process is repeated. Another alternative is to fracture the rock with a chipper dropped by a crane that is mounted on a barge or jack-up platform. After a sufficient volume of rock is adequately fractured, the chipper is removed and a clamshell bucket attached to the crane for removal of the fractured rock. An advantage of the crane and chipper is that they can operate in deeper water than the backhoe. The advantages of the backhoe and percussion drill are that the rock is fractured more quickly, the trench walls are more vertical, and the trench width can be more accurately controlled.

10.6 INSTALLATION OF MECHANICAL ANCHORS

It is sometimes necessary to install an outfall where it is not feasible to excavate a trench and the hydrodynamic oceanic forces exceed the holding capacity of concrete ballasts. In that case, mechanical anchors can be used to fasten the outfall to the seabed (see Section 9.4.2). For small to medium size outfalls these anchors can either supplement the concrete ballasts or be used alone. Driven or screw-in mechanical anchors are applicable for seabeds comprised of sand, silt, or clay. They are usually set in place from a barge or a floating platform. Expansion and grouted anchors are appropriate for solid rock seabed. They need a diver to drill holes into the rock seabed and set them in place. When mechanical anchors are used it is good practice to test them after installation to ensure they can withstand the maximum anticipated pullout forces.

10.7 DIFFUSER INSTALLATION

The method used to install the diffuser depends on many factors including, but not limited to, the diffuser configuration and length, the diameter, type, and class of pipe, the water depth, and the method of installation used for the outfall.

The diffuser may be a single section or multiple bifurcated branches. For a bifurcated diffuser, the main outfall pipe terminates in a wye or tee that serves a connection point for the diffuser branches. It is also common practice to install a cross or double wye at the end of the outfall pipe and close off the outer stem of the cross with a blind flange. This can later be opened briefly to clean out the diffusers and as an exit for a cleaning pig.

A straight diffuser on a cement-coated steel pipe outfall to be installed by bottom-pull will simply be included as an integral part of the outfall. If the diffuser is bifurcated, the outfall with a cross, wye, or tee at the end will be pulled into place and then each diffuser branch attached to it. The diffuser branches are usually positioned with supplemental flotation, lowered into place by winch and cables, and bolted or welded onto the cross, wye, or tee. Risers from the diffuser ports, or "duckbill" check valves may be installed prior to launching the diffuser or after it is set into its final position on the seabed.

For installation by mobile jack-up platform or floating crane barge the diffuser is installed in the same manner as the rest of the outfall, namely one or several lengths at a time regardless of whether or not the diffuser is bifurcated. If risers or check valves are used they may be installed prior to submersion or after positioning the diffuser on the seabed. If risers are attached after the outfall is on the seabed it should be done after any bedding material or armor rock has been placed.

If the outfall is installed by a lay barge or reel barge, the diffuser is usually welded to the end of the outfall on deck and then fed down the ramp into the ocean. Because it is important that risers and check valves are properly oriented, ballast weights should be affixed to the bottom of the pipe to ensure that the diffuser is properly rotated. This can be aided by use of cables and winches while lowering the end of the outfall (the diffuser) onto the seabed. If a bifurcated diffuser is installed, the diffuser branches are installed using supplemental flotation, lowering them into position where they are attached by divers to the wye or tee.

There are a number of options commonly used to install HDPE diffusers. For a straight diffuser that is a continuation of the outfall pipe, the simplest and most common options are:

1. For a constant diameter diffuser the outfall pipe is first placed on the seabed. Then divers drill ports (holes) of the designated diameter, spacing, and horizontal angle in the pipe wall using a hand drill with a hydraulic or air driven motor and a special bit that removes the coupon. This method also allows increasing the number of ports with time as flows increase to obtain the desired discharge velocity and to minimize seawater intrusion.

2. A second option is to fabricate a diffuser from the same pipe used for the outfall before submersion. This is accomplished by welding ports with the desired internal diameter at specified intervals to a shell which is later welded to the pipe wall. The ends of these short pipes are sealed by thermal fusion. The diffuser is fastened to the main outfall pipe by butt fusion welding, concrete ballasts are attached and the entire assembly (the outfall pipe and diffuser) is installed by flotation and submersion. After the outfall is resting on the seabed divers cut off the sealed ends. This also makes it possible to initially open only those ports needed for the low flow early years and then open others to accommodate future design flows. An alternative is to open install "duckbill" check valves that prevent seawater intrusion and distribute the discharge over the entire diffuser even during periods of low flow. These check valves have also been installed with some of them vulcanized closed. As flows increase, the vulcanized ends are cut off by divers.

3. A third option is to fabricate on land a tapered diffuser of pipes with sequentially smaller diameters connected by reducers. The purpose is to obtain adequate cleansing velocities throughout the diffuser. The ports can be formed by either of the previous two options. The diffuser can be installed in the ocean as an integral part of the outfall by flotation and submersion or can be installed separately after placement of the main outfall pipeline. This option requires fabrication of different ballasts for each pipe diameter used in the diffuser.

For a bifurcated diffuser, each branch is fabricated by one of the three previously described methods and then installed individually by flotation and submersion.

11
Outfall installation by trenchless techniques

11.1 DEVELOPMENTS IN THE UNDERGROUND INSTALLATION OF PIPELINES

The possibility of installing an underground channel by tunneling, rather than by digging a surface trench, goes back thousands of years to the Ghanats in the Middle East. A 1000 m tunnel was driven under the Euphrates River in 2500 BC.

Tunneling evolved in the nineteenth and twentieth centuries with the demands of industrial development. Applications were not only in the construction of road, rail and canal tunnels, but also in installing major sewer systems. The second half of the twentieth century saw major advances in the development of tunnel machines and the means of support and lining of tunnels. A limitation, which still exists today, is that the minimum diameter that can be economically driven for

© 2010 IWA Publishing. *Marine Wastewater Outfalls and Treatment Systems.* By Philip JW Roberts, Henry J Salas, Fred M Reiff, Menahem Libhaber, Alejandro Labbe, and James C Thomson. ISBN: 9781843391890. Published by IWA Publishing, London, UK.

a traditional tunnel is around 2.0 m. This limitation is created by the need to erect the lining using manual labor behind the advancing shield. An important development in the 1960s was that of pressure balance shields, which were capable of working in soils well below the water table. No longer was it necessary to use compressed air working, with all its dangers and drawbacks, to counterbalance the external soil and hydrostatic pressure.

As cities and towns developed, they demanded methods to install pipelines and other utility ducts less than 2.0 m in diameter at greater depths and by methods which did not disrupt the life of the city. For these smaller diameters, a new tunneling method was developed. Instead of erecting linings behind the advancing shield, pipe sections were jacked in from the working shaft behind the tunneling machine. These shields and tunneling machines were much the same as those used in traditional tunneling but smaller. Such pressure balance tunnel boring machines (TBMs) could be remotely operated. This new method, which became known as microtunneling, allowed installation of pipe from 300 to 3000 mm diameter.

A third method of trenchless installation was developed for crossing under rivers, roads and railways with pipelines and cables. Out of the experience of the vertical oil drilling industry, horizontal directional drilling (HDD) was borne. By using a slant-mounted drilling rig located at the surface it was possible to drill on a vertical curve to pass under the river and then pull back a pipe. Since the introduction of HDD in the 1970s we have seen major advances in the length and diameter of pipe that can be installed as well as the types of soil that can be successfully drilled.

These three trenchless approaches to installation are being increasingly applied to installation of marine outfalls. They can be used to install all or part of an outfall depending on economic and technical considerations that are discussed in this chapter.

11.2 ADVANTAGES OF INSTALLATION BY TRENCHLESS METHODS

Previous chapters have discussed the design and construction issues associated with traditional outfall pipeline installation methods. The three trenchless methods for installing pipes introduced above eliminate many of the risks and problems associated with these methods. Trenchless techniques have several inherent advantages:

- *Environmental impact is greatly reduced.* Installation takes place underground so there is no disturbance to the beach, the approaches, or the

seawater. A number of outfalls in environmentally sensitive areas have successfully used trenchless methods where installation from the surface would be unacceptable.
- *Impact on the community and the infrastructure is mostly eliminated.* Roads, beaches and shipping can continue in normal use without being aware of the work going on underground. Impact to the seabed is confined to a localized area where the tunnel or bore emerges and where the diffuser is to be installed.
- *Location of the launch site for the outfall is much more flexible.* For tunneling and microtunneling, the work is undertaken from a shaft which can be sited to minimize impact. The site can be back behind the foreshore and is often on a cliff top. Although HDD requires more space for the equipment than the other two methods, the equipment is set up on the surface with small shallow excavations and the time of occupation is relatively short.
- *Design issues are reduced.* It is no longer necessary to design for wave forces and currents, and seismic concerns are reduced as the pipeline is an integral part of the soil structure.
- *Concerns about erosion or accretion of seabed material around the pipeline leading to the diffuser are eliminated.* Scour is commonly found not only on outfalls laid directly on the seabed, but also to pipes laid on supports where, over time, support is lost and settlement occurs. Settlement can lead to excessive bending and failure of the pipe.
- *Pipe is installed well below the seabed which presents fewer risks.* There is no concern about the risk of losing surface protection to the outfall pipes or damage by natural forces or dragging anchors.
- *Trenchless methods are less affected by bad weather.* The only weather concerns are for the short period during the final breakthrough and when installing the diffuser section. The ability to continue working during gales and high seas greatly reduces cost and risk. For traditional installation, particularly in exposed locations, weather conditions can delay the program and accumulate significant cost while plant is standing idle.
- *Surf zone construction.* As discussed in Chapter 10, the most difficult, disruptive and expensive part of an outfall installation can be the surf zone where major temporary and permanent sheet piling may be needed. Trenchless methods can carry the line out to deeper water where offshore craft can operate and traditional installation can be used. This can be more economical and less disruptive.

Design issues for the various methods of trenchless construction are discussed in separate sections below.

11.3 THE SITE AND GEOTECHNICAL INVESTIGATION

All trenchless methods need the best possible information on the conditions likely to be encountered. The methods described in the following sections can deal with a wide range of soil conditions ranging from the hardest rock to the softest soil. Detailed knowledge of the conditions is required in advance in order to make the correct choice of methods.

The geotechnical investigations for the engineering characteristics, particularly the boreholes and laboratory testing of samples, are a key part of the design considerations and the selection of the best trenchless methods. The information required for trenchless methods is not concerned so much with the nature of the seabed but what lies below it.

The following are required for the preliminary site investigation:

- Carry out a desk study to collect all the known environmental, physical and historical information on the chosen location;
- Conduct a "walk-over" survey to identify key physical features, working areas, and environmental and other impacts on the public;
- Undertake a full geological assessment of the area, if possible using experts in the regional geology.

Having investigated the site, the engineer should have an emerging picture of the conditions likely to be encountered. A geotechnical investigation should be a confirmation of what is expected, not a venture into the unknown. If it does not confirm the anticipated conditions, re-evaluation of the data and further investigations may be required. For example, seabed profiling and seismic surveys can usefully supplement borehole samples for identifying the seabed and rock head profiles. Geotechnical investigation and laboratory testing can be undertaken in accordance with well-established guides and codes.

Drilling boreholes on the approaches and in the anticipated locations of the working shaft are normally not too difficult. Boreholes must be planned and undertaken, however, to provide a comprehensive understanding of the soil strata on and adjacent to the landward part of the outfall line including the shore and in the shallows if possible.

Marine borehole investigation is costly and may be difficult. The better the initial understanding of the essential locations and of the information required, the more effective will be the investigation. For tunnels that exit onto the ocean bed and in the area where the diffusers are to be installed, information is required on seabed conditions.

11.4 MARINE OUTFALLS INSTALLED BY TUNNELING

11.4.1 Background

Tunneling is today a highly skilled technical business. Sophisticated equipment is used to excavate the soil and remove it from the tunnel, to monitor and adjust the line and level, and to install a permanent tunnel lining.

Virtually any type of soil conditions, ranging from the hardest rock to the most unstable granular material under a high hydrostatic head, can be tunneled safely. The techniques and machines used must be matched to the anticipated soil conditions. This becomes a key issue for all forms of underground installation. The geological and geotechnical investigations are key elements for an engineer designing any tunnel. Ideally he would like to locate his tunnel in consistent conditions rather than not necessarily good conditions. Modern tunneling machines can deal with one set of conditions very well and with quite wide variations in an acceptable manner. However, some combinations of conditions can create greatly reduced progress and other difficulties. Once the TBM (Tunnel Boring Machine) has set off to drive a tunnel, there is no opportunity to change the head if unforeseen conditions arise.

11.4.2 The tunneling process

Installing a tunnel involves four key elements:

- The shield/TBM
- Removal of cuttings from the tunnel
- The tunnel lining
- The working shaft.

The shield and TBM

The shield and the technology used to cut the soil vary greatly depending on the soil condition and the length and diameter of drive. A shield can range from a simple steel cylinder in which men physically excavate the soil at the face, to a sophisticated TBM incorporating a wide range of controls for face stability and monitoring and control of line and level. Tunneling outfall work that is likely to involve significant drive lengths below the bed of the ocean and a high hydrostatic head except for the most impermeable soils will require the use of the more sophisticated pressure balance TBMs.

An exception might be in competent rock with low water inflows. The time honored technique of "drill and blast" from a simple open shield is frequently then the most cost effective approach. Because of technical advances, powerful TBMs with specially-designed rock heads are being increasingly used for rock strengths up to 300 MPa. The pressure balance facility is not required where water inflows can be contained. The actual cutting head of the TBM can be designed with cutters and bits appropriate to the rock strength and characteristics. This is illustrated in Figure 11.1 which shows a large diameter rock tunneling machine. It is possible to change worn cutters and bits from behind the cutting head during the drive.

Figure 11.1 Pressure balance TBM with a rock cutting head (Courtesy of Herrenknecht AG)

In alluvial and sedimentary soils, the development of pressure balance TBMs has been of great importance. These machines work on the principal of setting up a balancing pressure to the external soil and hydrostatic pressures. There are two main ways of achieving this counterbalance: Slurry and Earth Pressure.

The principles of a slurry pressure balance machine are shown in Figure 11.2. A feed line maintains the pressure in the slurry chamber and a return line carries away the slurry containing spoil to the separation plant. Slurry machines are the most widely used type; they work by having a chamber in which slurry is kept pressurized to the required level. Slurry flows into and out of this chamber through a continuous pipe loop. The outgoing line also carries away the cuttings arising from the cutting head. This line goes to a processing plant which removes the cuttings and returns the slurry back on the feed line. The slurry is most commonly based on a bentonite mix which can greatly assist in stabilizing the face and in keeping the cuttings in suspension. Slurry machines are suited to most granular materials, hard clays and cohesive soils like soft rock.

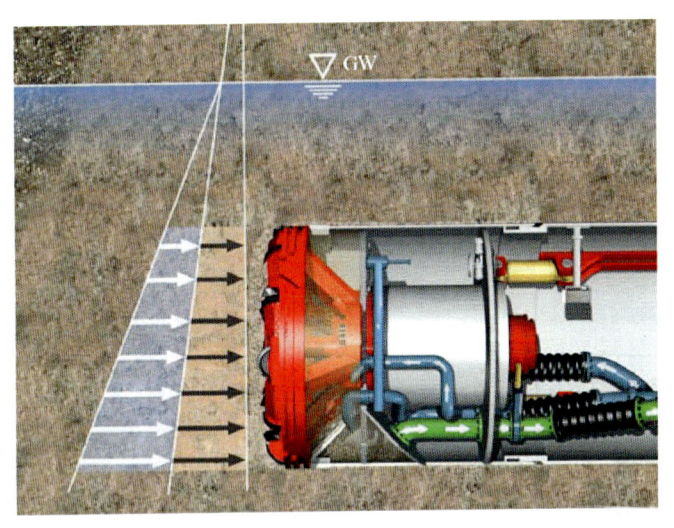

Figure 11.2 Principle of a slurry pressure balance TBM
(Courtesy of Herrenknecht AG)

Earth pressure balance machines (EPBMs, Figure 11.3) use the soil itself to set up the balancing pressure to counteract the external and soil and hydrostatic pressure. This is done by maintaining soil under appropriate pressure in the chamber behind the cutting head. Material is removed from this chamber in a controlled manner proportional to the new material being brought in from the cutting head in order to maintain the earth pressure balance. EPBMs are best suited to silts and plastic clays, although their range has been extended into granular soils and even rock by mixing in foams or other additives to the soil in the chamber to produce a plasticized material. The material is transported on conveyors, skips, or pneumatically pumped.

In many contracts, but not all, the choice of the TBM is the responsibility of the installation contractor.

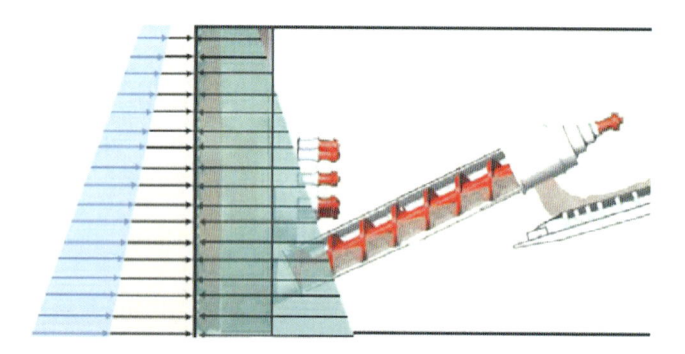

Figure 11.3 Principle of earth pressure balance TBM

Removal of cuttings from the tunnel

The method of spoil removal is normally left to the contractor. Spoil from excavation methods like drill and blast can be removed by simple wagons on rail tracks. For slurry machines, spoil is continuously removed by the slurry return flow line. Spoil is separated from the slurry by screens and centrifuges at a plant located adjacent to the shaft. The cleaned slurry is returned to the system. For a long drive it may be necessary to introduce booster pumps along the line. For earth pressure machines the use of positive displacement pumps is effective with plastic soils. An advantage of EPBM methods is that the spoil from the face is not slurried; it is in its natural state so no expensive separation plant is required.

Tunnel lining

A basic approach is to erect a temporary lining using steel ribs with lagging between them and then either forming an in-situ lining or placing pipe sections and filling the annular space to provide the finished pipeline. This approach is still employed in the USA but infrequently elsewhere.

For outfalls, a more common approach is to use segmental sections to line the tunnel. These are erected in the tail of the tunnel machine as it advances. Segments come in a variety of designs. Smooth bore versions with hydrophilic joint material provide a watertight pipe construction which can directly form the outfall pipe. Figures 11.4 and 11.5 illustrate segmental tunnel linings in a production yard and lining a tunnel.

As the tunnel lining forms part of the permanent works, it must be designed to withstand all permanent and temporary loads. For soft ground there are varying design approaches, but these are mainly based on the hoop stresses induced into the lining from the dead loads. Unlike pipe design in trenches, there is an arching relief which varies according to the nature of the soil and depth. For tunnels in rock, knowledge of the rock mass qualities is needed to ascertain the degree of support and lining which is required.

Figure 11.4 Tunnel segments in yard

Figure 11.5 Segment lined tunnel

As already noted, the minimum diameter for a segmental lined tunnel is about 2.0 m. Smaller diameters do exist but are only appropriate for short drives. The segments will usually be factory produced from reinforced concrete. For longer drives it may be more economical to set up a production unit on site.

A major advantage of tunneling is that vertical and horizontal curves can be incorporated into the design allowing the outfall to be located on a line and level that avoids problematic physical features and geotechnical conditions. A major consideration in determining the appropriate trenchless option is the outfall length. Tunneling is capable of longer installations than microtunnels or HDD.

The drive shaft

The drive shaft is placed in an appropriate location for the line of the outfall, most likely the onshore access for the outfall pipe. If the drive shaft is deep, a landward drive on an uphill grade can provide a better gradient for the outfall and a cost-effective installation. The depth of the shaft is determined by soil conditions, restraints on the outfall route, and the objective of locating the tunnel in consistent strata.

The shaft, as a permanent structure, is designed using established methods depending on the form of construction. The construction method is closely related to the nature of the anticipated ground conditions. The shaft can also be used as a means of providing permanent access to the outfall for maintenance such as pigging or diver inspection.

The reception shaft

In order to remove the TBM from the drive a requirement on soft ground drives is some form of reception shaft into which the machine is driven and recovered. This reception shaft can form part of the diffuser arrangements. Where the drive is in rock the design of such machines will allow the TBM to be backed out of the tunnel to the drive shaft.

11.4.3 Case studies

Aberdeen, UK

This tunnel, constructed in 1979, is an example of an outfall constructed in rock by traditional methods. Numerous other examples include earlier installations in Scandinavia where rock tunnelling is commonplace because of the geological conditions.

It became apparent to the Aberdeen City Council by 1970 that the existing outfall into the River Dee, constructed around 1900, was no longer adequate. Floating material was being washed up on the shore. It was decided to construct a larger and longer outfall. A detailed hydrographic survey was undertaken and a location was found for a sea outfall 1800 m long into the Bay of Nigg. Boreholes were carried out from a seagoing vessel using diamond drilling techniques. Cores were extracted and retained and packer tests were performed to determine the permeability of the rock. Standard laboratory tests were also undertaken.

The site investigations confirmed what was predicted from the geological study. The weathered granite had a compressive strength of 100 MPa to as little as 16 MPa for the gneiss. The investigation also showed that the rock head was lower than expected, and lowering the tunnel from 45 to 55 m would enable it to be located in more consistent rock strata.

Bids were taken and contractors offered both pulled pipelines and tunnel solutions. Two of the pulled pipeline bids were lower than the tunnel bid but heavily qualified. Aberdeen is exposed to the North Sea and often suffers from gales and high wave heights. A pulled pipeline construction could suffer from significant downtime and cost overruns.

The bid for tunnel construction was accepted. The contract works included the shaft, the driving and lining of the tunnel and the installation of the diffusers. The outfall length was 1800 m and the finished diameter 2.5 m. The shaft was about 66 m deep and passed through boulder clay, sands, and gravels before meeting the rock head. In the soft ground the shaft was lined by concrete segments which were sunk by the underpinning method. Because of the difficult interface between overburden and rock, the soil had to be frozen using the brine technique to provide a stable material. The relatively consistent nature of the rock allowed an oversize horseshoe section profile to be driven using drill and blast methods and temporary lining of steel ribs with lagging between ribs where necessary to retain the loose rock. Within this tunnel the 2.5 m diameter outfall pipe was constructed using in-situ construction. With recent advances in technology and rock cutting bits, it is likely that a full face TBM would be used today with a segmental liner or a sprayed concrete temporary liner.

The diffuser consists of ten risers that were installed using a jack-up barge. The risers are about 25 m long from the tunnel to the seabed, and hydraulic model testing showed that connecting the diffusers at the invert was preferable to soffit connections in order to purge seawater from the tunnel. The risers consist of a reinforced concrete head surrounded by four 200 mm diameter ports. The main shaft consisted of a 750 mm diameter GRP pipe surrounded by a 1016 mm diameter steel pipe. The annulus was grouted with a specially formulated mix.

The riser shafts were installed by reverse circulation drilling from the platform through a 1250 mm casing which extended from the barge down through the seabed deposits to the rock head. The drilling went down through the rock into the tunnel below. The casing was cut off just above the seabed and the diffuser placed and grouted. Finally a conical concrete deflector was placed on the diffuser head to protect it from damage by trawlers.

The final cost of the outfall at 1978 prices after settling all claims was £8.8 million (approximately $13.5 million at 1978 exchange rates). This is an average cost of about $7,500 per meter (at 1978 prices) which is comparable to prices of conventional outfalls discussed in Chapter 12.

Boston outfall

No reference to tunneled outfalls would be complete without including the largest outfall constructed to date by tunneling, the Boston outfall. To accommodate its peak design flow of 55 m³/s, this outfall, shown in Figures 4.6 and 11.6, has a tunnel 7.6 m diameter 16 km long. The tunnel was driven from a 9 m diameter shaft 130 m deep located at the Deer Island treatment plant. It was driven on an upward gradient to allow water inflows to drain back to the shaft for pumping out. The outfall tunnel is about 14 km long with an additional 2.1 km of diffuser tunnel. In the area of the diffuser, the tunnel is about 80 m below the ocean bed where the water depth is about 33 m. The diffuser consists of fifty five risers each containing eight ports (see Section 4.4). The risers were preinstalled under a separate contract from a jack-up platform. The risers are located about 8 m from the tunnel centerline, and connections were driven from the main tunnel to connect to the vertical riser shafts. The tunnel and outfall construction is reported to have cost $390 million, for an average of about $24,000 per meter.

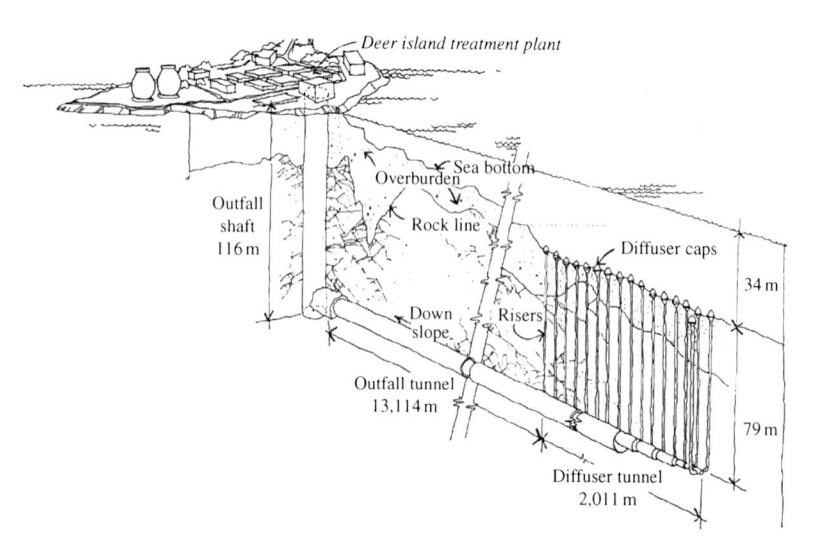

Figure 11.6 Cross section of the Boston outfall (Courtesy MWRA)

South Bay Ocean outfall

This outfall, shown in Figure 11.7, is located at San Diego near the Mexican border. It consists of a 5.8 km tunnel 3.3 m diameter installed 45 m beneath the seabed. In contrast to the Boston outfall, the diffuser consists of a single riser shaft 2.7 m inside diameter 4.2 km offshore. This riser conveys the effluent to a pipe on the seabed that is 1400 m long and 3.3 m diameter. This pipe connects to a "Y" structure, with two 600 m diffuser lines that discharge into a water depth of about 33 m.

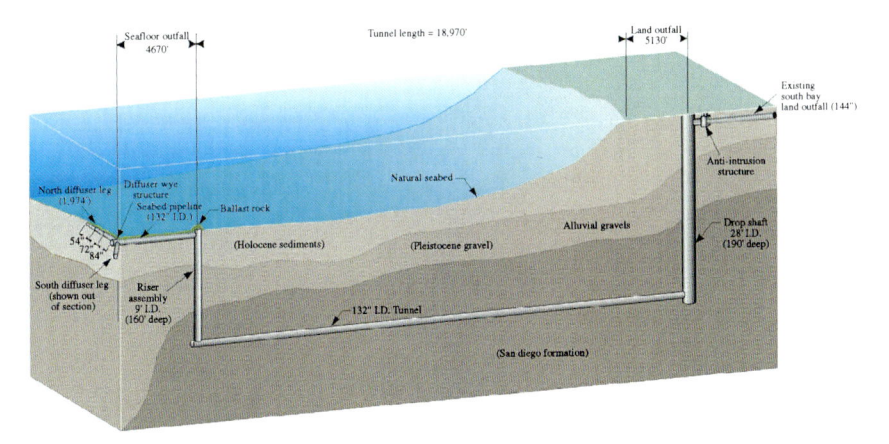

Figure 11.7 Cross section of South Bay outfall (Courtesy Parsons)

This project has a number of unique features. Unlike the two previous examples, the ground conditions were very difficult, consisting of sedimentary deposits of sands, silts, gravels, cobbles, boulders and over-consolidated marine clays. The alignment crosses 15 active faults and one dormant fault. Tunneling was selected because of environmental concerns and seismic conditions that could create significant displacement with the potential for liquefaction of marine sediments. Extensive geotechnical and geophysical investigations were undertaken to provide as comprehensive a picture of conditions as possible along the line of the tunnel. A probabilistic fault displacement analysis provided a basis for the anticipated movement that the tunnel liners must accommodate.

The tunnel liner design is unique in that single-pass concrete segments joined together by continuous hoop steel form a monolithic structure that was specifically developed for this tunnel. The segments provided the final outfall pipe without any form of secondary lining. Segments have an advantage over rigid outfall structures in seismic conditions. They have an inherent flexibility which comes from the articulation capability of the individual segment and the ductility of the connecting steel bolts. The segment lining is designed to withstand an internal operating pressure up to 3 bar and an external pressure of 7 bar.

A joint venture was awarded the contract to construct the drive shaft, the outfall tunnel and the riser shaft for the sum of $88 million (about $15,000 per meter). They chose to use an EPBM which incorporated air locks for accessing the cutter head and also freeze pipes to enable the ground at the face to be frozen. Injection ports were provided to allow foam and bentonite to be added to the material in the chamber.

Mumbai outfalls

Mumbai is the major industrial and commercial city of India and has a population estimated to be 12 million. The city has expanded rapidly but the infrastructure has not kept pace. Until recently, 23 m^3/s wastewater was dumped directly into the sea without treatment.

Brahnmumbai Municipal Corporation developed an integrated water management scheme with aid from the World Bank. Three of the seven areas into which the scheme is divided provides for disposal by outfall. Two of these, at Worli and Bandra, have now been installed (Figure 11.8). The wastewater receives preliminary treatment that includes coarse and fine screening and is pumped. Both of these outfalls extend about 3.5 km into the ocean and are located 30 m below the seabed.

The geotechnical investigation involved more than 20 boreholes for each outfall. Tuff, volcanic breccia, and basalt were indicated along the line of the tunnels. The boreholes produced a high RQD (Rock Quality Description), and permeability tests gave low results. This indicated a massive rock structure with few joints. Based on this data the contractor chose to use an open face hard rock machine.

The machine had a cutter head diameter of 4.05 m and 32 disc rock cutters distributed around the head which allowed cutting rock up to a strength of 300 MPa. The machine was fitted with an advanced probe drill which allowed investigation up to 120 m ahead of the face. The circular tunnel lining, which provides a finished diameter of 3.5 m, is formed from eight concrete segments. The rock was sufficiently stable to allow the segment erector assembly to be set back from 16 to 40 m behind the face.

Drive shafts eight and nine meters in diameter were sunk at Worli and Bandra respectively to a depth of 60 m, mainly through rock. They were lined with concrete. The tunnel at Worli had a short 600 m radius curve but was mostly straight. The Bandra tunnel had about a 45° change of direction again using a 600 m radius curve. Both tunnels were driven with an uphill incline of 1/250 to allow water to drain back to the shaft. The Bandra tunnel has a length of 3.7 km and Worli 3.4 km.

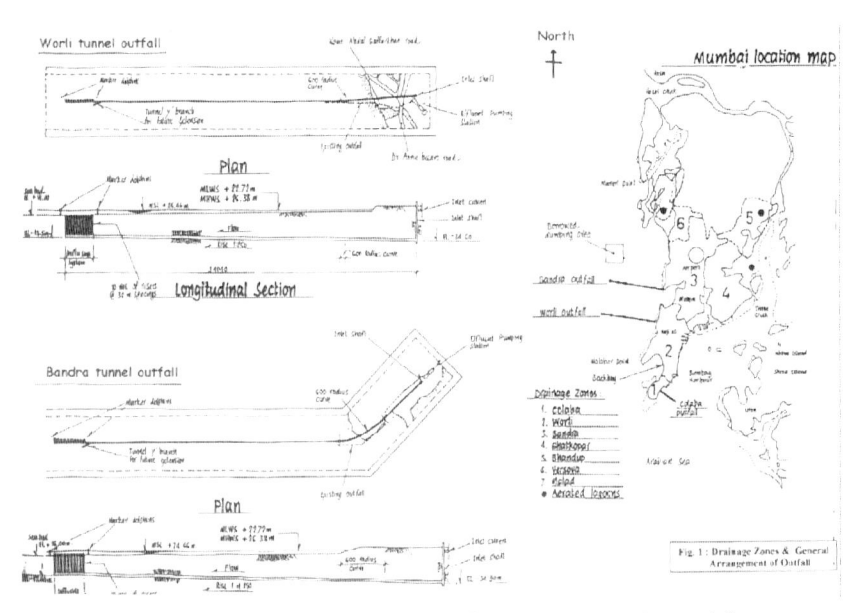

Figure 11.8 Location and layout of Worli and Bandra outfalls (Dykerhoff and Widman AG Mumbai)

The diffusers consist of ten 1600 mm riser shafts 12 m off the tunnel line. They were constructed in advance of tunneling by drilling into the seabed to a depth of 36 m, which is below the tunnel invert. The drilled shafts were lined with glass-reinforced plastic 1000 mm pipe and the annular space grouted. They were connected to the invert of the outfall tunnel by means of short adits.

Other examples

Table 11.1 shows a small fraction of the outfalls that have been constructed by traditional tunneling. Typically, tunneled outfalls are larger than 2.0 m diameter and where long lengths are needed. Most of those on Table 11.1 are where the geological conditions are competent rock and an open full face rock TBM could be employed.

Table 11.1 Tunneled outfalls

Location	Year	Diam (mm)	Length (m)	Max depth below MSL (m)	Geotechnical conditions	Tunnel lining	Installation methods	Comments
Brighton and Hove, UK	1971–1974	2100	1820	40	Chalk	Smoothbore segments	TBM	
Aberdeen, Scotland	1979	2500	1800	55	Weathered granite and gneiss	*In-situ* concrete	Drill and blast	
Vildenmald Helsinki, Finland	1985				Rock		Drill and blast	
Boulder Bay, Australia	1994	3000 × 3200	700		Rock	*In-situ*	Drill and blast	
Bondi, Sydney, Australia	1991	Inverted 2200	2200	120	Sandstones overlying shales and mudstone	Cast *in-situ* lining	Road header	
Malabar, Sydney, Australia	1990	3480 ID inverted	4200	120	Sandstones overlying shales and mudstone	Precast RC segments	Open full face rock TBM	36 risers in diffusion zone
North Head, Sydney, Australia	1990	3840 ID inverted	3100	120	Sandstones overlying shales and mudstone	Precast RC segments	Open full face rock TBM	36 risers in diffusion zone

(continued)

Table 11.1 (*Continued*)

Location	Year	Diam (mm)	Length (m)	Max depth below MSL (m)	Geotechnical conditions	Tunnel lining	Installation methods	Comments
Hong Kong, Stage 1	1999	5000	1000+ 2×620 pipelines		Granite overlain by seabed sediments	Gasketed segments	Open full face rock TBM	Two 2.75 m risers connect tunnel to seabed pipelines
Boston, USA	1999	7600	15000	125	Rock	Gasketed segments	Open full face rock TBM	Downward drilled and offset risers connecting at tunnel invert
South Bay, San Diego, USA	1998	3300	5800		Sedimentary deposits of sands, silts gravels, boulders and marine clays	Purpose designed segments	Earth Pressure TBM	Terminates in 2.86 m riser shaft connecting to 1600 mm seabed pipelines
Bandra, Mumbai, India	2000	3600	3700	41.5	Basalt/tuffs and tuff-breccias	Bolted concrete segments	Open full face rock TBM	10 risers of 1000 mm diameter each with 10 ports
Worli, Mumbai, India	2000	3600	3400	42.5	Basalt/tuffs and tuff-breccias	Bolted concrete segments	Open full face rock TBM	10 risers of 1000 mm diameter each with 10 ports

With the increasing capabilities of pressure balance TBMs, we are likely to see, even in sedimentary geological conditions, an increased number of large diameter outfalls constructed by tunneling. Several major outfalls in the USA are being currently designed as tunnels, including Los Angeles and Seattle.

11.5 MARINE OUTFALLS INSTALLED BY MICROTUNNELING

11.5.1 Background

A significant breakthrough in the installation of pipelines by trenchless methods has been the development of microtunneling. In some respects it is another form of tunneling, but is distinguished by the means in which the tunnel lining is installed and by the use of miniaturization and remote control methods. These allow tunnels less than 2.0 m diameter as well as larger ones to be constructed.

TBMs for microtunneling are designed to deal with a wide range of soil conditions from hard rock to the softest of marine clays. The same categories of machines described in the section on tunneling have models suited for microtunneling. In man-entry sizes, above 1200 mm, an open face shield might be used in cohesive soils. The capability of pressure balance machines to withstand high hydrostatic pressures makes microtunneling well suited to working under the seabed.

The initial impetus for developing these machines came from the need of many cities throughout the world to install or replace sewer systems. The technology began in Japan where a number of cities were without a sewer network. Today probably 3000 machines are in use around the world.

Sewers are designed to flow under gravity and must be installed to accurate gradients. For this reason they are laid at greater depths and with larger diameter pipes than other utilities. Sewers are usually located under the road to provide easy access via manholes for maintenance. Installing a sewer in a street often causes disruption; the work is slower and requires a greater width of working space than for other utility installations. In the last 20 years, microtunneling methods for installing sewer pipes in urban streets have become accepted practice for pipes ranging from 300 to 3000 mm and sometimes greater. Microtunneling has displaced traditional tunneling in diameters up to 2.5 m as it is safer, quicker, cheaper, and provides a better finished product.

The pipes are installed by pipe jacking (Figure 11.9) where hydraulic rams are used to push sections of pipe through the drive shaft behind the advancing TBM or shield. By installing a factory produced pipe with a flexible joint a high quality sewer can be installed. Pipe joints were initially developed for gravity pipes but there are now also pressure joint types.

Figure 11.9 Concrete pipes being jacked in behind a TBM
(Courtesy of Herrenknecht AG)

To allow longer drive lengths to be installed, a device called an Intermediate Jacking Station (IJS) or Interjack has been developed (Figure 11.10). This is a steel can mounted with hydraulic rams inserted between two pipes to enable sections of the line to be jacked forward successively and independently. In this way jacking loads on the pipe sections can be kept within safe limits. It is now common to find drive lengths of 1000 m between shafts. IJSs are not normally required on diameters less than 1000 mm as there is usually a limitation on lengths between access points of around 120 m so that utility maintenance equipment can be operated.

Another development has been the ability to install both vertical and horizontal curved drives. The curve radius is normally quite large so that the pipe joints accommodate the angular deflection.

For outfalls, microtunneling is normally employed in diameters between 1.2 and 2.5 m although smaller and larger diameters have been used. The economic length of drive has no lower limit and even short outfall lengths can be installed economically. The upper limit on the drive length is more a question of economics than technical limitation, although the nature of the soil conditions play a part. A practical maximum drive length would be around 1500 m. One major advantage is that depth of installation is not an important factor. Standard equipment is designed to work to depths of up to 100 m. Greater depths can be installed by modifying the machines for the higher pressure.

Figure 11.10 Intermediate jacking station

The first intakes and outfalls were installed by microtunneling in the early 1990s. Where conditions were favorable, however, a few outfalls had been installed by pipe jacking and open shield some years before.

For medium diameter outfall installation we are likely to see increased use of microtunneling because of the advantages if offers over alternative methods.

11.5.2 The microtunneling process

Microtunneling consists of the same four key components as for tunneling:

- The shield/TBM
- Removal of cuttings from the tunnel

- The tunnel lining
- The working shaft.

The shield and TBM

The same types of shield and TBMs used in tunneling are found in microtunneling but are generally smaller in diameter and remotely controlled. Slurry pressure balance machines are the most widely used. As these machines are smaller and do not have the power of some of their larger brothers they are not suitable for working in ground with numerous hard boulders. For diameters of 1.5 m and greater, rock machines and heads are available that cut rocks with strengths up to 300 MPa.

Removal of cuttings

The greater use of slurry pressure balance machines and the size limitations means that slurry systems are widely used for spoil removal. The same closed-loop system with separation is used with slurry installations and described in the section on tunneling. EPBMs are not available in diameters less than 1.5 m.

Tunnel lining

Tunnel linings are factory made pipe sections jacked in from the access shaft. Reinforced concrete pipes are most commonly selected for outfall work in diameters above 1.2 m. Alternatives are reinforced glass fiber polymer mortar and glass fiber reinforced resin products which are more corrosion resistant than concrete. Ductile iron and steel jacking pipes are also available. The jacked pipe differs from the normal products laid in a trench as the joint has to be incorporated into the pipe wall to maintain flush internal and external surfaces. Pipes must also be of a heavier and higher quality than pipes laid in trenches.

Pipes are designed to withstand the predicted live and dead loads they will experience in service. Probably the greatest loads that the pipe will experience are axial and occur during the jacked installation. Standards for jacked pipes exist in most countries.

The outside pipe diameter is slightly less than the diameter of the cutting head of the TBM. This means that the pipes are installed in close contact with the soil. To reduce frictional loads when jacking the pipe, a lubricant is injected into the small annular over cut. Not only does the lubricant reduce the frictional load, it also acts as filler to the annular space and reduces potential settlement at the surface.

The shaft

The shaft dimensions depend on the size of the TBM to be employed and the length of the individual pipe sections to be jacked. An additional feature of a shaft for jacking is that a wall has to be installed at the back of the shaft to take the reaction from the hydraulic rams pushing the pipe. The form of construction will be mainly determined by the nature of the soil in which the shaft will be sunk.

Some specific issues can arise when designing an outfall. Figure 11.11 illustrates some possible scenarios for installation of an outfall from a shaft located on a cliff top. It is assumed that the outfall will be installed with a vertical curve and terminate in a reception area where the diffusers are to be installed.

Figure 11.11 Schematic of a jacked outfall from a cliff top
(Courtesy of Herrenknecht AG)

To recover the TBM, it is necessary to install a bulkhead equipped with a valve in the end of the line. A reception pit is created by a dredged excavation in the bed of the ocean into which the TBM is driven. Closing the bulkhead allows the TBM to be jacked off the pipe and independent of the pipeline. Divers can connect the machine so it can be recovered onto the barge. The remainder of the work is completed using traditional ocean going craft and equipment. For deep installations, however, a reception shaft may be created in the ocean to allow the TBM to be recovered; it also acts as a riser shaft for the

Outfall installation by trenchless techniques 369

outfall. Figure 11.12 illustrates recovery from a seabed excavation. This ability to recover the TBM from a dredged excavation is a major advantage over traditional tunneling methods.

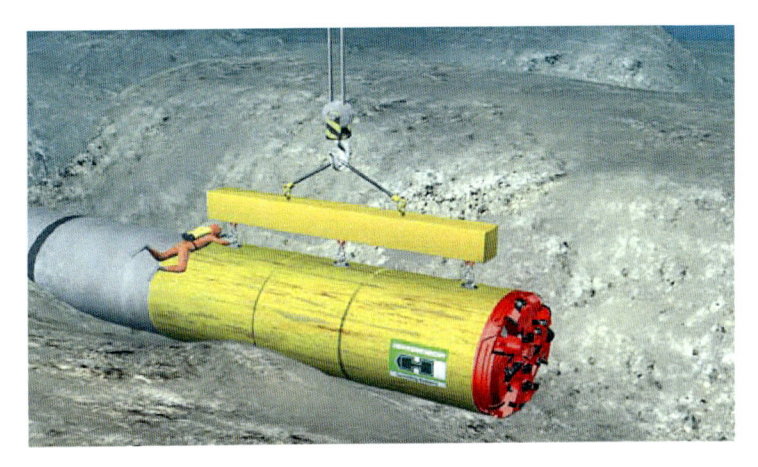

Figure 11.12 TBM recovery from an excavated exit area on the seabed (Courtesy of Herrenknecht AG)

11.5.3 Case studies

The following case histories illustrate a range of outfall constructions by microtunneling.

The Europipe

The German Europipe was completed in 1995 and was the longest single length of pipeline ever driven by pipe jacking. It was not an outfall but a landfall for gas pipelines coming from Norway's North Sea oilfields. The nature of the work as far as the tunneling was concerned is virtually the same as for an outfall, however. An underground solution was required as this was an environmentally sensitive shore area and a national park. Soil conditions along the alignment consisted of saturated sands, clays and peat. Figure 11.13 illustrates the variable soft ground marine deposits. There is a large tidal range, at high tide the hydrostatic pressure is a potential 1.5 bar.

The Europipe was driven from the shore out under the seabed to a reception shaft some 2600 m away. The pipes that were jacked to form the line were of 3 m internal diameter. The tunnel was located at some 6 to 8 m below the surface and was driven on a vertical curve. Working 24 hours a day and six days a week, it took just over 100 working days to drive the 2600 m, an average of 25 m per day.

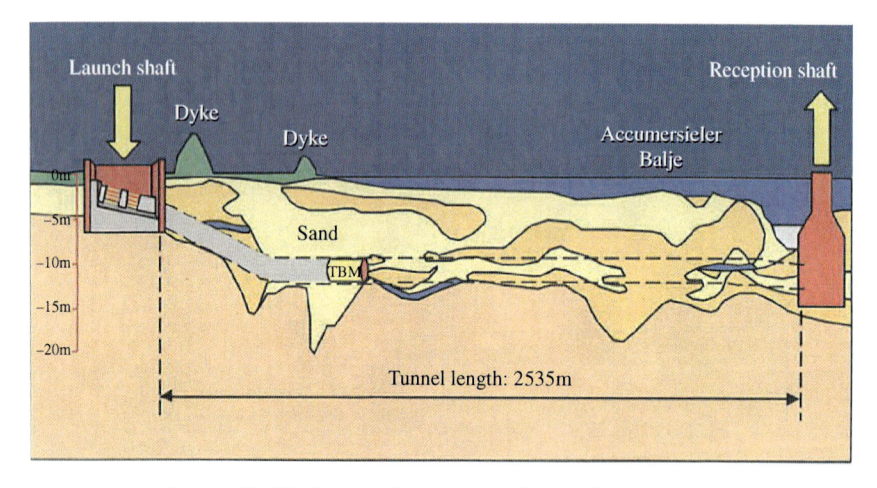

Figure 11.13 Schematic cross section of the Europipe
(Courtesy of Herrenknecht AG)

Horden outfall

The Horden outfall is an interesting example where the advantages of trenchless and traditional construction methods were combined to provide a cost effective solution.

Horden is located on the Northeast coast of England. To meet the strict European Community regulations on the discharge of wastewater to the sea, the local water authority, Northumbrian Water, designed a scheme involving a new treatment plant and a 1.8 m diameter outfall into the North Sea. As the site for the new treatment plant was located 42 m above the beach on a cliff top, and the adjacent land was a protected environmental area, trenchless installation was required.

Horden was one outfall out of four in the package. The design and contractual arrangements for the whole works were based on a three-way partnering arrangement of client, consultant, and contractor. A core team, with members from all three parties, was responsible for developing the overall design and determining the cost targets.

The solution accepted (Figure 11.14) was based on microtunneling a 1.8 m diameter concrete pipe. This involved a curved drive 550 m long. The first section out of the shaft was about 180 m on a downward gradient of 1:7. This was followed by a vertical curve of 1200 m radius to a point 300 m from the drive shaft. The final 250 m was on a 1:286 downward gradient terminating in a pre-dredged channel cut into the seabed.

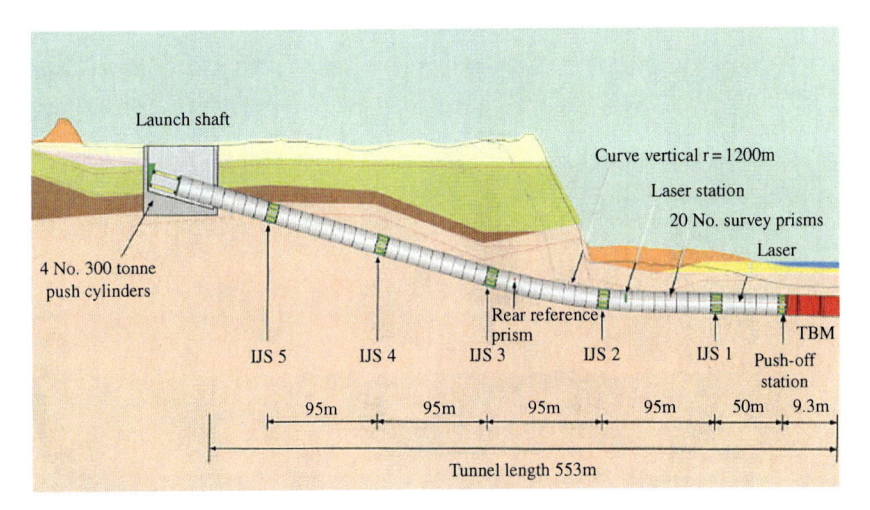

Figure 11.14 Cross section of the microtunneled section of the Horden outfall (Courtesy of Herrenknecht AG)

The TBM was driven into a dredged channel that had been excavated from the jack up dredger. All the flow and service lines to the TBM were removed and the machine was sealed using the in-line air locks. The use of two temporary pressure-equalizing pipes allowed the TBM to be pressurized to the high tide pressure calculated to be 1 bar. The rear of the machine and the end of the tunnel were sealed using an inflatable air bag. The tunnel was then flooded with a head of 10 m equaling the high tide level. The TBM was then jacked off the pipes. Divers attached straps to allow the crane on the lifting barge to raise it and suspend it below the craft. It was then taken in its suspended manner to a quay where a heavy crane was able to lift the 50 T machine on to a low loader.

The remaining section of the outfall was a 711 mm steel pipe extending 1300 m farther out to sea terminating with four risers each containing four ports. These were constructed in the traditional manner using offshore craft for dredging a trench and laying the pipe.

Marbella outfall Biarritz

Ageing outfalls with inadequate dispersion and a growing number of summertime visitors was causing pollution of the beaches of this famous French resort. The solution was a new outfall 1.6 m in diameter extending 780 m into Marbella Bay. There were a number of constraints, not least of which was the potentially negative impact of the works on tourism and residents. A second constraint was to maintain the existing outfalls during construction. The third constraint was the difficult weather conditions that prevail at certain times of the year.

The solution developed was microtunneling from a shaft set well back from the beach and the frontage properties. The aerial photograph in Figure 11.15 shows the outfall alignment and the drive shaft location. It was possible to locate this shaft so that the two existing pipelines carrying the effluent from the treatment plant, together with a line carrying storm water, could be intercepted and diverted into the shaft and the new outfall.

Figure 11.15 Aerial photo of the Biarritz scheme during construction (Smet Boring, Belgium)

The longitudinal profile is shown in Figure 11.16. The shaft is 12 m diameter and 20 m deep. Provision is made within the shaft for grit removal and odor control as well as energy dissipation for the high-elevation incoming pipelines.

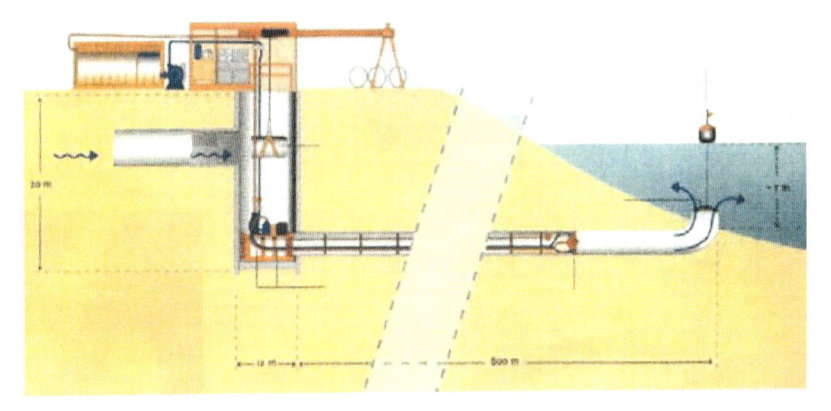

Figure 11.16 Cross-section of the Biarritz microtunneled outfall (Smet Boring, Belgium)

The geotechnical conditions were variable, with sand, limestone, and clay marl. A pressure balance slurry shield was adapted. The head combined both roller cutters and picks to deal with the heterogeneous conditions and with relatively large openings to allow the excavated material to pass into the shield. This shield is equipped with an air pressurized access chamber which allows man access for changing worn face tools. A unique feature was the provision of a rear section to the shield which was designed to seal off the tunnel and allow the front sections of the machine to be decoupled by divers and removed from the seabed exit point.

The shield was recovered and the diffuser installed by a jack-up platform equipped with a drilling rig and a crane with a grab to excavate down into the marl layers below the seabed. Divers recovered the microtunneller and placed the diffuser within the dredged channel excavated in the seabed.

Other examples

Table 11.2 shows examples of outfalls installed by microtunnelling. If one discounts the Eurotunnel, the length of drives is below 1000 m and the diameters 2.0 m or less. Ground conditions for these outfalls range from rock to sands and clays. They all used slurry pressure balance TBMs but the cutting heads would be matched to the soil conditions.

Table 11.2 Outfalls Installed by Microtunneling

Location	Year	Diam (mm)	Length (m)	Max depth below MSL (m)	Geotechnical conditions	Tunnel lining	Installation method	Comment
Europipe Germany	1994	3000	2530	13	Marine sands and clays	RC pipe	Pressure balance slurry TBM	Terminated in a shaft
Ramsgate UK	1994	1200	530	12	Sand, limestone	RC pipe	Pressure balance slurry TBM	
Horden UK	1997	1800	550	30	Sands/clay and limestone	RC pipe	Pressure balance slurry TBM	
Porto Portugal	1996	1600	520	18	Sand and granite	RC Pipe	Pressure balance slurry TBM	
San Sebastien Spain	1999	2000	410	35	Rock	RC pipe	Pressure balance slurry TBM	
Mompas Donostia Spain		2000	420		Sand and granite	RC Pipe	Pressure balance slurry TBM	
Honolulu Hawaii	2002	1200	2 × 400	5–14	Lagoon deposits, coral and sands	RC pipe	Pressure balance slurry TBM	Total length 3900 m Shaft located in ocean

(*continued*)

Outfall installation by trenchless techniques 375

Table 11.2 (*Continued*)

Location	Year	Diam (mm)	Length (m)	Max depth below MSL (m)	Geotechnical conditions	Tunnel lining	Installation method	Comment
Biarritz France	2003	1600	780	20	Rock and sand	RC pipe	Pressure balance slurry TBM	
San Pedro Spain	2005	1400	5000 (315 by micro)		Sand and clay and then calcareous sandstone	RC pipe	Pressure balance slurry TBM	First 325 m by micro then traditional
Sable d'Olonnes France	2006	1400	1563 (630 by micro)	8	Sand, clay and rock	RC pipe	Pressure balance slurry TBM	Vertical curve. First 630 m by micro then traditional
Valde-lentisco Spain	2006	2000	1452 (430 by micro)		Sand, sandy clay and limestone	Polymer concrete pipe	Pressure balance slurry TBM	First section micro then traditional with Pe
Berria Spain	2006–2007	2 × 2000	53		Limestone, silica with clay	RC pipe	Pressure balance slurry TBM	
Margate UK	2006–2007	1800	600		Chalk	RC pipe	TBM	
Dunedin New Zealand	2007	1500	1200 (250 by micro)			RC pipe	Pressure balance TBM	Beach and surf section then PE for deep water

11.6 MARINE OUTFALLS INSTALLED BY HORIZONTAL DIRECTIONAL DRILLING

11.6.1 Background

Horizontal Directional Drilling (HDD) evolved in the 1970s from a combination of technology from the directional drilling techniques used for drilling oil wells and the techniques used for boring road crossings. A very similar technique developed in the 1980s was guided drilling, which has some similarities to HDD but also has some key differences. Guided drilling is now widely used throughout the world for smaller diameter bores over shorter lengths. There has been cross-fertilization, however, between the two technologies. Today the larger guided drill rigs, known as midi-rigs, are installing drives with spans and diameters that a few years ago would have been undertaken by an HDD rig.

The HDD process is quite different from the previous two trenchless methods, tunneling and microtunneling, which evolved from tunneling technology. HDD is based on drilling technologies and the equipment and the skills needed to operate the rigs have little in common with tunnelling.

Unlike tunneling and microtunneling, HDD cannot install a pipeline to the close accuracy required for a gravity sewer. Rigs are positioned at the surface so the path of the bore is normally a vertical curve. The vertical curve can be designed and followed with an accuracy of around 1% of the drive length.

HDD is widely used for installing crossings for all kinds of pipelines and cables under rivers. The oil and gas transmission companies found that this was not only a cost-effective way to cross rivers for cross country pipelines but was environmentally acceptable as it did not disturb the river or the adjacent areas. The technology was also used to cross under highways and airports. HDD contractors modified their methods of working in the 1990s to enable installation of outfalls.

11.6.2 Horizontal directional drilling

There are three main stages to Horizontal Directional Drilling (HDD) as shown in Figure 11.17:

1. The pilot bore

A pilot hole is drilled along the prescribed path using a hydraulically driven "mud motor" which drives the cutting head. The cutting head is chosen to meet the anticipated geotechnical conditions. Behind the cutting head is a section of drill string which has a small dogleg configuration. The drill string is not rotated except to orient the dogleg; this provides a steering capability and allows a smooth path to be followed. Behind the dogleg is the survey tool that monitors the drill path position. This electronic package detects the relation of the drill string to the earth's magnetic field and its inclination. The data is transmitted back to the control where it is processed to give the drill head position.

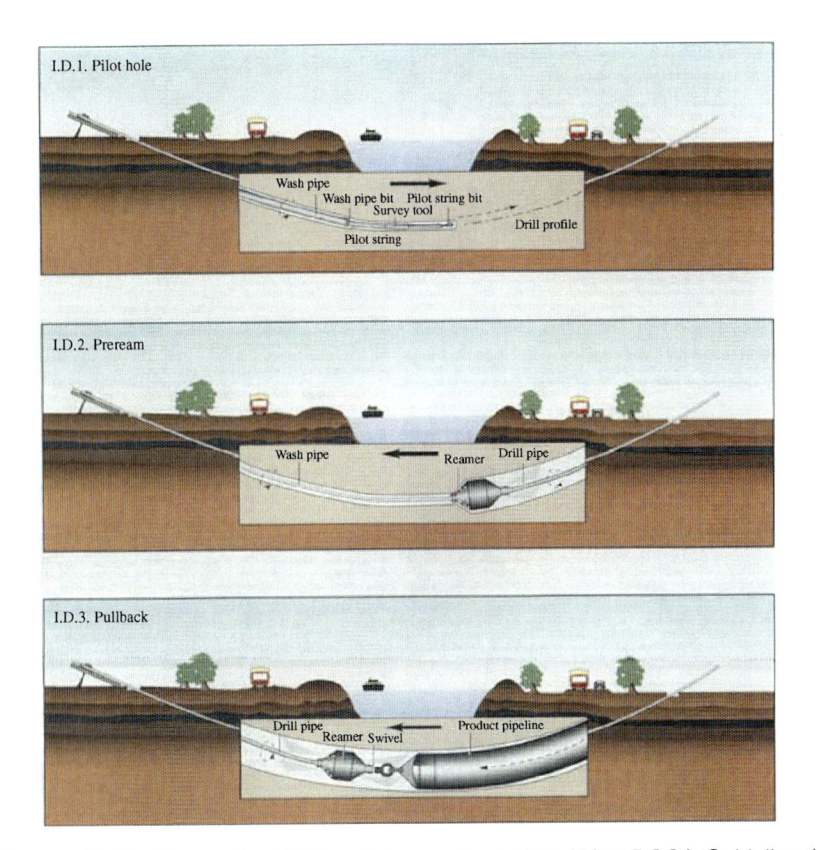

Figure 11.17 Stages of a HDD crossing under a river (After DCCA Guidelines)

The contractor often introduces a wash-over pipe, which is a larger diameter pipe that runs over the drill string and provides rigidity and hole stability if it is necessary to withdraw the drill line and drill head for a bit change. The cuttings from the pilot bore are carried back to the launch area in the drilling mud.

2. Pre-reaming

Having successfully run the pilot bore, the pre-reaming stage begins. The reamer tool is attached at the exit end and opens out the hole as it is pulled back by the rig. It may be necessary to carry out several reams in order to obtain the required hole size which is typically 50% greater than the pipe diameter to be installed. Large quantities of drilling mud are used in this operation to help maintain the stability of the hole as well as to carry cuttings to the surface. The drill rods are normally attached behind the reamer as it is pulled back to allow for additional reaming or pull back.

3. Pull-back

The pipe to be installed is attached to the final reamer by means of a swivel to avoid rotation. Again, large quantities of drill mud are used to pull back the drill-string, reamer and pipe to the launch area.

The HDD rig will depend on the length and diameter of pipe to be installed. Typically the launch area will be around 35 m wide by 50 m long behind the entry point. The angle of entry of the bore path lies between 8 and 20°. The rig is surface-mounted and has a number of support units. Figure 11.18 shows a typical launch arrangement.

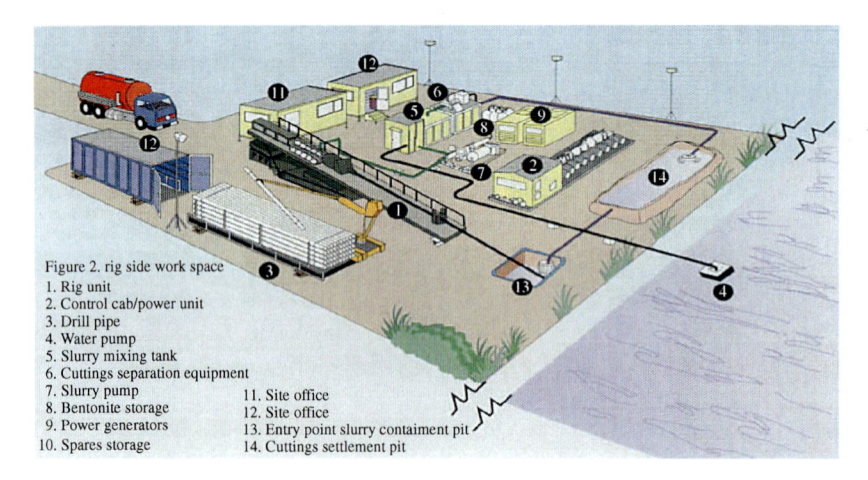

Figure 2. rig side work space
1. Rig unit
2. Control cab/power unit
3. Drill pipe
4. Water pump
5. Slurry mixing tank
6. Cuttings separation equipment
7. Slurry pump
8. Bentonite storage
9. Power generators
10. Spares storage
11. Site office
12. Site office
13. Entry point slurry containment pit
14. Cuttings settlement pit

Figure 11.18 Rig side layout (After DCCA Guidelines)

Typically, the pipe will be pulled in at the exit side, where the pipe to be pulled must be laid out on the surface. This can mean laying out the whole pipeline length behind the exit point or laying shorter lengths that are then jointed as the pull progresses. This is much less desirable.

Drilling mud and slurry

In HDD, drilling mud or slurry provides the following functions:

- Drives the cutting head;
- Provides hydraulic cutting of the soil;
- Lubricates the drill head and drill pipe;
- Protects and lubricates the product pipe during pull-back;
- Stabilizes the hole and builds up a "filter cake" barrier which prevents loss into the surrounding soil.

Most slurry used for HDD is based on bentonite, a naturally occurring clay with hydrophilic properties. As large quantities of slurry are used, the returns must be contained to avoid pollution. These returns are recycled or disposed of at an authorized landfill.

Design considerations must account for the stresses imposed on the pipe during its service life and during construction. The profile of the drill path must be designed not to impose unacceptable stresses on the drill pipe and the pipeline. The contractor must also calculate the required pull back based on frictional loads. This pull back loading will determine the rig size. Drilling mud is used at high pressures. In some circumstances, unless there is a sufficient cover, fracturing in the overburden can occur with loss of mud that causes pollution. For most work, a minimum of 8 m below the bed is recommended, but may need to be greater depending on conditions.

For HDD, the majority of pipes installed are either steel or polyethylene as they must have a degree of flexibility to follow the curved drill path. Steel is jointed by welding and coated with fusion-bonded epoxy. In some soil conditions external abrasion can occur during pull back and it may be necessary to provide a suitable sacrificial coating such as a polymer concrete. Polyethylene must be a high strength quality and a minimum SDR of 11 (SDR is the standard dimension ratio, the ratio of the outside diameter to the minimum wall thickness). Jointing is undertaken on site by fusion bonding, which is described in Chapter 10.

11.6.3 Marine outfall and intake installations

The HDD process has been adapted for applications such as outfalls, intakes, and landfalls for pipes and cables. There are two approaches for these types of installation:

- The drill is sited on the land side and the drilling operation follows the sequence set out above. The reason for the difference is that the drill head exits through the seabed so seagoing vessels are required for the reaming and product pipe pullback operations. A particular advantage is that if polyethylene is used for the pipe, it can be floated in the water and even brought to site by towing in a long length behind a tug (see Chapter 10). If the seabed floor is sand or soft clays, however, it may be difficult to maintain the bore open where it exits into the sea. This can create problems for installing the outfall pipe.
- Drill rigs and the support modules are sited at sea on a suitable jack-up platform. Drilling takes place through a moon pool and the drill pipe and wash-over pipe can be supported from the platform by sheaves. Succeeding stages of reaming and pullback of the pipe are done from the landward side. The problems of exiting into the seabed are avoided. This is illustrated in Figure 11.19.

Figure 11.19 A drill rig located at sea on a jack-up platform (Courtesy of Herrenknecht AG)

In either case, the installation of any outfall structure or diffusers is carried out using seagoing craft and equipment in the traditional manner.

11.6.4 Case studies

Castro Urdiales Spain

As part of the European Community requirements on wastewater discharges, the town of Castro in Cantabria, Northern Spain, installed a new wastewater treatment plant and two 900 mm outfalls 615 and 210 m long to carry treated effluent out into the Bay of Biscay.

The difficulties of the location can be seen in the aerial view of the site, Figure 11.20. The contractor opted to use HDD for both lines. A key consideration was that the outfall start on the top of a rocky cliff about 50 m above sea level. The total drop from inlet to outlet was 70 m. The geotechnical investigation showed that the bore would be located in limestone, some of which would be fractured. Because of the nature of the limestone, the final hole was opened up to 1320 mm OD. The pipe installed was 900 mm polyethylene pipe. The pipe was taken out to sea then pulled back to land with the aid of offshore craft. Pullback is shown in Figure 11.21.

Figure 11.20 Aerial view of the Castro Urdiales HDD installation.
(Haustadt & Timmermann)

Mear/Veg Rock, Isle of Man, UK

This is typical of a number of installations around the UK coast.

The Isle of Man is located off the west coast of England in the Irish Sea. An outfall was required to carry treated effluent from a new treatment plant out to sea. An 800 mm polyethylene outfall over a length of 400 m was required. The location of the outfall is in a very attractive natural coastline and environmentally sensitive area. The geology along the coastline consists of metamorphosed mudstones and silt/sandstones. By definition these are quartz rich, so are hard and abrasive. Because of the size of the outfall and the nature of the rock, the contractor opted to use a 100-tonne rig and four stages of hole

382   Marine Wastewater Outfalls and Treatment Systems

enlargement: 437, 650, 850 and 1000 mm. A 437 tungsten carbide insert tri-cone bit was used. The hole-opening reamer assembly is shown in Figure 11.22.

Figure 11.21 Pullback of 900 mm diameter HDPE pipe (Haustadt & Timmermann)

Figure 11.22 Hole opener assembly (LMR Drilling)

After the pilot hole exited on the seabed, a soft plug was created with swelling polymers and a cement pill at the end of the hole. Succeeding reaming stages stopped short of the end of the pilot exit point. The various sizes of reamers were then reintroduced one after the other to complete the bore. When the reaming operation broke through the seabed floor the bentonite drained away and from that point large volumes of fresh water were used which was environmentally more acceptable. To assist the divers in locating the end of the hole, the last reamer was left in place and air was pumped into the line. Once the head was located, an excavator on the barge dug down through the debris and the reamer was pulled back to the surface.

The polyethylene product pipe was welded on land and pulled out to sea ready for pullback as shown in Figure 11.23. The pipe was fitted with a pulling head at one end and a flange plate with valve and a pulling eye on the other. The valve allowed the buoyancy of the pipe to be regulated to assist the pull back. Pullback was by a rig on the landward side.

Figure 11.23 Towing the HDPE pipe out to sea (LMR Drilling)

The pipe was protected at the exit point by a concrete mattress. Prefabricated diffuser sections were then lowered and aligned so that they could be bolted to the end of the pipe, and finally the area around the diffusers was backfilled to provide protection. All this work was accomplished with seagoing craft and divers.

Rockaway Beach USA

The City of Rockaway Beach, Oregon was required to upgrade its wastewater. The new outfall was to extend 900 m offshore into a water depth of 15 m. The total length of the bore was 1340 m as it was set back and so went under 390 m of right away and 260 m of beach before reaching the ocean. Geotechnical field investigations showed mainly Quaternary – marsh, beach shelf, and marine terrace deposits overlying tertiary marine sedimentary bedrock. The bore was located in these deposits.

384   Marine Wastewater Outfalls and Treatment Systems

Various alternatives were considered including pulling the pipe from the seaward side and use of offshore vessels. This was estimated to cost around $3.3 million with a significant risk of downtime. An alternative approach, that was adopted, was to install a steel casing with a 300 mm HDPE pipe inside. The steel casing was installed by pushing in from the shore and using a mechanical press fit rather than welding of joints. The lowest bidder was awarded the contract for $1.6 million on the basis of using a 400 mm steel casing.

A 150 mm pilot bore was first drilled. It was found difficult to retain the drilling mud. To overcome this a secondary 250 mm casing was installed inside the 400 mm casing penetrating into the lower formations. Up to 330 m of pilot was installed in a day. The bore exited on the seabed about 8 m short of the design location. Barrel reamers were used to open the hole up. The 400 mm casing was inserted by pushing successive pipe sections in using the drill rig. All went well until a rapid increase of pushing force occurred when they were about 30 m from the end of the bore. After modifying the installation to increase the reaction resistance and applying about 80 tonnes of jacking load, the casing was pushed to the end of the bore. The HDPE pipe was then installed inside the casing. The only offshore work was the installation of the diffuser.

At a cost of around $1200 per meter for the outfall, this installation was substantially below the cost of a traditional pulled installation. Disruption to the community and the beach was also avoided.

Other examples

Examples of HDD outfalls are shown in Table 11.3. Installation of outfalls by HDD has been primarily for diameters less than 1000 mm. They have been made in a wide range of geotechnical conditions from sands to hard rock. Often the seabed completion and installation of diffusers was done with traditional offshore methods. Both onshore and offshore pulls of the pipe into the borehole have been done. In the case of Rockaway Beach the installation was all undertaken from shore except for the diffuser.

11.7 NEW DEVELOPMENTS

Trenchless techniques continue to advance. Notable recent developments are:

Pipe Thruster. This could impact HDD outfall construction. It can be used to install pipelines of up to 1.2 m diameter and lengths of 3000 m. The device is introduced into the line at the point of pullback and at the exit side. It can simultaneously push the pipe being installed as the rig pulls back.

Outfall installation by trenchless techniques 385

Table 11.3 Outfalls Installed by Microtunneling

Location	Year	Diam (mm)	Length (m)	Max depth below MSL (m)	Geotechnical conditions	Pipe type	Comments
Santander, Spain		900	615		Limestone	HDPE	400T Rig
Santander, Spain		900	210		Limestone	HDPE	
Grangemouth Scotland	1998	710	1400				
St Andrews, Scotland	1999	450	614	10	Sandstone limestone	HDPE	Drilling from an offshore platform
Crail, Scotland	1999	315	700	10	Sandstone limestone	HDPE	Drilling from an offshore platform
Buckhaven, Scotland	1998	250					
Fraserburgh, Scotland	2000	355	710				
Macduff, Scotland	2001	355	250		Slate quartzite	HDPE	
Cornborough England	2002	710	540		Mudstone	HDPE	
Largs, Scotland	In design						
Kirkudbright, Scotland	In design					HDPE	

(continued)

Table 11.3 (Continued)

Location	Year	Diam (mm)	Length (m)	Max depth below MSL (m)	Geotechnical conditions	Pipe type	Comments
Holyhead, Wales	2004	560	300		Rock	HDPE	
Mear Veg, Isle of Man	2004	800	400		Metamorphic mudstones	HDPE	
Severn, UK	2004	400	400		Fractured limestone	HDPE	
Barton on Sea, UK	2004	560	560			HDPE	
Centerville, USA		102	1300		Mudstone siltstone	Steel	
Rockaway Beach, Oregon USA	2004	400	1370		Sand and marine deposits	HDPE in steel casing	
Twin Rocks, Oregon, USA	2004	200	1450			Steel Permalok	
Agingan Saipan, USA	2003		350		Limestone		
Port Orford Oregon, USA	2004	300	2250	15	Rock	HDPE	600 mm casing installed by ramming for initial 80 m through sand

(continued)

Table 11.3 (Continued)

Location	Year	Diam (mm)	Length (m)	Max depth below MSL (m)	Geotechnical conditions	Pipe type	Comments
Otway-shore crossing for gas line, Australia	2005	725	800	20	Limestone	Steel casing	1200 mm casing 210 m long installed using a retractable microtunneler
Otway-shore crossing for gas line, Australia	2005	268	800	20	Limestone	Steel casing	A 500 mm casing 24 m long installed using a boring machine
Warrenton, Oregon, USA	2006	450	1600	Steel			

Direct Pipe: This combines microtunneling and HDD and allows the direct installation of steel pipelines in diameters up to 1.2 m. The bore excavation is undertaken with a microtunneler and the pipe is pushed in behind using the "Pipe Thruster". This technique is illustrated in Figure 11.24. Unlike HDD it is a single stage operation with full bore excavation and pipe installation simultaneously. Spoil removal is by a slurry circuit. The method has obvious applications to installing steel pipe sea outfalls.

Figure 11.24 The 'Direct Pipe' installation technique
(Courtesy of Herrenknecht AG)

11.8 SUMMARY OF APPLICATIONS

Table 11.4 compares typical applications for the three methods for different site conditions. This is only a broad guide, however, and every site will have unique characteristics that must be fully evaluated before determining the most effective method of installation. It has become apparent in a number of recent projects that the combination of trenchless with traditional methods and the combination of two or more trenchless methods can achieve the best solutions.

Table 11.4 Suitability of trenchless installation methods (After Lutz Meyer)

		Tunnelling	Micro-tunnelling	HDD
Soil conditions	Cohesive	√	√	√
	Sand	√	√	√
	Gravel	√	√	x
	Blocks/boulders	√	x	x
	Mixed	√	o	o
	Rock	√	√	√
Pipe/lining material	PE	x	x	√
	Steel	o	o	√
	Precast concrete pipe	x	√	x
	Tunnel segments	√	x	x
	In-situ lining	√	x	x
Pipe diameter	≤ DN 1000	x	o	√
	⩾ DN 1200	x	√	o
	⩾ DN 2000	√	√	x
	> DN 3500	√	x	x
Laying length	≤ 100 m	x	√	√
	100–500 m	o	√	√
	500–1200 m	√	√	√
	> 1200 m	√	o	o

√ particularly suitable; o conditionally suitable; x unsuitable

11.9 BIBLIOGRAPHY

"Boring under the Holyhead SSI." NATM Magazine February 2005.

Britch, M.J. and Murray, D.S. *"Directional drilling an ocean outfall for Rockaway Beach, Oregon."* No-Dig 2005 Conference Orlando 2005.

Brooks, P. and Perronne, J. *"Marine Construction of the Boston Outfall."* ASCE 1995 Construction Congress San Diego.

Dave Watson, Montgomery Watson Garza, UK. Personal communication.

Directional Drilling Contractors Association. *"Guidelines for a Successful Directional Crossing Bid Package, 1995.*

Eloy Pita *"Design and Construction of a sealine in the Spanish coast – Berria Outfall."* MWWD Antalyla 2006.

Eloy Pita *"Design and Construction of sealines in the Spanish coast."* MWWD Antalyla 2006.

Henderson, J.B. *"Aberdeen Outfall."* Institution of Civil Engineers Conference on Long Sea Outfalls 1988.

Ian Clarke. *"Sewage Treatment for the North East."* No-Dig International August 1997.

Jack Burke. *"South Bay Outfall."* World Tunnelling, March 1997.

Jas, E.P. and McPhee, A.T. *"A state of the art shore crossing."* APPEA Journal 2005-1.

Jez Seamans, LMR Drilling UK Ltd, Birkenhead, UK. Personal communication.

Lutz Meyer. "Methods of construction of tunnelling outfalls, advantages and applications of various techniques." MWWD 3rd International Conference on Marine Waste Water Disposal and Environment 2004.

Martin Cherrington, Cherrington Corporation, Sacramento. Personal communication.

Munz, K. and Haridas, G. *"Marine outfalls project- issues and challenges."* Tunnelling Asia Conference 2000 New Delhi.Philip Hattersley, Montgomery Watson Harza, Singapore. Personal communication.

Shani Wallis. *"Sydney's ocean outfalls keep Bondi Beach crystal clear."* Tunnels and Tunnelling September 1987.

Shani Wallis. *"Europipe sets the record"* World Tunnelling Sept 1994.

Sid French, Burns Dr. and Roe Worley, Sydney, Australia. Personal communication.

Verbecke, R. and Marlier, P. *"Construction de l'emissaire en mer de Marbella a Biarritz."* Tunnels et Ouvrages Souterrain.

12
Operation and maintenance

12.1 INTRODUCTION

This chapter covers operation and maintenance of outfall pipelines and diffusers. It does not cover pumping stations or treatment processes other than milliscreens and grease and oil removal. In comparison to sewage treatment plants, ocean outfalls that are well designed and constructed require little operational effort and only routine maintenance. Although repairs can be quite difficult and costly, they can be precluded or at least minimized by good planning, appropriate design, and careful quality control during construction.

Routine operational requirements for ocean outfalls mainly consists of measuring and recording the flow and head loss in the outfall at predefined intervals, and checking if there has been any significant increase or decrease in headloss. The response to either occurrence should be prompt inspection and/or testing to determine the cause.

© 2010 IWA Publishing. *Marine Wastewater Outfalls and Treatment Systems.* By Philip JW Roberts, Henry J Salas, Fred M Reiff, Menahem Libhaber, Alejandro Labbe, and James C Thomson. ISBN: 9781843391890. Published by IWA Publishing, London, UK.

A significant increase in head could indicate either blockage of orifices in the diffuser or an obstruction in the bore of the outfall pipe. If the increase is sudden it usually indicates damage to the outfall pipe such as from a heavy boat anchor crushing a polyethylene or steel pipe or possibly burial of diffuser ports during a storm that has deposited seabed sediments or moved armor rock over the diffuser orifices. If the increase is gradual it could signify accumulation of grease and/or sediment in the outfall, the gradual burial of orifices due to movement of seabed material, or accumulation of marine organisms around orifices.

A sudden decrease in head would most probably indicate a rupture in the pipeline. This could be caused by displacement of a pipe section and pullout of O-ring pipe joints due to a dragging ship anchor, snagged fishing equipment, liquefaction of the sea bed during an earthquake, or a similar catastrophic event. A gradual decrease in head could signify development of leaks such as might occur with corrosion of ferric materials or deterioration of pipe joints. Both circumstances call for underwater inspection by divers or remote operated vehicles (ROV), to determine the cause. This should be carried out promptly, especially if a sudden decrease occurs.

Maintenance requirements are somewhat different for the different types of pipe materials and are discussed below.

12.2 MAINTAINING A CLEAN PIPE BORE

Outfalls with low internal velocity may require cleaning up to the diffuser to remove accumulated deposits of sediment and/or grease. Figure 12.1 illustrates long term accumulation of deposits in an outfall diffuser and the penalty for ignoring them.

Figure 12.1 Deposits in an outfall diffuser (Pedro Campos, Chile)

Pretreatment of the effluent before discharge will remove most, but not all, grease and sediment. Because grease is less dense than water the deposits are usually thickest on the crown of the pipe, and sediments that are denser than water tend to be primarily deposited on the invert. Grease buildup occurs in great part due to the temperature difference between effluent and ocean water; sediment buildup is primarily due to flow velocities that are inadequate to keep the particles in suspension.

Cleaning grease and sediment from the outfall can be accomplished in two ways:

- The easiest, and usually least costly, method is to periodically increase the flow in the pipe, by pumping or other means, so that scouring velocities are obtained. This is usually effective for removal of grease that has not become tightly adhered to the pipe wall and hardened over a long period of time, and sediment deposits on the invert that have not cemented together. In general, it is easier to dislodge and remove grease and sediment from outfalls constructed of polypropylene pipe than from steel, ductile iron, or concrete pipe. A sufficient velocity is also often able to remove other materials that can accumulate in an outfall;
- More stubborn deposits can be removed by sending a "PIG" through the pipeline. A PIG commonly used in ocean outfalls is a bullet shaped swab with an abrasive outer coating that is usually manufactured of flexible plastic. Figure 12.2 illustrates some of the types of PIGS that are available, each for a special purpose.

Figure 12.2 Various type of PIGs available for specific cleansing purposes (courtesy of Girard Industries)

When using pigs for cleaning an outfall it is necessary to install an entrance for insertion of the PIG on land and an exit hatch through which the accumulated scrapings and the PIG can be removed. The hatch is usually located on the seabed just prior to the diffuser and it is only opened during the "pigging" process. The PIG is propelled through the outfall pipe by increasing the water pressure behind the PIG.

These devices should be compatible with the pipe material and they should be designed to scrape grease and sediment from the pipe walls. A PIG designed for concrete or steel pipe is too abrasive for HDPE and could damage the pipe wall; a PIG that is designed for HDPE pipe may not be abrasive enough to clean steel or concrete pipe. These devices also must be specially manufactured for the inside diameter of the pipe to be cleaned. If the diameter is too small it will not remove all of the accumulated material; if is too large it can abrade the pipe wall and may become stuck. Another precaution in using a PIG is to avoid entrapment of large volumes of air between the water in the outfall and the PIG so as to avoid the possibility of flotation of the outfall.

A unique method of cleaning an outfall was recently used in Gisborne, New Zealand. The flow was first stopped for a while, and then pumped at maximum flow rate for preliminary removal of debris and accumulated material. Then large blocks of ice with ping-pong balls frozen into them were released into the outfall. The ice blocks act as pigs to scour the pipe wall. They cannot jam inside the pipe, because if they do the warm wastewater will melt them. The balls serve as a marker, showing that the pig has traveled through the outfall when they surface.

Since sediment and grease buildup are related to low pipe velocities, their accumulation tends to be more of a problem during the initial years of operation when flow velocities are lowest. For this reason, during the first four or five years of operation it is recommended that preventive maintenance be carried out periodically by increasing the flow for a set length of time to assure cleansing velocities. Recommended flow velocities to ensure cleaning during periodic flushing depend on the pipe diameter as shown in Figure 12.3. If it is not feasible to increase the flow sufficiently to reach these velocities then it is recommended that a PIG be used to clean the pipe after the first few years of operation. The pipe should be cleaned by a PIG or flushing if significant increase in headloss has been observed.

12.3 CORROSION PROTECTION SYSTEMS

Maintenance requirements depend on the pipe material. Outfalls that are constructed of ferrous materials require protection against galvanic corrosion.

Cathodic corrosion protection is used to control surface metal corrosion by making the metal surface the cathode in an electric cell. This is usually accomplished by either the galvanic anode system or the impressed current anode system. Both require routine maintenance, and the impressed current anode system also has some operational requirements.

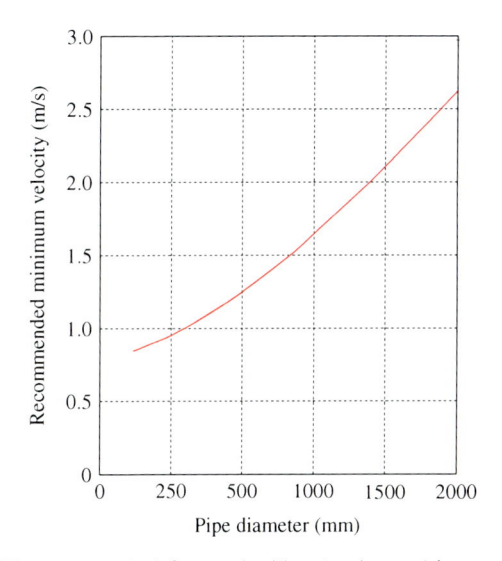

Figure 12.3 Recommended flow velocities to cleanse by periodic flushing

The galvanic anode system consists of a sacrificial anode such as aluminum, zinc, magnesium, or a special alloy, attached to the surfaces being protected. It is attached directly or by an electrical conduit. A ferrous outfall usually has multiple galvanic anode systems placed at intervals along its entire length. The anode material, size, and spacing are selected to yield the current required for protection over a period ranging from 10 to 25 years. For ferrous pipe with rubber O-ring joints, each pipe section should be electrically connected to the adjacent sections or have its own galvanic anode system.

The galvanic anode systems and interconnecting electrical cables between pipe sections must be routinely inspected at established intervals to ensure that electrical continuity has not been broken and that no sacrificial anode has been depleted. Any defects or deficiencies should be promptly corrected because galvanic corrosion of ferrous material in the marine environment can be quite rapid. Divers that have been trained in inspection and repair and corrosion protection, or technicians trained as divers, are required to carry out the inspection and any necessary repairs.

Galvanic anode systems are also frequently utilized to protect ferrous metals such as bolts, washers, anchors, etc. that are used in the construction of non-ferrous outfalls.

The impressed current anode system requires routine operation in addition to routine submarine inspections. An operator must monitor the impressed voltage and current at pre-set intervals and make adjustments to maintain the current within the determined range to provide optimum protection. If there is a failure, such as broken electrical continuity, it will show up as an anomalous readout and a diver should be promptly sent to inspect the system to determine the cause and make necessary repairs. If galvanic caps for corrosion protection are placed on the bolts fastening concrete anchors to HDPE pipe, they should be inspected at least annually and replaced as they approach depletion.

12.4 DIFFUSER MAINTENANCE

Because of nutrients in the effluent, colonies of marine organisms can flourish on the diffuser around the orifices. Organisms rarely invade the orifice itself because of the low effluent salinity. However, if there are extended periods of very low flow or no-flow conditions, organisms can invade the diffuser. If this occurs, it is necessary for divers to remove them. A giant octopus entered a small diameter outfall in Alaska during a no-flow period of several weeks when repairs were being made to a sewage treatment plant. It died from asphyxiation when the flow of anaerobic effluent resumed, but its suction cups were firmly affixed to the pipe wall and the creature could not be removed by increasing the flow. It was necessary to cut open the pipe to remove it. Accumulation of barnacles, sponges, soft coral, and other marine life can accumulate at a surprising rate, so regularly scheduled inspection by divers or ROVs are recommended.

A properly designed diffuser should not trap sediment contained in the effluent; it should be expelled through the diffuser ports. Nevertheless, unanticipated changes in effluent quality and flow may result in sediment buildup within the diffuser, necessitating its removal. It is not possible to send a PIG through a diffuser due to the ports relieving the pressure needed to propel it, and a tapered diffuser precludes use of a PIG because of the decreasing bore diameter. In most cases it is possible to remove accumulated sediment by temporarily increasing the flow entering the outfall. If this does not work, cleaning can be carried out by divers inserting high pressure water jets into the ports.

There have been situations where mass movement of seabed material during storms or gradual transport by strong currents resulted in blocked orifices. This problem is usually considered to be a repair rather than maintenance because the permanent solution necessitates modification of the structure after the deposited material has been removed. This usually entails installation or extension of port risers to an elevation above anticipated future levels of deposits.

If the size and density of bedding rocks used to support a diffuser are not adequate, they can shift during a hurricane or strong storm causing blockage or partial blocking of orifices. If the blockage is not too severe removal of the rock can be handled as a maintenance item. If the problem occurs repeatedly it should be corrected by modifying the ports.

12.5 ROUTINE VISUAL INSPECTION

It is good practice to visually inspect an outfall immediately after it is placed in operation, one year later, and then every two years. In no case should inspection intervals be greater than five years. Divers can usually carry out this inspection up to a depth of 40 m; for greater depths a remote operated vehicle (ROV) or piloted submersible will be required.

It is highly recommended that the inspection be recorded by video to enable viewing by non-diver personnel responsible for upkeep of the outfall. The video and dive log should relate problems that are encountered to the station number on the as-built drawings. Ideally, station numbers should have been permanently and clearly marked on the outfall pipe or concrete ballast weights during outfall construction to facilitate this correlation.

Before this inspection is carried out, the dive team should be provided with a list of items on which to focus. The inspection should include, but not be limited to, any corrosion protection devices, structural or corrosion damage of the pipe, damage to or loss of protective coatings, condition of pipe joints, bolts and other fastening devices, as well as concrete ballasts and mechanical anchors. It should also check for signs of pipe deterioration, misalignment of the outfall pipe, dislocation of armor rock, erosion of seabed under the pipe, accumulation of biological growth on the diffuser, and especially the condition of the diffuser.

The team should also inspect for leaks at each joint, particularly O-ring compression joints. Even small leaks are usually visually evident due to the density and color difference between effluent and seawater. If leaks are suspected in the buried outfall section, dye studies may be used to determine their locations as shown in Figure 12.4.

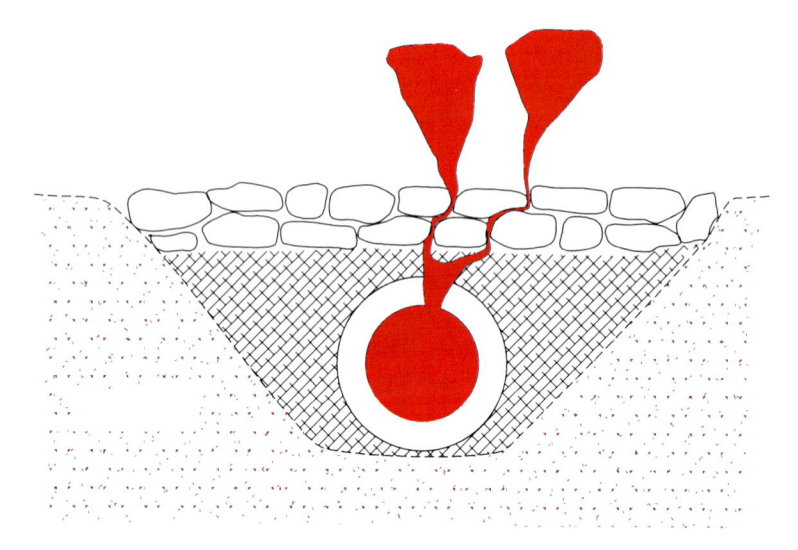

Figure 12.4 Dye studies may for leak detection in buried sections of an ocean outfall

The inspectors should also look for seabed scour that could cause vertical pipe displacement. This is especially significant for outfalls made of concrete or GRP pipe that utilize O-ring compression insert type joints. HDPE pipe can withstand considerable displacement without damage because of its inherent flexibility, but seabed erosion should be noted nonetheless.

Visual inspection should ideally be carried out when visibility in the water is good. If this is not possible, ultrasound imaging may be necessary. This emerging technology has been successfully used to detect defects in zero visibility conditions.

Internal visual inspection of an outfall can also be carried out by divers if the diameter is large enough and entrances have been designed and installed in the outfall at suitable intervals for that purpose. It can also be carried out by a special PIG fitted with a video camera and devices to record its location along the outfall.

12.6 RECORDING DISCHARGE FLOW

It is recommended that flow be continuously recorded during the first year of operation to establish peak and minimum hourly flows and times that are approximately representative of the average daily flow rate. After the first year the schedule for measuring flow can be adjusted according to these data and the results of the recorded headlosses to enable the recording of flow at strategically important hours of the day or for certain days of the week or months of the year.

A flow-measuring device such as a Parshall Flume or Ultrasonic device should be placed at the outfall entrance. The Parshall Flume is only useful if open channel flow occurs at the outfall entrance. Ultrasonic flow measuring devices can be installed on full flow pressurized pipes. The measuring device should be equipped for continuous recording of flow or some flow parameter such as voltage or current.

12.7 RECORDING TOTAL HEAD

The total headloss in the outfall should be recorded for different flows during the initial years of operation to establish a baseline for routine comparisons. The sensors should preferably be located in a place and manner that enable the readings to be translated to meters of head above sea level. During the initial operational period the recordings should be made hourly to establish daily, weekly, and seasonal trends. The reading schedule can then be adjusted to obtain representative readings for peak, minimum, and average flows. A graph of total head versus flow should be prepared for future comparisons to determine if the internal hydraulics of the outfall has changed.

12.8 MISCELLANEOUS MAINTENANCE AND REPAIR

It is good idea to have buoys with signs anchored to the seabed in the vicinity of the outfall. The signs should direct boaters to not anchor within a stated distance from the sign, meaning the outfall. They should also prohibit fishing in the area, especially fishing with bottom trawling nets. Fishing of any kind within a stated distance from the sign should be prohibited. The signs and buoys should be maintained in good condition and the agency responsible for the outfall should report them to the proper authorities to have them entered on nautical charts.

Submerged logs can act as battering rams during storms and should therefore be removed from the vicinity of the outfall if they are observed during visual inspection.

Because HDPE is a soft material, it cannot withstand abrasion against a sharp rock. If a visual inspection shows the pipe between ballast weights is pressing against a sharp rock such as coral, the sharp rock should either be hammered down or the pipe removed from contact with it. During storms or heavy seas, vibration of the pipe against sharp rocks or coral formations can wear a hole completely through the pipe wall in a matter of days or even hours.

If routine inspection reveals that seabed erosion is occurring around the outfall because of turbulence from currents or waves, the eroded seabed should be replaced with a material of greater size and/or density such as coarse gravel, cobble stone, armor rock, or bags filled with a dry sand and cement mix, whichever is capable of resisting further erosion.

12.9 COSTS

The costs associated with operation and maintenance (O&M) vary greatly and depend on many factors. The most important is the appropriateness and quality of the planning, design, and construction. Another important factor is whether or not pumping is required and if so under what hydraulic conditions. Other factors include the outfall depth and length, how much of the outfall is buried and how much is resting on the seabed, the pipe material, the severity of the hydrodynamic sea conditions, the prevailing marine organisms, proximity of shipping corridors and commercial fishing, and the proximity of sediment sources such as large rivers. An additional factor that can greatly affect operational costs is whether or not the responsibility for environmental monitoring, sampling, and testing rests with the entity that is responsible for operation and maintenance.

Annual O&M costs are often expressed as a percentage of the initial investment cost. They can be as low as 0.5% for an appropriately designed and constructed, relatively short gravity flow outfall with a depth less than 30 m located in dependably calm waters. A more typical range is probably 1 to 2%, but under unfavorable circumstances it could be as high as 4%.

12.10 SUMMARY

A summary of the minimum typical activities and responsibilities for the operation and maintenance of marine outfalls is presented in Table 12.1.

Table 12.1 Typical activities and responsibilities for operation and maintenance of marine outfalls

Activity or responsibility	Suggested frequency
Removal of material accumulated on the milliscreen	Daily with adjustment of frequency with experience
Removal of grease, oil, and floatable material from the grease trap	Daily with adjustment of frequency with experience
Measurement of wastewater flowrate	Continuously
Ensure the flow meter is functioning correctly	Daily
Record flow in the operating log	Daily
Measure hydraulic head above sea level at peak hourly flow and record it in the operational log	Weekly
Inspect the surge chamber or other outfall entrance and clean as necessary	Daily
During the first years of operation, if the flow is not sufficient to guarantee self-cleansing of the pipe, increase the flow rate by supplemental means to obtain sufficient velocities. With time, after the flow increases sufficiently to guarantee self cleansing, this activity is not necessary.	Weekly
Visually inspect the entire outfall, and record the inspection with videos and photos utilizing divers or ROV.	Immediately after the outfall is placed in operation, 1 year later, and every 2 years thereafter.
If flushing the outfall does not successfully remove the accumulated sediments and grease, clean the outfall with a PIG.	As required
If there is a high point in the outfall equipped with an air release valve have a diver check to see that it is operating properly	As required
Observe the sea level and conditions (waves) of the sea and the weather and record this in the operators' log. This information is especially important during storms and hurricanes if possible.	Daily
Visually inspect entire outfall and carry out necessary repairs to outfall pipe and diffuser as well as replacement of bedding and armor rock, etc.	As soon as possible after a major storm, exceptionally rough seas, earthquake, etc.

13
Monitoring

13.1 INTRODUCTION

A monitoring program should be implemented with any outfall scheme to verify the performance of the system and its compliance with water quality criteria and to determine the need for changes in wastewater treatment and the marine management plan. The program will establish baseline levels of potential contaminants before discharge begins, measure any impacts of the discharge, and monitor beaches to ensure compliance with bathing water standards.

Plans for two types of monitoring programs are presented in this chapter. The first is specifically related to the proposed outfalls, and consists of both pre- and post-discharge monitoring. The second is a beach monitoring program to ensure compliance with recreational water quality standards. We are here discussing only routine monitoring. Other types of studies may sometimes be required by the regulatory agency, such as direct measurements of near field dilution and wastefield behavior. An example is that required for the discharge

© 2010 IWA Publishing. *Marine Wastewater Outfalls and Treatment Systems.* By Philip JW Roberts, Henry J Salas, Fred M Reiff, Menahem Libhaber, Alejandro Labbe, and James C Thomson. ISBN: 9781843391890. Published by IWA Publishing, London, UK.

permit of the Boston outfall (Roberts *et al.*, 2002). These studies require the addition of a known quantity of a tracer to the effluent, such as Rhodamine WT dye, measurement of the dye *in situ* by an instrument such as a submersible fluorometer to measure dilution, and simultaneous measurements of wastewater flow rate, current speed and direction, and instrument depth. The logistics of such an operation are complex and require careful planning to ensure that useful information is obtained. Because of this, they are not discussed further here; for an example, see Roberts *et al.* (2002).

The best monitoring programs are those designed to answer specific scientific and regulatory questions, such as:

- Do effluent contaminant concentrations exceed permit limits?
- What are the concentrations of contaminants and characteristic tracers of sewage in the influent and effluent and their associated variability?
- Are the near and far field model predictions of short-term (less than 1 day) effluent dilution and transport reasonably accurate?
- Do contaminant levels outside the mixing zone exceed water quality standards?
- Are bacteria transported to shellfish beds at levels that might affect shellfish consumer health?
- Are bacteria transported to beaches at levels that might affect swimmer health?
- Is there any evidence of eutrophication or other nutrient impacts of the discharge?

As the questions are answered, new issues may arise, and the monitoring program and sampling plan should be continuously updated. For an excellent example of this process, see MWRA (2004). Monitoring programs for a major outfall can be quite detailed and complex. Because they are site-specific and depend on many variables, we can here only give some general guidelines to aid in planning a particular study.

13.2 MONITORING PARAMETERS

The main parameter of concern is often indicator bacteria because all other wastewater constituents are subject to rapid dilution and dispersion, resulting in low concentrations around the outfalls. Nutrients are sometimes monitored to address potential concerns about eutrophication and potential effects on coral reefs.

The main purpose of indicator bacterial monitoring is to ensure and demonstrate that the beaches meet relevant bathing water quality standards for protection of human health. As discussed in Chapter 2, many standards or guidelines have been proposed by various agencies around the world, including the California State Water Resources Control Board (Ocean Plan), the California Department of Health Services (DHS), the World Health Organization (WHO), the European Union (EU), and the U.S. Environmental Protection Agency (USEPA). These standards are mainly based on total and fecal coliforms, and enterococcus/fecal streptococcus. Epidemiological investigations have shown that the intestinal enterococcus has the best relationship with disease incidence, and is also more able to survive in salt water.

Bacterial monitoring therefore often consists of measuring total and fecal coliforms and enterococci. The detection methods can be those presented in the most recent edition of "Standard Methods for the Examination of Water and Wastewater, 21st Edition" (APHA, 2005), those of the International Standards Organization (ISO), or any other approved method. Detection methods for enterococci are also presented in USEPA (1997). Standard and approved sample collection procedures should be followed and the samples analyzed in a certified laboratory.

13.3 OUTFALL-RELATED MONITORING

13.3.1 Introduction

Monitoring specifically related to the outfalls should be conducted both before and after discharge commences. The purpose of pre-discharge monitoring is to assess environmental conditions prior to discharge to establish baseline conditions to assess any changes that may result from the discharge. The purpose of post-discharge monitoring is to confirm the system effectiveness and plume behavior, to show compliance with permit requirements, and to measure any environmental effects of the discharges. The parameters monitored depend on environmental considerations and permit requirements and should be measured *in-situ* near the diffusers and at nearby beaches. The protocols for sampling in the marine environment must include quality assurance and quality control (QA/QC).

13.3.2 Pre-discharge monitoring

Pre-discharge monitoring should generally include at least two intensive surveys during different seasons. The purpose of this is to assess seasonal variations in oceanic and environmental conditions.

Temperature, salinity, dissolved oxygen, and transmissivity profiles through the water column should be made using a multi-parameter probe. Currents should also be measured from the survey boat by a downward-looking ADCP, if available. Depending on the local water depth, measurements should be made at one to five meter increments through the water column. Water samples should be obtained at least at three depths: the water surface, mid-depth, and as near to the seabed as possible. The samples shall be analyzed for indicator bacteria, biochemical oxygen demand (BOD$_5$), suspended solids, and grease and oil. Samples should be taken along the outfall alignment at, for example, the diffuser location, at 2/3 and 1/3 of the distance from shore to the diffuser, and at 300 m from shore. The proposed diffuser locations should be verified before commencing monitoring in case changes have occurred as a result of the ongoing design process.

To address potential eutrophication concerns, the following nutrients could also be measured:

- Nitrogen series (organic N, NH$_4$, NO$_2$, NO$_3$);
- Total phosphorus and ortho-phosphate;
- Chlorophyll *a* (euphotic zone).

Any recreational beach at potential risk due to the outfall discharge should be sampled for indicator bacteria. Samples should be taken weekly at each beach in the morning before 10 am. The samples should be taken according to standard protocols.

Wastewater quality and quantity should also be measured along with surface runoff quality and quantity. This will allow an evaluation of possible impacts of non-point sources of contamination that would continue uncontrolled after outfall construction. This will also provide a basis for a comparison of non-point sources and sewage discharges.

13.3.3 Post-discharge monitoring

13.3.3.1 Physical and chemical sampling

Post-discharge monitoring also consists of offshore water and beach measurements. Samples would typically be obtained four times per year at different seasons at the pre-discharge stations. Vertical profiles can be taken at five points located approximately 100 m from the diffuser. If the plume can be detected from the salinity or transmissivity profiles, additional water quality samples should be taken in the middle of the plume. Otherwise, mathematical models should be run using the measured data to predict the plume rise height and samples then taken at the predicted height.

In addition, water samples should be taken at stations in a grid that includes the diffuser and extends about 500 m up and down coast from the diffuser. Samples should be taken between the diffuser and the shoreline, at distances of 300 m, 1/3 and 2/3 of the distance to shore. The purpose is to measure any offshore gradients of contaminant properties to assess whether any effluent is approaching or reaching the shore. Reverse gradients (shore to outfall) would indicate shore-based sources such as runoff.

The parameters sampled and their spatial and temporal sampling frequencies can be adjusted as time proceeds, but should at minimum be those measured in the pre-discharge program, Section 13.3.2.

13.3.3.2 Biological sampling

Biological sampling will not generally be necessary, except for very large discharges or discharges suspected of impacting coral reefs. It will depend on any particular local concerns such as fisheries or nearby areas of particular ecological significance such as coral reefs. Following are some suggestions for parameters to be measured in these cases.

Water samples can be analyzed for phytoplankton biomass, as chlorophyll *a*. Zooplankton can be sampled in triplicate using a conical net having a 50 cm mouth opening and 200 μm mesh. Vertical hauls of the water column should be made near each diffuser.

The health of any local coral reefs should be monitored by SCUBA divers to determine whether chronic effects are occurring. Sampling should be done along established fixed transect lines whose start and end points are marked with permanent stakes with known GPS coordinates. The transects are monitored for bottom substrate composition and fish abundance and diversity and the abundance of floral and faunal macrobenthic organisms associated with the coral reef is also categorized. Reef fish should be visually identified by divers, and enumerated and sized.

Reef fish are a vital part of the reef community and may be caught for consumption. Thus, the levels of potential contaminants in fish should be measured. To do this, individual species of fish are captured, muscle and liver tissues resected, and samples analyzed to determine their chemical profile. The samples are analyzed for moisture and lipid content and chemicals such as listed in Table 13.1.

Table 13.1 Marine fish tissue monitoring parameters

Physical	Chemicals
Lipid content	Mercury
Moisture content	PCBs

408   Marine Wastewater Outfalls and Treatment Systems

Indicator species should be selected for specific monitoring to provide focused long-term results. Some suggested organisms are:

- algae, sponges and bivalves encrusting mangrove roots;
- soft bottom fauna such as polychaetes, bivalves, burrowing crustaceans;
- holothurians;
- hard-bottomed species such as coralline algae, sponges and sea urchins;
- herbivorous fish.

Bioassays might also be considered.

Benthic invertebrate samples can be collected to assess changes to community composition. Retained benthic organisms should be analyzed for taxonomic composition and abundance.

The location and frequency of sampling are based on the spatial and temporal resolutions over which measurable changes are expected. Monitoring should be conducted near the diffusers and beyond to provide information on spatial gradients of any environmental impacts, and the sampling locations should be spaced to determine these gradients. In addition, reference locations should be sampled to control for non-outfall related environmental changes. Sampling will also be done at key coral reef locations.

The monitoring results obtained at different times and different stations are compared to evaluate spatial and temporal changes. They are also compared to changes occurring at reference locations outside the project area to account for natural changes to the environment.

13.3.3.3 Wastewater characteristics

Samples of wastewater should be obtained after leaving the treatment plant just before entering the outfall pipeline. The samples should be collected using strict and consistent quality assurance and quality control (QA/QC) procedures, including clean sampling methods, replicated sampling, field and travel blanks. At a minimum, the parameters analyzed will include the physical and microbiological parameters listed in Table 13.2. PCBs and selected heavy metals may also be measured if thought to be present.

13.3.4 Sample monitoring program

The following monitoring example is considerably more extensive than would generally be required and much more than we would recommend. We provide it only to indicate an upper extreme of parameters and their sampling frequencies from which the reader may extract a more reasonable program. It is for the outfall to be built at Cartagena, Colombia (Figure 13.1). The initial monitoring program is for two years, one before construction, and one after. Three campaigns will be conducted each year.

Table 13.2 Wastewater monitoring parameters

Physical/Biological/Chemical	Microbiological
Flow	Total coliform
pH	Fecal coliform
Total suspended solids (TSS)	Enterococci
Turbidity	
Grease and oil	
Nutrients	
BOD_5	

The following water quality parameters will be measured at stations A, B, C, and D: Temperature; transparency; salinity; pH; dissolved oxygen; nutrients; oils and fats; BOD_5; suspended solids; phenols and detergents; total and fecal coliforms.

Oils and fats are measured only at the surface, and temperature, salinity, pH, and dissolved oxygen, will be profiled through the water column. Parameters that require laboratory analysis will be obtained at three depths: surface (0.5 m), mid-depth (about 5 m), and bottom (greater than 10 m) and will be collected once in each sampling campaign.

Bottom sediments will be collected and analyzed at Stations O, A2, B2, C2, A4, B4, C4, D1, D2, D3, and D4. The samples will be analyzed for: Organic Nitrogen; organic Carbon; Copper; Lead; Zinc; Cadmium; Arsenic; Nickel; Chromium; Silver; Manganese; Aluminum; Chlorinated organic pesticides; Polycyclic aromatic hydrocarbons; grain size sorting; and polychlorinated biphenyl (PCB).

Phyto and zooplankton will be obtained and measured at Stations O, B2, B3, B4, and D1. This will consist of three sampling campaigns each year for two years. The parameters to be measured are: Composition; abundance-biomass; density; and Chlorophyll concentration.

Benthic communities will be sampled at stations O, A2, B2, C2, A4, B4, C4, D1, D2, D3, and D4.

Fish communities will be sampled for: Composition; abundance-biomass; spatial distribution; and diversity.

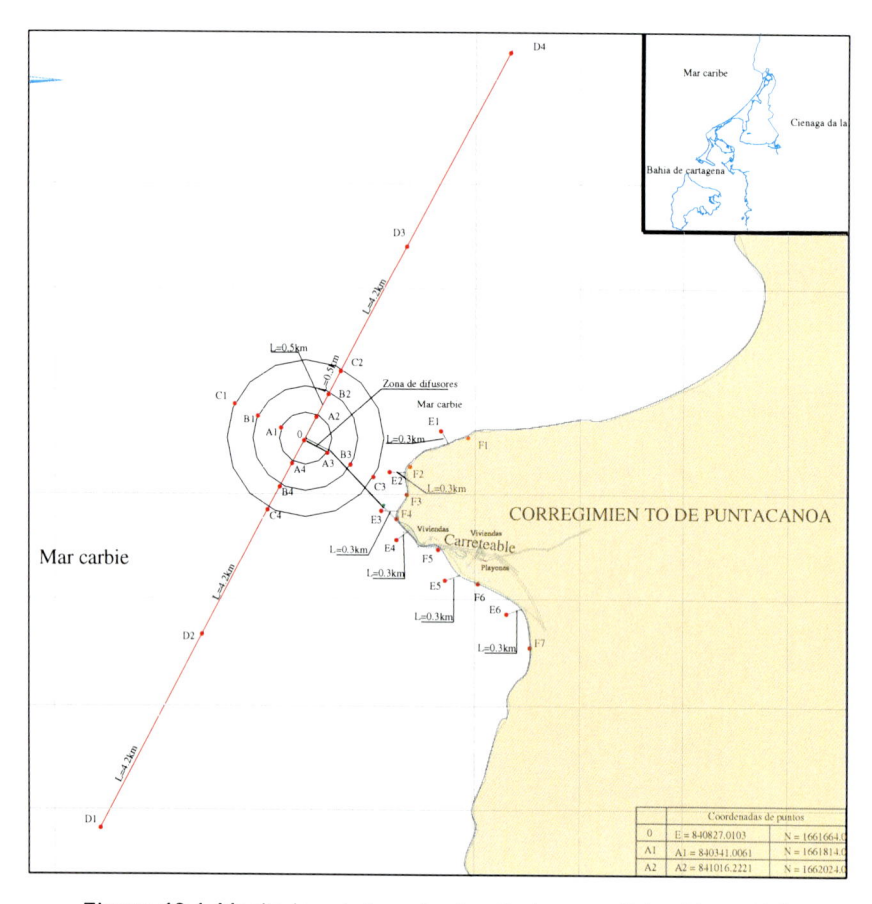

Figure 13.1 Monitoring stations for the Cartagena, Colombia, outfall

Microbiological quality at the beaches at Stations E, F will be monitored monthly and analyzed for total and fecal coliforms.

13.4 ROUTINE BEACH MONITORING

The purpose of beach monitoring is to ensure and demonstrate compliance with bathing water standards and the safety of the bathing beaches. This is especially important for beaches frequented by tourists.

Routine national and/or local beach surveillance programs are conducted in most countries at specified locations and frequencies, for example, once per week or fortnightly, to comply with water quality standards and classify and manage bathing beaches. Should a more detailed evaluation of the performance of an outfall be required by a regulatory agency, a site-specific monitoring program complementary to the routine surveillance programs should be implemented. See Reeves and Bartram (1999) for details on bathing beach monitoring.

As part of the WHO development of the Guidelines of Safe Recreation Water Environments (WHO, 2003), an Experts Consultation was held in 1999 in Annapolis, Maryland, USA (WHO, 2000) co-sponsored by the WHO and the USEPA. The *Annapolis Protocol* that resulted from this meeting proposed an innovative approach to bathing beach classification based on risk evaluation using long term water quality data combined with sanitary inspections and the application of beach management practices.

Present day bathing beach classification systems are generally based on mandated periodic vigilance monitoring (5 samples per month is common) or single measurements of indicator organisms consistent with water quality standards. Such approaches may or may not capture transient events such as storms and the single measurement systems have the inherent flaw of after-the-fact bathing beach classifications due to the 24 to 48 hour delay of microbiological indicator measurement results (although new rapid methodologies are presently being tested by the USEPA, Haugland *et al.*, 2005; Wade *et al.*, 2006; and Wade *et al.*, 2008). Furthermore, a monitoring program of fixed periodic measurements applied indiscriminately to all beaches would result in the same repeated values and beach classification for those beaches that are either heavily contaminated or pristine and, as such, would be an inefficient use of monitoring resources.

The reader is referred to WHO (1999, 2003) for details of the *Annapolis Protocol*. This protocol encompasses a combination of sanitary inspection, water quality measurements, and risk categories to monitor and classify beaches. The proposed monitoring frequency is presented in Table 13.3.

13.5 SUMMARY

A monitoring program should be established to determine any impacts of the discharge, to verify predicted effects, and to provide a rational, scientific basis for changes in wastewater treatment. Biological, physical, and chemical parameters should be monitored in the marine environment over suitable spatial and temporal scales to evaluate short and long term impacts. The monitoring program should ideally be designed to answer specific scientific questions rather than the simple performance of routine sampling.

Table 13.3 Recommended monitoring schedule (WHO, 2003)

Risk category identified by sanitary inspection	Microbial water quality assessment	Sanitary inspection
Very low	Minimum of 5 samples per year	Annual
Low	Minimum of 5 samples per year	Annual
Moderate	Annual low-level sampling 4 samples × 5 occasions during swimming season Annual verification of management effectiveness Additional sampling if abnormal results obtained	Annual
High	Annual low-level sampling 4 samples × 5 occasions during swimming season Annual verification of management effectiveness Additional sampling if abnormal results obtained	Annual
Very high	Minimum of 5 samples per year	Annual

Monitoring should begin one year before outfall commissioning to obtain background data at the same locations intended for future monitoring in order to assess changes due to the outfall. After commissioning, monitoring frequency will vary for different parameters. After the first year of operation, the monitoring frequency could be decreased unless specific effects have been identified, in which case monitoring of specific parameters may be increased. A suggested monitoring program is outlined in Table 13.4.

Table 13.4 Sampling summary

Activity	Pre-discharge	Post-discharge	
	One year before commissioning	First year of operation	After first year of operation
Effluent characteristics: pH, turbidity, temperature, conductivity	Historical data	Monthly Bi-weekly	Monthly Bi-weekly
Water column physical structure: Temperature, salinity, dissolved oxygen, and transmissivity	Biannually	Quarterly	Biannually
Ambient water quality	Biannually	Quarterly	Biannually
Plankton	Biannually	Quarterly	Biannually
Coral reefs and fish tissue	Biannually	Biannually	Annually
Benthic invertebrates	Biannually	Quarterly	Annually

14
Case studies

14.1 INTRODUCTION

This chapter discusses three outfalls in Latin America. The first refers to the water and sanitation sector in Cartagena, a vacation resort city on the Caribbean coast of Colombia, which has a current population of about one million. It describes the water sector reform that has been undertaken with private sector participation, and the World Bank financed project. It explains the wastewater disposal strategy, consisting of discharge of preliminary treated effluent via an outfall into the Caribbean Sea. The results of outfall modeling, based on two years of oceanographic measurements, demonstrate the environmental benefits of the project and served as the basis for the outfall design. Difficulties in obtaining the environmental license for the outfall due to opposition of various interest groups are described, along with the strategy to gain consensus through the aid of a high level panel of international experts that provided support in clarifying technical issues.

© 2010 IWA Publishing. *Marine Wastewater Outfalls and Treatment Systems.* By Philip JW Roberts, Henry J Salas, Fred M Reiff, Menahem Libhaber, Alejandro Labbe, and James C Thomson. ISBN: 9781843391890. Published by IWA Publishing, London, UK.

For coastal cities in developing countries, wastewater disposal by preliminary treatment followed by an appropriate outfall is affordable, effective, simple to operate and free of negative impacts, with a proven track record in many coastal cities throughout the world. However, objections to this concept often, in developing countries, results in no action towards wastewater disposal being taken, since treatment above and beyond preliminary would render most projects financially non-viable. Avoidance of actions towards solving wastewater disposal problems is the worst option of all, which usually leaves the vulnerable population (mostly the poor) under the worst conditions. In Cartagena, for instance, if the outfall project were derailed on the basis that higher than preliminary treatment is required, nothing related to wastewater disposal would be done for many years, since any other sustainable project would be much more expensive, and additional financing is not available. That would leave the poor neighborhoods with unacceptable sanitation conditions for a long time.

14.2 CARTAGENA, COLOMBIA

14.2.1 Background

Cartagena de Indias, the fifth largest city of Colombia, has a population of about one million and has been growing in recent years at about 2.5% per year. It is surrounded by water on almost all sides – the Caribbean Sea to the north and west, Cartagena Bay to the south and the Cienaga de la Virgen lagoon (an in-city large coastal lagoon hereafter called the Cienaga) to the northeast. It is also traversed by several interconnected water courses (caños y lagos). Thanks to its historical landmarks, spectacular natural scenery, and tropical climate, Cartagena is Colombia's largest tourism area, as manifested by its name: Distrito Turistico y Cultural (Turistic and Cultural District), with a large annual influx of about 700,000 national and foreign visitors. A floating population of thousands of tourists can be encountered in the city all year long, with peak numbers during vacation seasons and holidays. UNESCO declared in 1984 the Old City, Fortresses and Group of Monuments as a Cultural Heritage Site of Humanity.

The city's economy is heavily dependent on tourism, which generates an estimated US$315 million annually. Cartagena has also a thriving industrial sector with important petrochemical, beverage and seafood processing industries, most of which are located in the Mamonal industrial estate, and is an important port city and a gateway for supplies, goods, lumber, and general merchandise imports to the country The port moves over 10 million tons of cargo per year.

The city and its surroundings is a complex area, particularly from socio-economic and ecological perspectives. Cartagena is a destination for rural Colombians displaced by violence in the countryside looking for better economic opportunities, making it one of Colombia's most rapidly growing cities; Cartagena has doubled its population in the past 20 years. As a result of high immigration of the poor from other parts of Colombia, 84% of the fixed population is of low and medium-low income, 31% of which is extremely poor. About 14% of the population is of medium and medium-high income and less than 2% is high income.

The city's water resources contribute significantly to the quality of life of its residents by providing venues for both leisure (swimming and water sports) and commercial activities (fishing and commerce). Unfortunately, the water bodies surrounding Cartagena are severely fouled by wastewater as they are repositories of the untreated municipal and industrial liquid wastes of the city. The total flow is 145,000 m^3/day, 30% of which is discharged into Cartagena Bay, 60% into the Cienaga, and 10% to the in-city water courses. Overflows from the overloaded sewage collectors in the wealthy neighborhoods of the city contaminate the beaches and the in-city water courses, and industrial wastes from the Mamonal industrial estate contribute additional contamination loads to the Cartagena Bay. The population residing in the poor neighborhoods around the Cienaga suffers from the worst sanitation conditions, since, as a result of lack of sewerage in these neighborhoods, raw sewage flows in its streets. As the city grew rapidly, water and sanitation infrastructure lagged behind. The rich parts of Cartagena, including the hotel areas, have high coverage of water and sewerage, but the poorer parts have low coverage. This situation has generated severe public health, sanitation, and environmental problems, which, in addition to their direct impact on the population in deteriorating quality of life, pose a serious impediment to the sustainable economic development of Cartagena, especially in the important tourism sector.

Ocean sewage discharge is not an unknown technology in Cartagena. A 38 year-old outfall (800 m long, 30 inch diameter, 22 m deep, with no diffuser) discharges raw sewage (without even preliminary treatment) into the Cartagena Bay. This outfall, an important factor in preventing even worse sanitary conditions in Cartagena during the past three decades, has outlived its design capacity. The sewage plume is usually trapped below the surface because of thermal stratification. Bacteriological pollution in the inner bay is high, however, and floating materials are visible when the plume surfaces.

14.2.2 Institutional aspects of the Cartagena water and sanitation sector

For many years prior to June 1995, the water and sewerage services in Cartagena were the responsibility of Empresas Publicas Distritales (EPD), an overstaffed and inefficient public utility. This utility suffered from continually increasing financial problems, resulting in poor and deteriorating performance and a low level of service delivery. During 1994 and early 1995, a major initiative was taken by the Municipality, supported by the National Government and the World Bank, to incorporate an experienced private operator in the management and operation of Cartagena's water and sewerage services. This initiative was brought to a successful culmination by liquidating EPD and transferring the responsibility for providing water and sewerage services, as of June 25, 1995, to ACUACAR (Aguas de Cartagena). ACUACAR is a mixed capital company constituted as an association between the Municipality of Cartagena and Aguas de Barcelona (AGBAR), an experienced European private operator. The District of Cartagena was the first municipality in Colombia to incorporate private sector participation into the water sector.

Since the takeover of responsibility for water and sewerage services by ACUACAR, major efforts were invested in developing and implementing the water supply and sewerage master plan of the city. Major investments have been carried out, financed by the District of Cartagena and ACUACAR, and with Government support. For that purpose, ACUACAR took on major loans from local commercial banks (for an amount equivalent to US$41 million at the time the loans were received) and from the Interamerican Development Bank, IDB, for an amount of US$24 million, used exclusively for sewerage works in the Cartagena Bay drainage basin.

Since its constitution, ACUACAR has been a well-managed water and sewerage service company, the operational performance of which compares well with the highest international standards. Corporate operation has developed along modern partnering lines. The ACUACAR model brings together a private operator and a public authority in a service company under transparent corporate governance rules, providing an opportunity for disciplined activity and accountability before customers and shareholders. The partnership of the two major stock holders of the company, the District of Cartagena and Aguas de Barcelona, works well. The privatization of water and sanitation services in Cartagena is considered the most successful in Colombia and has served as a catalyst for similar privatization processes in other Colombian cities.

During its first seven years of operation, ACUACAR has achieved significant improvements in operational performance. These include putting the business on a sound commercial footing, increasing efficiency and staff commitment, creating a credit-worthy public/private company, providing customer satisfaction and dialogue, and increasingly reliable audited information. Improvements include: (i) increasing water supply, in production, treatment and coverage; (ii) increasing sewerage coverage; (iii) improving water quality; (iv) introducing modern commercial and recognizable management information systems; (v) installing water meters on practically all its customers connections; (vi) improving continuity of service; (vii) rationalizing labor force; (viii) putting in place a staff training program and creating an enabling and encouraging working environment; (ix) investing heavily in Information Technology/Information Systems; (x) introducing an environment of maintenance of assets; (xi) introducing automated control of processes and operation; (xii) improving efficiency and financial performance; (xiii) undertaking significant investments for expansion of services by borrowing from local banks and multilateral donors, although such investments are not mandatory under the ACUACAR contract, (xiv) improving relations with customers; and (xv) creating relations of trust with suppliers. These efforts resulted in: (i) ISO 9002 accreditation for its quality of operations and laboratory (the first water utility in Latin America which obtained such accreditation); and (ii) high customer satisfaction rating achieved in customer surveys.

The achievements of ACUACAR are demonstrated by the improvement of its performance indicators (Table 14.1 and Figure 14.1).

Table 14.1 Performance indicators of the water utility of Cartagena before and after takeover by the private sector

Indicators	Before PSP (1994)	After PSP (2004)	Common values in industrialized countries
Number of employees	1,300	272	
Number of employees per 1,000 connections	15	2.3	2
Water coverage	68%	99%	100%
Sewerage coverage	56%	78%	100%
Domestic metering	30%	99%	100%
Number of connections	84,143	117,194	
Unaccounted for water	60%	41%	25%
Production capacity (m^3/s)	1.6	3.1	
Continuity of service (hrs/day)	7	24	24
Response to complaints (days)	6	0.5	0.5
Connections in poor areas (strata 1 and 2) as percentage of new connections installed in 1995–99		92%	

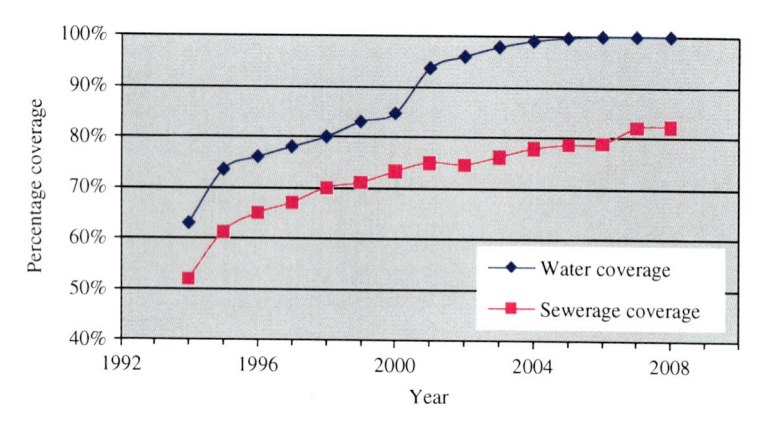

Figure 14.1 Evolution of the water and sewerage coverage in Cartagena since incorporation of the private sector

14.2.3 The World Bank's support to Cartagena

In spite of this progress, the water and sanitation sector in Cartagena still faces the following challenges:

- *Insufficient water supply coverage:* Despite continuous major efforts and investments by ACUACAR and the District of Cartagena to expand the water coverage and improve service, the current coverage of about 90% is still insufficient and needs to be increased. Most of the areas lacking water distribution networks are the poor neighborhoods in the southwest part of the city, near the Cienaga.
- *Insufficient sewerage services:* The sewerage coverage level in the city is currently about 73%, leaving about a third of the population, especially in the poorest areas around the Cienaga and in part of the southwestern zone adjacent to the Cartagena Bay, without sewerage services. In these areas raw sewage flows in open channels in the streets. Since the Cienaga itself is heavily polluted and open canals carrying sewage from other parts of the city cross poor neighborhoods on their way to the Cienaga, Cartagena's poor population is exposed to about 60% of the city's untreated sewage. The sewage collection system is particularly deficient in the southeast and southwest sections of the city. The network in other parts of Cartagena is very old and many sewers are hydraulically constrained, particularly during the peak tourism seasons. In view of deficient sanitation services, city authorities are forced to restrict construction of new buildings in order

to avoid a further deterioration of services. Sewage collection and disposal problems represent a major obstacle to Cartagena's overall economic development.
- *Inadequate wastewater management system:* Cartagena has no sewage treatment facilities, and liquid wastes from municipal and industrial areas are discharged without treatment to the Cartagena Bay (receiving 30% of the discharge), the Cienaga (60%), and the in-city water courses that eventually reach the bay and the Cienaga (10%). Projections for 2025 indicate that Cartagena will generate about 305,000 m^3/d of sewage. The discharge of raw wastewater is creating serious environmental and health problems and is hampering activities such as swimming, fishing, and shellfish harvesting.
- *Industrial wastewater discharges in Mamonal:* The industrial zone of Mamonal is located on the outskirts of the District. It consumes an estimated 20,000 m^3/d of potable quality water and 61,800 m^3/d of raw water for process and cooling requirements. The industrial zone is not connected to the municipal sewerage network but rather discharges its wastewater to the Cartagena Bay, after only partial or minimal treatment.

Institutional weakness of public utilities is usually the main impediment to sustainable development and mobilization of investment funds by such utilities. In the case of Cartagena, institutional problems have been dealt with up-front through the creation of ACUACAR. Consequently, the World Bank agreed to provide, as of 2000, partial financing to support an investment project for the water and sanitation sector of Cartagena, aimed at addressing the mentioned issues.

Following is a description of the project components:

- *Expansion of the water supply system:* This includes: (i) expansion and improvement of the water production system; (ii) increasing water coverage in the city; (iii) replacement of primary distribution mains; (iv) mitigation of environmental impact of water treatment sludge; (v) remote control systems; and (vi) implementation of an Unaccounted For Water (UFW) reduction plan.
- *Expansion of sewerage system in the Cienaga Basin:* This includes: (i) enhancement of conveyance capacity of existing sewage collectors in the southwest, southeast, and central parts of the city that currently drain to the Cienaga; (ii) expansion of secondary sewerage network in the southwest, southeast, and central parts of the city, as well as the Boquilla area, that currently drain to the Cienaga; (iii) construction of new pressure

lines and pumping stations; and (v) construction of new gravity collectors in residential areas.
- *Construction of the main wastewater conveyance system:* This includes: (i) upgrading the Paraiso pumping station; (ii) construction of the pipeline from Paraiso pumping station to the treatment plant site; and (iii) construction of the effluent pipeline from the treatment plant to the submarine outfall at the Caribbean shoreline. The conveyance system would consist of a 72″ reinforced concrete cylindrical pressure pipe (RCCP) with a total length of 23.85 km.
- *Construction of wastewater treatment facilities:* The preliminary treatment installation will remove floatable materials, grease, oil, sand, and grit. It includes six rotary screens (0.6 mm clearance) followed by two vortex-type grit chambers. The rotary screens will remove rags, floatable material, and large solids. The expected volume of screenings generated by the rotary screens is about 8.5 m^3/d, and the vortex type grit chambers will remove about 5.1 m^3/d of sand and grit.
- *Construction of an ocean outfall:* This consists of construction of an outfall for the safe discharge of the treated effluent to the Caribbean Sea near Punta Canoa. A main conveyance system will connect the treatment plant with the submarine outfall. The outfall would be constructed using a 72″ reinforced concrete (subsequently changed to a two meter diameter HDPE) pipe. The total outfall length would be 2,850 m and the diffuser will be submerged at a depth of 20 m. The diffuser will have a length of 540 m, with a riser spacing of 20 m, i.e., a total of 27 risers. Each riser will be made of a 12″ diameter pipe, 2 m long. The upper end of each riser will be sealed by a welded plate, with 2 openings of 8″ diameter close to the plate, i.e., the total number of discharge ports will be 54. In the surf zone between the shoreline and the 3 m-depth contour, outfall pipes will probably be laid using a trestle built over the water.
- *Industrial wastewater discharge control:* This would address issues related to industrial wastes discharged to the municipal sewerage network, and includes the following activities: (i) a survey to identify key sources of industrial pollution; (ii) establishing a system for regulating the discharge of industrial wastes to the sewerage system or to receiving bodies; (iii) establishing a system for auditing the status of industrial waste discharges; (iv) defining strategies to control small and dispersed sources of industrial pollution discharging to the sewerage networks (gasoline stations and mechanical repair shops); and (v) providing technical assistance in selection and design of pretreatment processes.

Case studies   421

- *Environmental and social component:* This part of the project seeks to implement mitigation measures of the environmental and social impacts of the project. The environmental management program includes: (i) environmental supervision during construction; (ii) restoration and conservation of the Cienaga de la Virgen nature reserve; (iii) carrying out a monitoring program before and after construction of the marine outfall to study the fate of pathogenic coliforms and other contaminants discharged through the outfall; and (iv) an environmental institutional strengthening program. The social management program includes: (i) organization and strengthening of the community; (ii) construction rehabilitation and equipping of community centers; (iii) support to in-house basic sanitation in La Boquilla; and (iv) strengthening of the Community Relations Unit of ACUACAR.
- *Project management, technical assistance, studies, design and supervision:* This component would support and partially finance the following activities: (i) project management; (ii) design and supervision of the water supply systems works; (iii) design and supervision of the sewerage systems works; (iv) design of the main wastewater conveyance system, treatment installations and submarine outfall; (v) supervision of the main conveyance system works; (vi) supervision of the treatment installation and submarine outfall works; and (vi) procurement audits.

Cost details are provided in Table 14.2. The total project cost is US$117.2 million, of which US$85 million is a World Bank loan.[1]

14.2.4 The proposed wastewater disposal scheme

A feasibility study for treatment and disposal of the wastewater of Cartagena[2] was carried out during project preparation. This study identified and analyzed a comprehensive set of alternatives related to the final location for the sewerage collection system, the level of wastewater treatment, the alignment and exact location for an ocean outfall, sites of major pumping stations, sewerage pipe material, and the size and depth of the outfall diffuser pipe. The main options for the final disposal of domestic sewage analyzed were (i) the Cartagena Bay; (ii) the Cienaga de La Virgen; (iii) the Caribbean Sea; and (iv) reuse and irrigation.

---

[1] The World Bank, Report No. 18989 CO, Project Appraisal Document, Cartagena Water Supply, Sewerage and Environmental Management Project, June 28, 1999.
[2] Hazen and Sawyer, Feasibility Study for the Treatment of the Wastewater of Cartagena and Disposal of the Effluent to the Adjacent Sea via a Submarine Outfall, October 1998, in Spanish.

Table 14.2 Cartagena project cost estimate (US$millions)

Component	Cost (including contingencies)	% of total	Bank financing
Expansion of the water supply system	9.9	8.4	7.3
Expansion of the sewerage system in the Cienaga basin	35.7	30.5	26.2
Construction of the main wastewater conveyance system	28.1	24.0	20.6
Construction of the wastewater treatment installations	6.8	5.8	5.0
Construction of the submarine outfall	22.8	19.5	16.6
Industrial wastewater discharge control	0.6	0.5	0.6
Environmental and social component	3.3	2.8	3.3
Project management, technical assistance, studies, design and supervision	10.0	8.5	5.4
Totals	117.2	100.0	85.0

Local and regional environmental authorities have established cleanup of the Bay and the Cienaga as a top regional priority. Continuing the discharge of sewage to the Bay was considered to be unsustainable because of its limited assimilative capacity. Discharge of effluent to the Cienaga would be very expensive, as it would require nutrient removal. Wastewater reuse for irrigation is not presently feasible given the lack of local agricultural areas able to use such high flows. In any case, any wastewater reuse scheme must be coupled with the capability for wastewater disposal when irrigation is not required and in emergency situations. An outfall could serve as such an alternative discharge option. The project does not exclude future partial reuse of the Cartagena wastewater, and the proposed conveyance and submarine outfall systems could well support such initiatives in the future. Alternatives for the final disposal and treatment levels of wastewater included combinations of the following:

Disposal options:

- Continuing wastewater discharges into the Cartagena Bay and the Cienaga in the same current proportion (40% to the Bay, 60% to the Cienaga);
- Elimination of the Bay discharge and disposal of all wastewater into the Cienaga;
- Elimination of both Bay and Cienaga discharges and disposal of all wastewater into the Caribbean Sea;
- Reuse for irrigation.

Treatment options:

- Preliminary treatment: microstrainers and grit chambers for the removal of coarse solids and floatables;
- Primary treatment: settling tanks, and anaerobic digestion and drying of sludge;
- Conventional secondary treatment: activated sludge plants, anaerobic digestion and drying of sludge;
- Secondary treatment in oxidation ponds;
- Advanced treatment: biological and chemical removal of nutrients.

Over fifteen combinations of different treatment levels and final disposal sites were considered and four different outfall alignment and discharge sites into the Caribbean Sea were identified. All alternatives were evaluated from technical, economic, environmental and social perspectives. Water quality, initial investment costs, operation and maintenance costs, and land uptake were the main comparison criteria, although other environmental and social criteria were also used.

The comparisons indicated that disposal of all wastewaters to the Caribbean Sea is the most feasible option. Having identified the receiving water body, intensive oceanographic studies were carried out for more than a year in the coastal waters of Cartagena. These measurements constituted the main inputs to hydrodynamic and water quality mathematical models which were used to determine the best discharge site. Consideration of the patterns of wind, current speed and direction, wastewater coliform loadings, and bacterial decay rates, indicated that not all of the outfall alternatives would meet the standards adopted. The best discharge site was at Punta Canoa (Figure 14.2), about 20 km north of Cartagena. Although located at the greatest distance from the city, this site was the least cost alternative because the sea bottom slope there is quite steep, so the length of an outfall that reaches comparably deep water at that site is only 2.85 km. The other sites nearer the city have very mild bottom slopes, requiring a very long outfall (of about 9 km) to reach deep enough water. Therefore, the combined cost of the onshore and offshore pipes was lowest for the Punta Canoa site.

An additional reason for selecting the Punta Canoa site is that the water there is highly turbid due to a current which carries huge amounts of suspended solids and silt discharged to the sea by the Magdalena River. This river flows into the sea near Barranquilla, about 100 km to the east of Cartagena, at an average flowrate of 10,000 m^3/s. Due to this turbidity, the sea is essentially void of fish and other life forms near the planned outfall discharge site. The discharge of about 3 m^3/s of effluent with a suspended solids content after dilution that will be much lower than the natural sea environment, will have no negative effect on the marine environment (if any, it will have a positive effect), so there is no logic to providing a higher than preliminary level of treatment to the wastewater.

Figure 14.2 The proposed Cartagena wastewater conveyance, treatment, and disposal system

Detailed oceanographic surveys were then carried out at the Punta Canoa site over a period of two years and simulation models were prepared using the data from these surveys. The surveys are continuing into the third year. Details of the modeling results are available in a report prepared by Roberts and Carvalho.[3]

Based on the feasibility study and the model results, the alternative for wastewater conveyance, treatment, and disposal shown in Figure 14.2 was chosen. The system consists of:

- Expansion and improvement of the existing Paraiso pumping station;
- On land conveyance system, a 23.85 km long pipe of 72″ diameter;

---

[3] Philip J.W. Roberts, Joao Luiz Carvalho, Modeling of Ocean Outfall for Cartagena, Colombia, Report Prepared for Aguas de Cartagena, June 23, 2000.

Case studies 425

- Preliminary treatment works consisting of micro-strainers and vortex-type grit chambers;
- Submarine outfall, 2,850 m long, two meter diameter at Punta Canoa, discharging at a seawater depth of 20 m, with a total discharge capacity of about 4 m^3/s in 2025;
- A diffuser 500 m long with 27 risers, each continuing 2 ports each for a total of 54 ports.

Near field dilutions were predicted using the model NRFIELD (Section 3.3.2) and are presented in Figure 14.3. The plume usually surfaces due to the lack of density stratification in the water column. The dilution is highly variable with a mean value of about 300 and a maximum about 1000. Low dilutions are infrequent; the minimum dilution was predicted to be greater than 100 for 87% of the time.

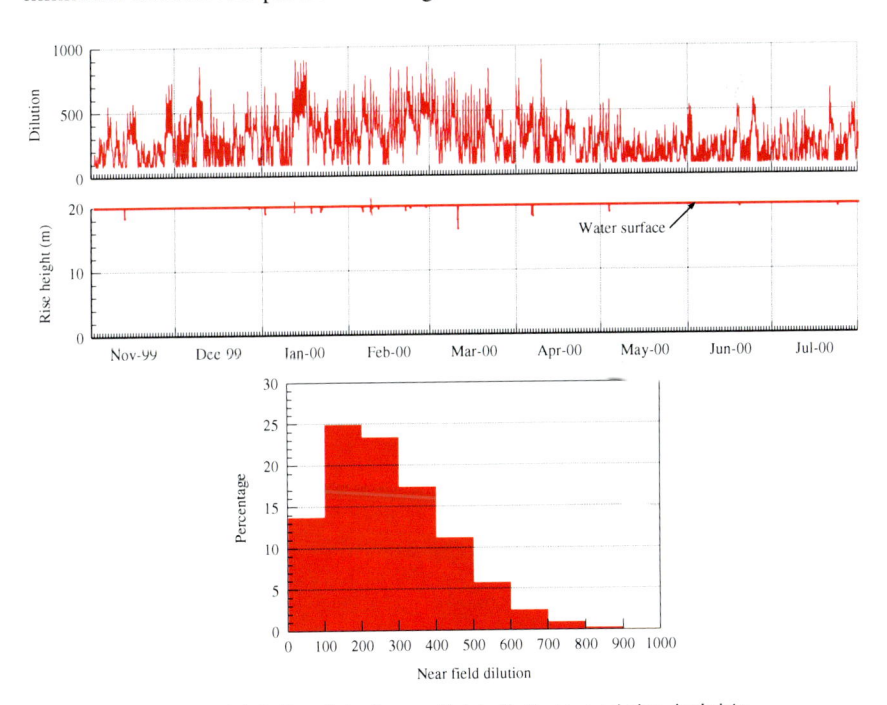

Figure 14.3 Predicted near field dilutions and rise heights

Spatial variations of bacterial concentrations around the diffuser were predicted using measured currents and the methods of Section 3.3.3.3. The results are shown in Figure 14.4 as frequency of exceedance of bacterial levels, chosen to correspond to the bathing water standards of the California ocean plan. Figure 14.4(a) shows the frequency with which the total coliform concentration exceeds 1000 per 100 ml (which is the same as fecal coliform level of 200 per 100 ml). Figure 14.4(b) shows the frequency with which the fecal coliforms exceed 400 per 100 ml. The area where the bathing water standards of the California Ocean Plan were exceeded is indicated.

(a) Total coliforms = 1000 per 100 ml
(Fecal coliforms = 200 per 100 ml)

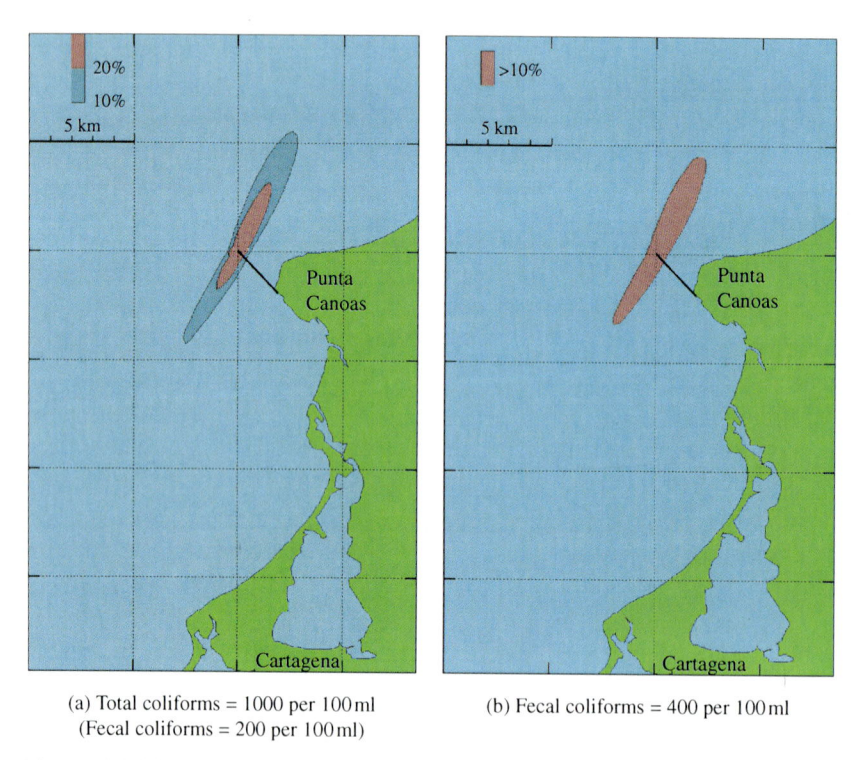

(b) Fecal coliforms = 400 per 100 ml

Figure 14.4 Predicted coliform exceedance frequencies (area where California water contact standard may be exceeded is shown in red)

The contours elongate considerably along the first principal current axis which is approximately NE/SW due to the predominance of currents in this direction. Little onshore flow is predicted due to the slow speeds and random directions of the onshore currents. Bacterial concentrations at four points on the shore near Punta Canoa were computed. No bacteria were predicted to be transported to these points. In other words, the fecal coliform levels at the beaches were predicted to be always zero, so the median value should be well below the water quality standard. It should be noted that estimates of onshore transport are generally conservative, i.e. they overestimate it, as the currents do not actually continue moving onshore as the model assumes.

The strategy adopted is to provide minimal level of wastewater treatment and an outfall long enough to ensure that no contamination reaches bathing beaches. Any environmental impacts should therefore be minimal. The validity of this strategy will be tested by a monitoring program (Section 14.3.4) begun prior to and continuing after construction of the outfall. The monitoring results will indicate whether the outfall is performing as designed and predicted. In the unlikely event that the outfall's performance does not conform to the planned environmental standards, additional treatment capacity could be added. This strategy prevents unnecessary investments in costly treatment processes that cannot be afforded by developing countries.

As shown in Table 14.2, the estimated cost of the proposed wastewater disposal system (including the main conveyance system, the treatment plant and the outfall) amounts to US$57.7 million. For the population forecasted for 2025, this is equivalent to an investment of about US$44 per inhabitant. This is a small investment for a wastewater disposal system (including conveyance, treatment and final disposal) of such a large city.

The bidding documents for the construction of the submarine outfall allowed proponents to select the pipe material from a list of specified materials. The winning bidder proposed a HDPE pipe of internal diameter 1,800 mm (about 72″ as required) and an external diameter of 2,000 mm. This is the largest HDPE pipe manufactured to date (2006) for submarine outfalls and one of the first cases that such a diameter HDPE pipe is used for construction of an effluent outfall. The cost of construction of the Cartagena submarine outfall proposed by the lowest bidder was close to the cost estimated by the design engineer.

The project will have enormous benefits in providing water and sewerage services to all the poor neighborhoods of Cartagena, cleaning up the water bodies surrounding the city (Cartagena Bay, the Cienaga, the Caribbean beaches, and the in-city water canals) while avoiding generation of environmental nuisances. It is the least costly alternative and generates many benefits; no other wastewater disposal alternative is less expensive or more effective.

14.2.5 Social aspects of the outfall

Detailed environmental and social analyses were prepared as part of the appraisal activities of the project. An Environmental Impact Assessment study (EIA) was carried out by an independent joint venture of national and local environmental Non Government Organizations (NGOs) reinforced by a group of national and international consultants),[4] and a Social Assessment of the project was prepared by national consultants.[5] The main conclusion of the EIA was that the proposed disposal solution, preliminary treatment followed by the 2.85 km long outfall at Punta Canoa, will not result in negative environmental impacts.

The most sensitive project component from environmental and social standpoints would be the construction and operation of the ocean outfall. A project component was designed to address these issues by providing and disseminating to the community and other stakeholders project information and study results, especially the environmental studies, to inform them about the project benefits and absence of negative environmental impacts. In addition, a Social Component was included in the project, aimed at mitigating negative social impact. This component included a set of action in benefit of communities which felt that they would be adversely affected by the project. Additional details on the social aspects related to the Cartagena submarine outfall are discussed in Chapter 14.

14.3 CONCEPCION, CHILE

Two outfalls were constructed in 1994 in Concepcion Bay, Chile, and have been in operation since. One serves the town of Tome (population about 50,000), and the other, the Penco outfall, serves the towns of Penco and Lirquen (with a total population of about 40,000). The locations of the outfalls in the Bay of Concepcion are shown in Figure 14.5.

Each disposal system consists of preliminary treatment followed by an effective outfall. The Tome outfall has an internal diameter of 450 mm and is 1200 m long, equipped with a 25 meter long diffuser zone discharging the effluent at a depth of 25 m. The Penco outfall has an internal diameter of 580 mm and is 1300 m long, equipped with a 25 m long diffuser zone discharging the effluent at a depth of 22 m.

A rigorous monitoring program of the performance of both outfalls has been in effect since they began discharging. The results for the first 5 years of operation have been reported by Leppe and Padilla (1999)[6] and are summarized below. According to more recent information, the performance of these outfalls continued

---

[4] Environmental Impact Assessment – Ocean Outfall in Cartagena, 1999 (prepared by a joint venture of national and local NGOs, Fundacion Neotropicos and Fundacion Vida.

[5] Social Assessment of the Cartagena Sanitation Project, 1998.

[6] Leppe, A.Z. and Padilla L.B., Emisarios Submarinos de Penco y Tome, 5 anos de Vigilancia y una Evaluacion Global de sus Efectos Ambientales, XIII Conreso Chileno de Ingenieria Sanitaria y ambiental, AIDIS Chile, Antofagasta, October 1999.

Case studies 429

with similar results during the first 10 years of operation. The locations of the sampling points around the Tome outfall are shown in Figure 14.6.

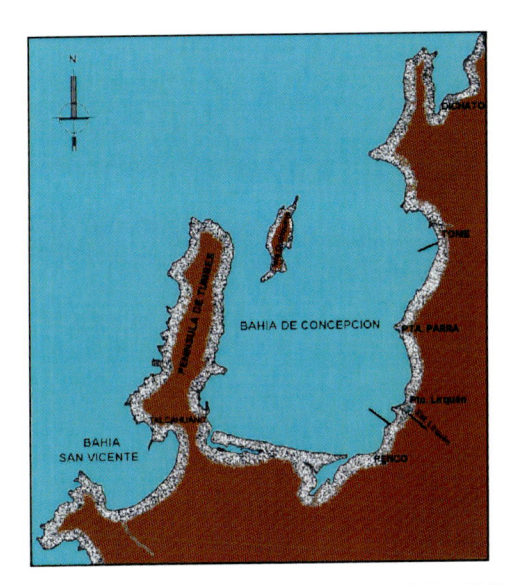

Figure 14.5 The Penco and Tome outfalls, Chile

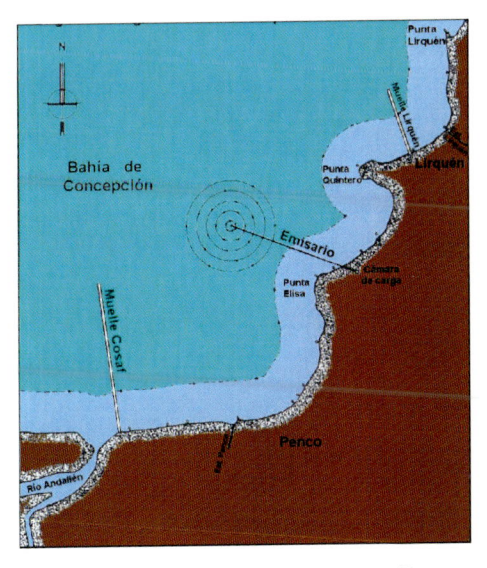

Figure 14.6 Sampling points around the Tome outfall

The uniqueness of the Chilean outfalls is that: (i) a comprehensive monitoring program of their performance has been in effect for a long period of 10 years; and (ii) water quality was measured not only near shore, which is the common practice, but also near the discharges at several points located on circles with a radius of 100 m centered on the discharge.

The results are presented in Tables 14.3 and 14.4, based on information provided by Leppe and Padilla (1999). The marine water quality composition at 100 m from the discharge points refers to averages of six measuring points around the outfalls. The results are averages of five years of measurements.

Table 14.3 Marine water quality near the Tome Outfall, Concepción, Chile (Average of 5 years monitoring)

Parameter	Effluent discharged into outfall	Maximum according to local standards	At the discharge point	At 100 m from the discharge Point	Typical back ground values
pH	7.3	5.5–9.0	7.6	7.7	7.8
Temperature °C	17.8	–	13.1	14.1	14.2
Oil and grease mg/l	46.3	150	4.6	9.2	2.4
TSS (total suspended solids) mg/l	206	300	7.6	1.8	3.0
BOD5 (Biochemical oxygen demand) mg/l	348	–	2.3	2.2	2.4
TOC (Total organic carbon) mg/l	203	–	2.1	3.4	2.0
DO (Dissolved oxygen) mg/l	0.5	–	4.8	7.5	7.2
Detergents mg/l	9.2	15	0.07	0.06	0.05
Nitrogen Kjeldal mg/l	72	–	0.52	0.41	0.57
Nitrites mg/l	0.01	–	0.02	0.02	0.02
Nitrates mg/l	0.17	–	0.25	0.28	0.18
Total Phosphorus, mg/l	12.8	–	0.20	0.21	0.20
Phosphate, mg/l	38	–	1.00	1.00	1.00
Sulfur, mg/l	1.0	5	0.20	0.15	0.24
Phenol	<0.002	1	<0.002	<0.002	<0.002
Fecal Coliforms, MPN/100 ml	3.5×10^7	–	2546	30	11
Total Coliforms, MPN/100 ml	5×10^7	–	3,450	41	18

Source: Arodys Leppe Z., 1999.

For all quality parameters, except fecal and total coliforms, concentrations at a distance of 100 m from the discharge points are the same as background levels, which demonstrates the high treatment capacity of the outfall systems.

The concentrations of coliforms in the raw sewage and preliminary treated effluents are extremely high. Even so, their concentrations are markedly reduced at a distance of 100 m from the discharges, to levels that meet the most stringent bathing water standards, although they are still higher than background. With decay, the concentrations reduce to background levels within a short additional distance.

Table 14.4 Marine water quality near the Penco outfall, Concepción, Chile (Average of 5 years monitoring)

Parameter	Effluent discharged into outfall	Maximum according to local standards	At the discharge point	At 100 m from the discharge point	Typical background values
pH	7.4	5.5–9.0	7.6	7.6	8.0
Temperature °C	18.4	–	11.5	13.2	13.8
Oil and grease mg/l	48.8	150	4.0	5.6	3.2
TSS (Total suspended solids) mg/l	216	300	3.5	3.7	3.8
BOD5 (Biochemical oxygen demand) mg/l	236	–	2.5	2.8	2.5
TOC (Total organic carbon) mg/l	142	–	2.5	3.5	1.4
DO (Dissolved oxygen) mg/l	1.1	–	5.1	7.9	7.9
Detergents mg/l	15.4	15	0.05	0.07	0.06
Nitrogen Kjeldal mg/l	49.3	–	0.5	0.36	0.4
Nitrites mg/l	0.01	–	0.02	0.01	0.01
Nitrates mg/l	0.15	–	0.18	0.21	0.16
Total Phosphorus, mg/l	12.9	–	0.2	0.25	0.2
Phosphate, mg/l	39.6	–	1.0	1.0	1.0
Sulfur, mg/l	0.8	5	0.18	0.25	0.22
Phenol	<0.002	1	<0.002	<0.002	<0.002
Fecal Coliforms, MPN/100 ml	1.1×10^8	–	5420	172	3.0
Total Coliforms, MPN/100 ml	1.4×10^8	–	7,317	223	5

The results of the studies, as reported by Leppe and Padilla, can be summarized as:

- The water quality on the shoreline after commissioning the outfalls improved and is independent of the discharge flows;
- Water quality at 100 m from the diffusers is similar to that at distances of 200, 300, 400, and 500 m;

- Water quality at 100 m from the diffusers complied with the Chilean standard for bathing water that was in effect at the time;
- The only detectable effects near the discharges were an increase in coliforms, reduction in dissolved oxygen, and occasional variations in total suspended solids or oil and grease;
- Studies carried out with a Rhodamine dye tracer showed that the plume moves to the northeast and out of the Bay;
- The initial dilution of the Penco outfall, measured by Rhodamine dye, was 375:1;
- Studies carried out with radioactive tracers showed that the effluent aligns itself with the direction of the main current with dimensions of about 550 by 200 m;
- No bacteria potentially pathogenic for fish or other marine organisms were found in depths over 10 m;
- Measurements of Colifages showed that viruses were detected only in the raw effluent and at the points of discharge, but not at any other points in the sea;
- The wastewater was classified as medium to low strength. The physico-chemical composition remained constant for the duration of the studies;
- The ratio of COD to BOD indicated that the outfalls received only domestic wastewater;
- Toxicity tests of the raw wastewater using *Bacillus Subtillis* indicated that toxic materials, which cause enzymatic inhibition, were not present;
- The sea water quality did not change over the five year monitoring period;
- Physico-chemical water quality parameters did not show perceptible effects in the sea water distant from the discharge points;
- Bacteria reduced by 99.9% as they traveled within the outfall;
- Gram Negative bacteria were reduced by 98.6% at the discharge point and were not detected at other points of the receiving body;
- The Vibrios sp., which are a potential cause of gastrointestinal diseases, were reduced by 98.7% at the discharge points;
- No bacteria potentially pathogenic to fauna were found;
- Epiflourescence microscopy studies showed that the biomass and volume of bacteria is smaller in the marine water than in the raw wastewater;
- The *in-situ* value of T_{90} for bacterial decay in the bay of Concepcion, is in the range 0.67–0.74 hours, i.e. less than one hour;
- Studies with Chondria sp., an algae which grows in the vicinity of outfalls, demonstrated an intensive antibiotic activity of this alga on *Eschericia coli, Aeromonas, Pseudomonas,* and *Vibrios,* inhibiting the growth of these organisms and killing them;

- During the first year there was a disruption of the fauna, probably due to outfall construction;
- The number of species increased from three in 1994 to 10 in 1995, 15 in 1996, and 17 in 1998;
- The fauna diversity index increased from 0.4 bits in 1994 to 2.54 bits in 1997;
- The fauna density increased from 520 individuals/m^2 in 1994 to 9,100 in 1997;
- Sediment studies demonstrated a 12.1% organic matter content in 1994, increasing to 16.3% in 1995 and stabilizing at 13.9% thereafter.

Conclusions:

- Measurable impacts of the outfalls on the marine environment extend a distance less than 100 m from the discharge points;
- Even at the discharge points the dilution is so high that background environmental conditions are almost reached there;
- The only exception is coliforms, which are higher than background values at the discharge point. At a distance of 25 to 30 m from the discharge, however, they fall to values lower than that permitted by the bathing water standard;
- The effect of the outfalls on the marine water quality is insignificant;
- Heavy metals and micro-pollutants were below detection limits;
- Benthic activity increased after outfall installation, indicating that the outfalls did not cause any adverse effects on benthic organisms;
- The two outfalls did not generate risks to public health or to the marine environment;
- Considering that the raw sewage does not contain industrial wastes or toxic materials, it is reasonable to assume that the impact of the outfalls will not change;
- The small area affected by the outfall and the water quality in this area suggest that there will be no future affects on water quality in the Bay of Concepcion;
- The organic nature of the compounds discharged indicate that, after dilution and dispersion, they will be easily incorporated into the trophic chains and will not accumulate in the water column or on the seabed;
- The principal factor in the purification process is dilution. Increasing the wastewater flow or organic matter should not significantly change the effects on the marine environment;

- Changes in effluent composition, such as an increase in toxic, persistent or bio-accumulative materials, in spite of their rapid attenuation due to dilution, may generate long-term negative effects due to bio-accumulation and bio-concentration.

In summary, Leppe and Padilla conclude that (i) the effect of the outfalls on the water quality in the sea is insignificant; (ii) heavy metals and micro-pollutants were below detection limits around the outfalls; (iii) no negative effects on benthic communities were observed. Note that these conclusions refer to typical municipal sewage which does not contain toxic materials.

An interesting finding is that bacteria are reduced by 99.9% during the travel through the outfall pipe. This has been reported in other studies and is probably caused by antibacterial or antibiotic activity in the pipeline.

14.4 SANTA MARTA, COLOMBIA

Santa Marta, located on the Caribbean coast of Colombia, is the capital of the Department of Magdalena with a population of about 300,000. It is an important port city but the main economic activity is tourism; it has attractive beaches and hotels and is one of the most popular vacation resorts in Colombia.

In 1996, like Cartagena, the City's water sector underwent a reform that incorporated a private operator to manage water and sanitation services (a reform being implemented in many Colombian cities). Since then, the sector performance and the level of service delivery have improved significantly. The wastewater was discharged on the beach without treatment and constituted a major nuisance to a city whose economy depends on resort hotels and tourism. The new operator moved quickly to expand coverage of the sewerage network and to provide adequate wastewater disposal. A study of alternatives was undertaken and from the outset it was clear that the final effluent disposal must be to the sea. The level of treatment required and the length of the outfall were the only issues.

The seabed bathymetry and profile along the potential outfall alignment are shown in Figure 14.7. As a result of the Sierra Nevada Mountains meeting the sea, deep water is reached close to shore.

Four alternatives were developed and compared:

(1) Secondary treatment followed by discharge on the beach without an outfall;
(2) Secondary treatment followed by an outfall discharging at a depth of 15 m;

Case studies   435

(3) Primary treatment followed by an outfall discharging at a depth of 30 m;
(4) Preliminary treatment followed by an outfall discharging at a depth of 30 m.

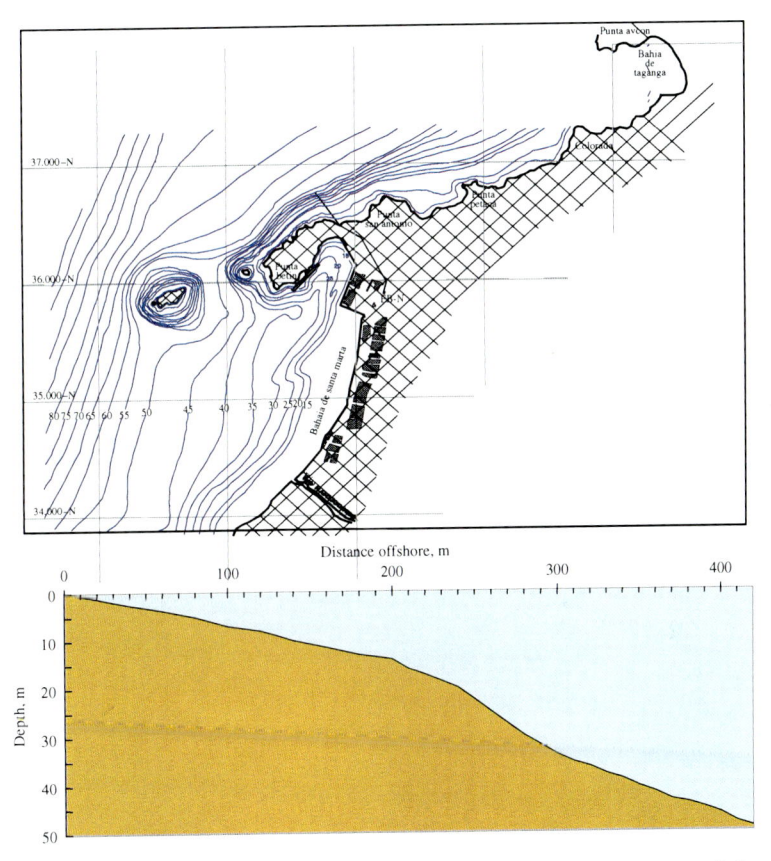

Figure 14.7 Bathymetry near Santa Marta Bay and profile along the outfall route

Selection of the outfall lengths and treatment levels was arbitrary and did not conform to the current WHO 2003 guidelines (which were not published at that time). Alternative (iv) was finally selected, not based on oceanographic measurements or modeling but on experience with other outfalls, although the outfall length was extended to discharge at a depth of 56 m. The slope of the seabed in Santa Marta is steep so the length of the outfall required to discharge at this depth is only 428 m. The reasons provided by the operator for selecting alternative (iv) were that it was cheapest, the level of treatment was highest (when considering the outfall as a treatment system, accounting for dilution and dispersion in the sea), and operation and maintenance requirements and costs are minimal.

The outfall was inaugurated in April 2000 and was designed to handle an initial flow of 650 l/s rising to 2,500 l/s in the year 2050. The diameter is 1 meter and it is an HDPE pipe of high molecular weight. To prevent floatation, 72 concrete blocks weighing 3 tons each were attached to the pipe. The diffuser is located at the final 120 m of the pipe and consists of 30 ports, each with an orifice of 15 cm diameter, spaced 4 m apart on opposite sides of the pipe as shown in Figure 14.8.

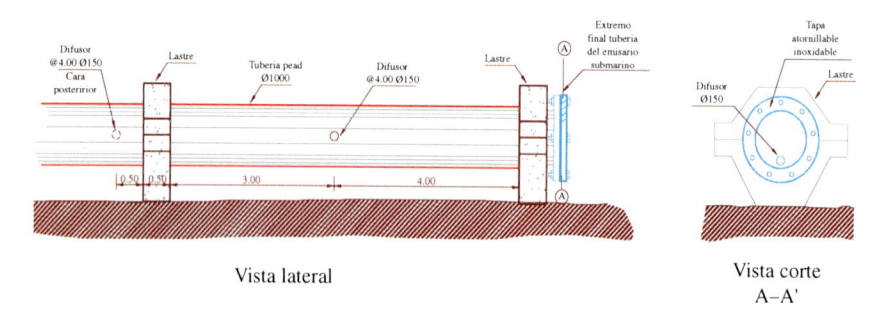

Figure 14.8 Design of the concrete blocks

The outfall was constructed in the port of Santa Marta, a few kilometers from the final outfall location. It consisted of three sections, two 150 m long and the third one was the diffuser section, about 120 m long. The concrete blocks were mounted on the pipes which were then floated on the sea surface (they were filled with air) and towed to the installation site. They were then sunk to the sea bottom by filling with water and connected by flanges.

The wastewater treatment is preliminary and quite elementary. There is no single treatment plant; rather the wastewater passes through pumping stations prior to reaching the outfall. Each station is equipped with bar screens of 1.5 cm opening in front of the pumps. These screens remove 90% of the coarse and floating material and this is the only treatment provided.

Bacterial levels are low around the outfall as shown by the monitoring results in Figures 14.9 and 14.10. The daily average T_{90} was measured to be 1.2 hour and the initial dilution was estimated to be 100. As stated by the operator, the seawater is thermally stratified so the plume is trapped and does not reach the surface.

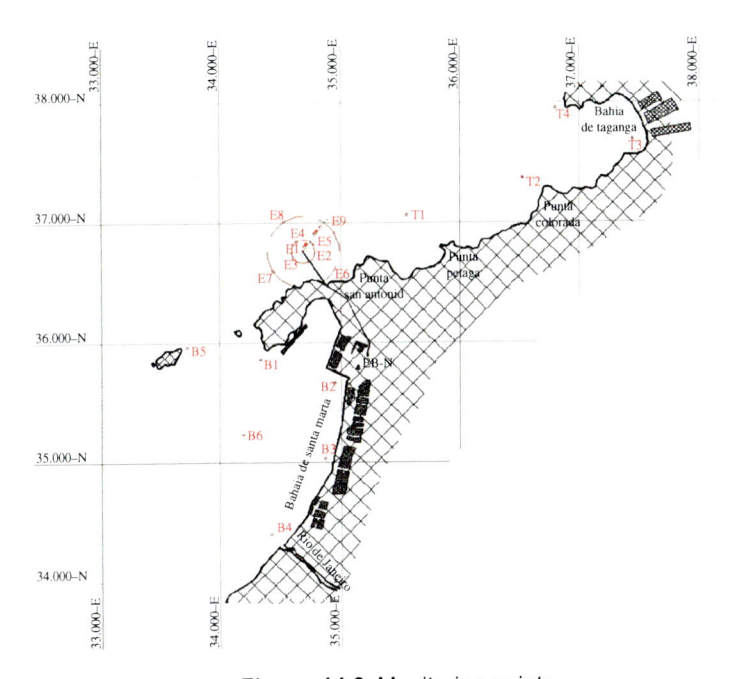

Figure 14.9 Monitoring points

This solution proved to be extremely successful and resolved the city's wastewater problem. The conclusion of the operator was that there is no other solution that can achieve consistently comparable results to those of the outfall.

The Santa Marta outfall was the first submarine outfall in Colombia for discharge of a municipal effluent to the sea. The operator was instrumental in obtaining the environmental license despite the fact that Colombian Law requires secondary treatment of municipal wastewater prior to its discharge to any kind of receiving body. The construction of the outfall, from the date of awarding the contract to completion, was 4 months and cost about US$2 million.

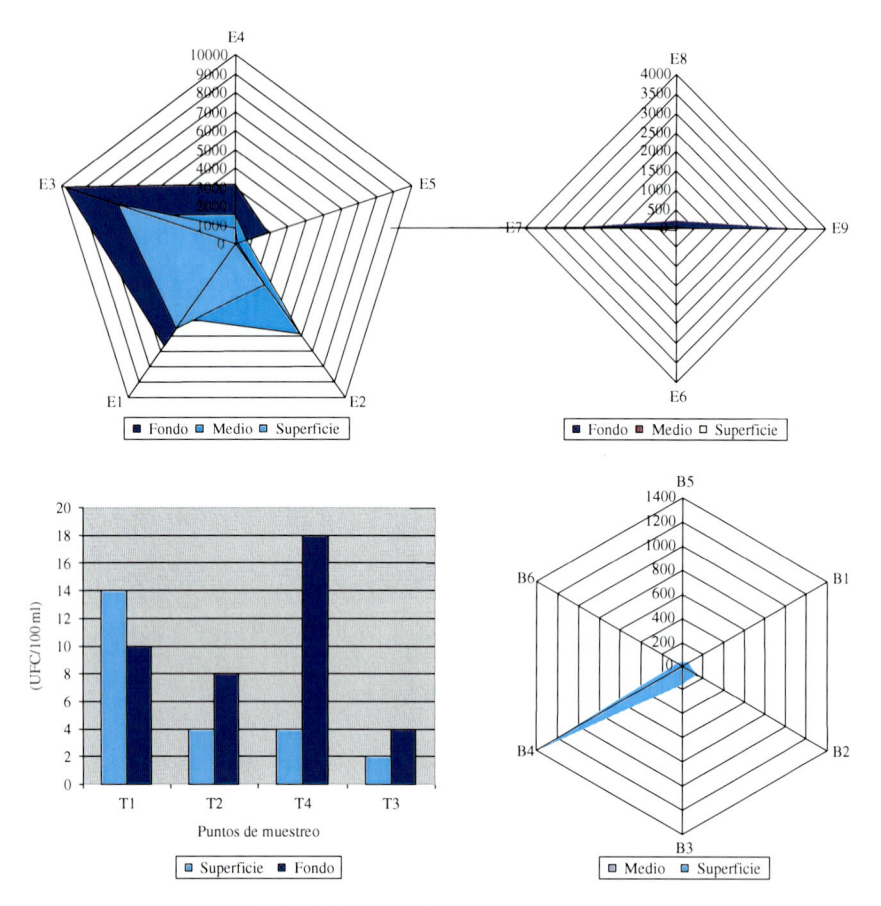

438   Marine Wastewater Outfalls and Treatment Systems

Figure 14.10 Total coliforms at monitoring points

The Santa Marta outfall is an example of a simple and effective solution to dispose the wastewater of a medium-size coastal city at low cost and with a short construction time. The low cost results from the favorable conditions at this location, i.e. a very steep seabed slope.

15
Gaining public acceptance

15.1 INTRODUCTION

Most wastewater projects of large cities, and even smaller ones, not specifically related to effluent disposal in the sea, generate some opposition in spite of their benefit to the public. This is usually due to the NIMBY (Not In My Back Yard) effect, but there are also other reasons, such as opposition by environmental Non-Governmental Organizations (NGOs) and by private interest groups that fear economic damage. Effluent disposal to the sea is even more sensitive, especially in developing countries where the first phase of a project can financially sustain only preliminary treatment. This level of treatment is conceptually unacceptable to many environmental NGOs, even though, as we have demonstrated in this book, an effective outfall can discharge preliminary effluent without adverse environmental effects.

Consequently, opposition should be expected to any wastewater disposal project, and a program to gain public acceptance must be prepared and implemented.

© 2010 IWA Publishing. *Marine Wastewater Outfalls and Treatment Systems.* By Philip JW Roberts, Henry J Salas, Fred M Reiff, Menahem Libhaber, Alejandro Labbe, and James C Thomson. ISBN: 9781843391890. Published by IWA Publishing, London, UK.

First and foremost, the proposed project should provide a sound solution that does not cause adverse environmental effects or any other type of damage. An effective outfall can comply with these requirements, even when discharging preliminary effluent. It is advisable to locate the treatment plant and outfall away from residential and industrial areas, so as not to face the NIMBY problem. This is usually difficult to achieve, especially in developing countries, since it increases investment costs. Strong political support and commitment at the municipal and central government level should be a condition for project implementation. Satisfying this requirement can be tricky, however, since implementation of a large wastewater project takes a long time, often exceeding 10 years from the initiation of studies to completion. Such a period can span the terms of several political administrations. At the concept and initiation stages political administrations are usually supportive, but subsequent administrations, especially in developing countries, usually take the position of not objecting, but not supporting, because of the political risk.

Recent trends encourage disclosure of information to the public and consultation and public participation activities, especially in large projects. Because of the sensitive nature of outfall projects, they should have a participatory approach including a series of stakeholders workshops. Social and environmental impact assessment studies should be implemented that include consultations with the main stakeholders. Seminars and community meetings, including meetings with community leaders and members, courses, and training events, should continue during project preparation and implementation. Environmental laws in many countries require public hearings in each community that may be affected by construction. The initial consultations should be held concurrently with presentation of the first draft of the project feasibility study. It is advisable that the terms of reference for the Environmental Impact Assessment (EIA) of the project be discussed in a public workshop and agreed on with the workshop participants. This should include representatives of the environmental authorities, national and local environmental NGOs, municipal government agencies, community organizations, professional associations, and representatives of the community and private sector.

During preparation of the EIA, public consultation meetings should be held in the communities closest to the outfall site. Community concerns should be registered and taken into account in the Environmental Management Plan, which should be prepared to mitigate potential adverse effects. A final public consultation workshop at the project preparation stage (with the same range of participants as the first one) should be held to review and discuss the final draft of the feasibility study and the EIA report. The EIA report should be publicly available. It should include an annex that summarizes the consultation process, including participants, a record of concerns, and details on the agreements reached. This participatory approach should continue during project implementation. It is also advisable to implement, especially in developing countries, a Social Impact Mitigation program and a Community Development Program.

A problem related to the participatory approach in developing countries is that communities near the outfall do not oppose it in the early stages, possibly because they do not see any potential problems. With time, however, perhaps due to incitement by other groups, their opposition builds, and this happens when project implementation is quite advanced and it is neither feasible nor practical to change course and come up with a different solution.

Attention should be paid to social issues, and, if possible, compensation should be provided to communities that feel damaged. This is not easy to carry out in an outfall project, because the damage is only perceived and usually does not materialize. Legislation in most countries does not allow compensation for perceived damage, only for actual damage. This can be resolved by the use of damage insurance. The project sponsor, usually the water and sanitation utility, can contract an insurance policy for which it pays annually, so that the insurance company will compensate for any damage. The concept of insurance against potential environmental damages has gained ground recently.

It is important to explain to the community, both the benefiting and the opposing factions, that opposition to wastewater disposal through an outfall preceded by preliminary treatment will, in most cases, prevent any progress towards improving sanitation, i.e. it will maintain the "No Action" alternative. This is because treatment above and beyond preliminary would render most projects financially non-viable. Avoidance of action towards solving wastewater disposal problems, i.e., selection of the "No Action" alternative, is the worst option of all, and usually leaves the most vulnerable population (mostly the poor) under the worst conditions as well as an adverse environmental impact.

Wastewater project sponsors, usually water and sanitation utilities, should be aware that in spite of all efforts devoted to gaining public acceptance, they will probably encounter opposition that can in many cases derail good and worthy projects. Wastewater disposal projects with an effective outfall usually bring high benefits to the community and do not cause any damage, except imaginary damage in the minds of specific interest groups. The many beneficiaries are usually a silent majority that does not actively express its support, whereas the opposition groups that represent a minority, sometimes very few people, can be determined, fierce, and noisy. They use various strategies to fight a project, including public campaigns in the media that present incorrect and misleading information, legal procedures, and sometimes physical opposition. The presentation of scientific evidence and experience from similar projects and reasoning usually do not have good outcomes with such groups and the project fate is often determined by litigation and legal actions.

15.2 CARTAGENA CASE STUDY

The case of Cartagena, a vacation resort city on the Caribbean coast of Colombia, is presented as an example of gaining public acceptance for a wastewater project involving a marine outfall. Cartagena had a population of about one million in 2005, and details of the project and the proposed wastewater disposal scheme were presented in Chapter 14.

The water bodies surrounding Cartagena (Cartagena Bay, the Caribbean Sea, and the coastal lagoon, Cienaga de La Virgen) contribute significantly to the quality of life of its residents by providing venues for both leisure (swimming and water sports) and commercial activities (fishing and commerce). These water bodies are also a main attraction for tourists and are the basis of the local tourism industry. Unfortunately, these water bodies are severely fouled by the city's untreated municipal liquid wastes. These have a total flow of about 145,000 m^3/day in 2005, 30% of which is discharged into the Cartagena Bay, 60% into the Cienaga de La Virgen, and 10% to the in-city water courses. Overflows from the overloaded collectors in the wealthy neighborhoods contaminate the beaches and the in-city water courses, while industrial wastes from the Mamonal industrial estate area (which is not connected to the city's sewerage network) contribute additional contamination loads to the Cartagena Bay. The poor neighborhoods around the Cienaga suffer the worst sanitation conditions, since, as a result of lack of sewerage, raw sewage flows in its streets.

As the city rapidly grew, water and sanitation infrastructure lagged. The rich parts of Cartagena, including the hotel areas, have high coverage of water and sewerage, but the poorer parts have low coverage. This has generated severe public health, sanitation, and environmental problems that, in addition to their direct impact on the population in deteriorating quality of life, pose a serious restriction to sustainable economic development, especially in tourism which is the city's main source of income. The project is designed to provide universal water and sewerage coverage, which means providing water and sewerage networks in the poor neighborhoods and cleaning up the surrounding water bodies. The project benefits are enormous, the most important of which is eliminating inundation of the poor neighborhoods (the southeastern zone adjacent to the Cienaga where about 400,000 residents live) with raw wastewater.

Detailed environmental and social analyses were prepared as part of the appraisal activities of the Cartagena project. An EIA study was carried out by an independent joint venture of national and local environmental NGOs reinforced by national and international consultants,[1] and a Social Assessment was prepared by national consultants.[2] The main conclusion of the EIA was that the proposed disposal solution, consisting of preliminary treatment followed by the 2.85 km long outfall at Punta Canoa (see Figure 14.2), will not result in any adverse environmental impacts.

The most sensitive project component from the environmental and social standpoints would be construction and operation of the ocean outfall. A project component was therefore designed to address these issues by providing and disseminating to the community and other stakeholders project information and results of studies, especially the environmental studies, to inform them about the project benefits and the absence of negative environmental impacts.

Because of the sensitive nature of the project, it was prepared with a participatory approach that began in February 1998 with a stakeholders workshop and concluded in February 1999 with a second workshop. In between, the social and environmental impact assessments also included consultations with the main stakeholders. Seminars and community meetings continued during project implementation, and about 250 meetings with community leaders and members, seminars, courses and other training events were held. As required under Colombia's environmental law, public hearings were held in each community that would be affected by the construction works. The initial consultations were held in February 1998, concurrent with the presentation of the first draft of the project feasibility study. At this workshop, the terms of reference for the EIA Study were discussed and agreed on with the participants, which included representatives of the Ministry of Environment, Aguas de Cartagena (ACUACAR, the water and sewerage utility of Cartagena), CARDIQUE (the regional environmental authority in charge of issuing the environmental license), DAMARENA (the Municipal Environmental Department), national and local environmental NGOs, municipal government agencies, community organizations, professional associations, community representatives, and private sector representatives.

During preparation of the EIA, public consultations were held in the communities closest to the outfall site. These were Punta Canoa, Arroyo de Piedra, Manzanillo del Mar, La Boquilla and the Southeastern Zone. Community concerns were registered and taken into account in the Environmental Management Plan. A final public consultation workshop at the project preparation stage to review and discuss the final draft of the feasibility study

---

[1] Environmental Impact Assessment – Ocean Outfall in Cartagena, 1999 (prepared by a joint venture of national and local NGOs, Fundacion Neotropicos and Fundacion Vida).

[2] Social Assessment of the Cartagena Sanitation Project, 1998.

and the EIA report was held in February 1999. The EIA report was made publicly available. This report included an annex that summarized the consultation process including participants, a record of concerns, and agreements reached during the meetings. This participatory approach continued during project implementation through the Social Impact Mitigation and Community Development Programs.

As in any wastewater disposal project of a large city, the Cartagena project generated opposition from various interest groups. ACUACAR used the following strategy for gaining public acceptance, which provided information to all interested parties and facilitated obtaining consensus which led to issuing the environmental license by the regional authority:

- Expansion of the participatory approach and working with the community to provide information regarding the impact of the outfall and its benefits (about 250 events were been carried out);
- Execution of a publicity campaign regarding the outfall, including publicity in the media (articles in newspapers and advertisements on radio and TV), and preparation and distribution of brochures, etc.;
- Implementation of the social community development program including: (i) support for urban rehabilitation, improvement of sanitary conditions and cleanup activities; (ii) strengthening and development of community organizations to promote participation and social control; and (iii) promotion of community development to consolidate communities, to avoid or reduce conflicts and to recover cultural heritage, mainly by rehabilitating the Cienaga;
- Organization of a study tour for community leaders and representatives of the media, the municipality, the environmental authorities, and other stakeholders to similar outfall sites in Latin America. The group included about 30 persons that visited outfalls in Chile (Viña del Mar, Valparaiso, and Concepción), Montevideo in Uruguay, and Guaruja in Brazil. All these outfalls are of comparable size to that proposed for Cartagena and have the same type of preliminary treatment. Unlike Cartagena, they are located in front of the most desirable residential areas and beach resorts, whereas in Cartagena the outfall will be about 20 km north of the city. In all sites the outfalls are functioning successfully, to the complete satisfaction of all the local stakeholders;
- Utilization of a panel of five international experts (hired to review the project), with broad experience in wastewater management, design and construction of ocean outfalls, water quality and oceanographic modeling, environmental impact assessment and private sector participation. The

panel provided valuable support in clarifying the technical issues to the various stakeholders;
- A series of workshops with the opposition groups to explain the scientific, technical and engineering aspects of the selected alternative and its advantages over all others; and
- Financing the participation of representatives of the key stakeholders in an international course on marine outfalls for sewage disposal in the Caribbean, organized recently by PAHO/WHO in Barbados.

Several opposition groups mounted a campaign against the project and invested enormous efforts to derail it. They presented a wide range of reasons for their opposition, some of which were not relevant to the issues presented in this book. The fundamental issue was based on the NIMBY argument. Some NGOs and other special interest groups argued that discharge of preliminary treated wastewater is unacceptable under any conditions since it would impose severe health risks on the residents of the village of Punta Cano, a small village of about 500 residents near the outfall. They also argued that fishing is the main source of income for the residents and it will be reduced as a result of the outfall. To fight the project, the opposition groups strongly argued that the village residents are a poor and socially vulnerable group.

The social component of the project took account of the social aspects related to Punta Canoa and other neighboring communities known as the communities of the North Zone. The project was planned to benefit these communities in many ways:

- Providing piped water and sanitation through a sewerage network to Punta Canoa, Arroyo de Piedra and Manzanillo, and two additional communities which were not involved in opposition to the project (Portezuela and Bayuca);
- Providing sanitation through a sewerage network to La Boquilla, which already had piped water services;
- Providing in-house sanitation to all households in Punta Canoa, Manzanillo and La Boquilla;
- Construction of two community centers, one in La Boquilla and another in Punta Canoa, to complement the District of Cartagena's urban rehabilitation program which is aimed at stabilizing urban growth and consolidating these communities;
- Carrying out a study to optimize fishing activity in Punta Canoa and providing funds to implement the recommendations, including improvements in commercial practices;

- Financing a plant nursery, managed cooperatively by the residents of Punta Canoa, which will be used to support the project's post-construction reforestation activities, as well as provide an additional revenue source to the residents;
- During project construction, the North Zone residents may benefit from work opportunities in construction; and
- In general, the improvement in the environment as a result of better wastewater disposal will help bolster tourism in the region, creating jobs and economic opportunities for North Zone residents.

The opposition groups were not satisfied by the explanations and scientific evidence that showed that the outfall will not cause adverse effects or damage to anyone, or by the actions taken by ACUACAR to support the communities that felt impacted by the project. They took a series of actions in an attempt to derail the project, ignoring its huge benefits, especially to the low income communities of the southeastern zone of Cartagena. During the entire period of project preparation and execution they maintained a media campaign against it that was based on inaccurate and false information. They submitted an appeal to the environmental license issued for the outfall by CARDIQUE, the regional environmental authority. This appeal was rejected, and they then submitted an appeal to the highest Colombian authority, the Ministry of the Environment, which was also rejected. Meanwhile ACUACAR carried out the social component of the project which was designed to alleviate social problems and address the opposition issues. Implementing the social strategies and ratifying the environmental license took a long time, 4.5 years, which delayed project implementation by three years. In parallel, the opposition groups also submitted a series of complaints to the World Bank, which participated in financing the Cartagena project. These complaints were thoroughly investigated over several years. In spite of the repeated complaints, the World Bank maintained its position and continued to participate in the project. Construction of the Cartagena outfall is anticipated to be completed in October 2009.

It is expected that, as has happened in many projects involving outfalls for effluent discharge, after constructing the project and putting it in operation, the opposition will fade away. This is because it will be recognized that there are no significant adverse impacts, only benefits, and even the opposition groups realize that their fears were not justified.

Appendix A
Outfall costs

1. INTRODUCTION

Costs of ocean outfalls vary over a huge range, from several hundred thousand dollars for a short, small diameter HPDE pipe, to $390 million for the Boston tunneled outfall. And others, such as the new tunneled outfall proposed for Los Angeles may approach $1 billion. Every job is site-specific and depends strongly on local conditions, so the costs of even seemingly similar outfalls at different locations can differ significantly. Furthermore, the many technologies associated with marine outfalls, including alternative installation methods, material fabrication, construction, and dredging, are all advancing rapidly, so cost estimation is extremely difficult. The real cost is only known when the construction bids are received and implemented, and even these can vary widely. Therefore, we can only give guidance that may be useful for feasibility planning. In this chapter, some cost data on actual outfalls are presented along with a general discussion of the main factors that affect the construction costs to enable

© 2010 IWA Publishing. *Marine Wastewater Outfalls and Treatment Systems.* By Philip JW Roberts, Henry J Salas, Fred M Reiff, Menahem Libhaber, Alejandro Labbe, and James C Thomson. ISBN: 9781843391890. Published by IWA Publishing, London, UK.

preliminary cost estimation and its possible range. Of course, a complete analysis of the cost of a marine disposal scheme would include the costs of treatment, pumping, maintenance, etc., which are beyond the scope of this book.

2. OUTFALL COSTS

Cost data was gathered from the literature and from personal experience and contacts. Wallis (1979) was the first major publication to report on ocean outfall construction costs. He presented 40 actual costs and 8 estimated costs of projects built from 1947 to 1977. The costs were converted to a common Engineering News Record (ENR[1]) index of 3200. Correlation equations were presented between the unit construction cost in $/foot and the inside diameter in inches for expensive, ordinary, and inexpensive conditions.

Gunnerson and French (1996) summarize costs of 65 outfalls of all construction types built worldwide from 1920 to 1995 with diameters 0.4 to 7.6 m. They included 24 of the outfalls in Wallis (1979). Costs were converted to 1995 USD at an ENR index of 5500 and ranged from about 500 to 30,000 $/m. Most outfalls less than about 2.4 m diameter are 2,000 to 10,000 $/m (in 1995 USD), and most of those greater than 2.4 m diameter are 10,000 to 20,000 $/m. A few, mostly with diameters greater than about 3.5 m, cost more than 30,000 $/m. The most expensive is the Sydney, Australia tunneled outfall at Bondi which is 2300 m long, 2.5 m diameter with a cost of 30,610 $/m.

More data on 56 outfalls built from 1979 to 2009 were obtained from personal experience and contacts among which 42 were HDPE outfalls (17 in Chile), four steel, three glass-reinforced plastic (GRP), two reinforced concrete and five tunneled outfalls. A recent large proposed tunneled outfall for Buenos Aires, Argentina, which is 7 km long, 4 m diameter, has an estimated cost of 28,000 $/m. The Cartagena, Colombia, outfall is HDPE pipe 2.0 m in diameter with an estimated cost of 8,750 $/m.

Data for these 145 outfalls are summarized in Table A.1, which is updated to an April 2009 USD at an ENR of 8528. The unit cost per meter is plotted as a function of diameter (in meters) in Figure A.1. The data are quite scattered and generally increase with diameter, as would be expected.

---

[1] http://enr.construction.com/features/conEco/default.asp

Table A.1 Example outfall costs at ENR = 8528 (April 2009)

	Country	City and Project	Year	Length (m)	Diameter (m)	Original cost (USD millions)	ENR Index	ENR Factor	Cost (USD) Total (millions)	Cost (USD) Per meter	Source/ Material
1	Chile	Mejillones	2000	959	0.35	3.22	7,959	1.07	3.45	3,602	HDPE
2	Chile	Huasco	2003	850	0.36	0.86	7,959	1.07	0.92	1,080	HDPE
3	Israel	Soreq	1984	5,200	0.40	2.70	4,146	2.06	5.55	1,068	HDPE
4	USA	Rockaway Beach	–	900	0.40	1.60			1.60	1,778	Tunnel
5	DK	Kolta	1995	150	0.40	0.45	5,500	1.55	0.70	4,652	G&F
6	Brazil	Itaquanduba Ilhabela, SP	2008	923	0.40	5.40	7,959	1.07	5.79	6,269	HDPE
7	Chile	Lebu	2005	838	0.45	1.78	7,959	1.07	1.91	2,280	HDPE
8	Yugoslavia	Dubrovnik	1974	2@1,500	0.45	5.50	2,020	4.22	23.22	7,740	G&F
9	Bermuda	Bermuda	1992	180	0.49	0.50	5,000	1.71	0.85	4,738	G&F
10	DK	Vallo	1985	800	0.50	0.30	4,220	2.02	0.61	758	G&F
11	Turkey	Trabzon, Akçaabat	2002	1,075	0.50	1.88	6,538	1.30	2.45	2,280	HDPE
12	Turkey	Ordu, Fatsa	2004	1,820	0.50	3.54	7,115	1.20	4.24	2,330	HDPE
13	Turkey	Trabzon, Söğütlü	2001	948	0.50	1.72	6,343	1.34	2.32	2,445	HDPE
14	Chile	Higuerillas	2002	500	0.50	3.11	7,959	1.07	3.33	6,667	HDPE
15	USA	Coos Bay, OR	1972	1,440	0.52	1.80	1,753	4.86	8.76	6,081	G&F
16	USA	Toledo, OR	1965	1,130	0.53	0.96	1,000	8.53	8.19	7,245	G&F
17	Chile	Coronel (Norte)	2004	2,130	0.56	2.20	7,959	1.07	2.35	1,104	HDPE
18	Chile	Lota	2000	1,527	0.56	2.34	7,959	1.07	2.51	1,641	HDPE
19	Turkey	Trabzon, Havaalan	2001	948	0.56	1.88	6,343	1.34	2.52	2,663	HDPE
20	USA	Los Angeles, CA	1957	11,260	0.56	2.60	724	11.78	30.63	2,720	G&F
21	USA	Passaic, NJ	1920	460	0.60	0.10	357	23.89	2.39	5,193	G&F
22	Dom. Rep.	Sosua	2007	787	0.61	2.00	7,966	1.07	2.14	2,721	HDPE
23	USA	Carmel, CA	1971	272	0.61	0.41	1,581	5.39	2.21	8,131	G&F
24	Chile	El Tabo y El Quisco	2003	1,729	0.63	1.84	7,959	1.07	1.97	1,142	HDPE
25	Turkey	Guzelbahce, Izmir Bay	2002	600	0.63	2.20	6538	1.30	2.87	4,783	HDPE
26	DK	Ketamindc	1989	650	0.71	0.50	4,800	1.78	0.89	1,367	G&F
27	Chile	San Antonio	2006	1,050	0.71	2.00	7,959	1.07	2.14	2,041	HDPE
28	Turkey	Rize, Centre	2004	1,206	0.71	3.54	7,115	1.20	4.25	3,522	HDPE

(continued)

Table A.1 Continued

	Country	City and Project	Year	Length (m)	Diameter (m)	Original cost (USD millions)	ENR Index	ENR Factor	Cost (USD) Total (millions)	Cost (USD) Per meter	Source/ Material
29	USA	Salmon Creek, WA	1974	312	0.76	0.28	2,100	4.06	1.14	3,644	Wallis
30	USA	Seattle, WA	1962	640	0.76	0.27	872	9.78	2.64	4,126	G&F
31	USA	San Eiijo, CA	1965	820	0.76	0.96	1,000	8.53	8.19	9,984	G&F
32	Chile	Concón, Con Con Oriente	2002	2,500	0.80	1.50	7,959	1.07	1.61	642	HDPE
33	DK	Guldborg	1976	700	0.80	0.20	2,401	3.55	0.71	1,015	G&F
34	Chile	Punta Arenas (Bahia Catalina)	2003	1,244	0.80	1.50	7,959	1.07	1.61	1,292	HDPE
35	Turkey	Trabzon, Yomra	2001	957	0.80	2.90	6,343	1.34	3.89	4,069	HDPE
36	Turkey	Trabzon, Değirmendere	2001	1,094	0.80	3.32	6,343	1.34	4.46	4,081	HDPE
37	Australia	Victoria, Latrobe Valley	1992	1,300	0.80	5.00	7,959	1.07	5.36	4,121	HDPE
38	Perú	Lima, Playa Venecia	2002	900	0.80	3.57	7,959	1.07	3.82	4,248	HDPE
39	USA	Seattle, WA	1962	490	0.84	0.15	872	9.78	1.47	2,994	G&F
40	USA	N. San Mateo CSD, CA	1962	760	0.84	0.41	945	9.02	3.70	4,868	Wallis
41	USA	San Mateo, CA	1962	490	0.84	0.84	872	9.78	8.22	16,765	G&F
42	Chile	San Pedro de la Paz	2006	2,140	0.90	1.92	7,959	1.07	2.06	963	HDPE
43	Chile	San Vicente	2000	1,800	0.90	3.27	7,959	1.07	3.50	2,086	HDPE
44	Chile	Antofagasta	1998	985	0.90	3.49	7,959	1.07	3.74	3,801	HDPE
45	Turkey	Trabzon, Moloz	2001	856	0.90	2.99	6,343	1.34	4.02	4,701	HDPE
46	Turkey	Mersin, Mezitli	2007	1,805	0.90	8.04	7,966	1.07	8.60	4,766	HDPE
47	Turkey	Mersin	2007	1,805	0.90	8.034	7,959	1.07	8.61	4,770	HDPE
48	USA	Oceanside, CA	1972	2,500	0.91	1.90	1,753	4.86	9.24	3,697	G&F
49	Sweden	Monsteras	1999	5,000	1.00	3.00	6,059	1.41	4.22	844	HDPE
50	USA	Watsonville, CA	1959	1,170	1.00	0.47	797	10.70	5.03	4,298	G&F
51	Colombia	Santa Marta	2003	500	1.00	2.00	6,694	1.27	2.55	5,096	HDPE
52	Australia	Coffs Harbour, NSW	2005	1,550	1.00	10.00	7,959	1.07	10.71	6,913	HDPE
53	USA	Eureka, CA	1977	1,940	1.10	9.40	3,200	2.67	25.05	12,913	Wallis
54	Canada	Vancouver, BC	1993	250	1.17	1.26	5,300	1.61	2.03	8,110	G&F
55	Spain	Albufereta	2007	4,393	1.20	8.11	7,959	1.07	8.69	1,846	HDPE

(*continued*)

Table A.1 Continued

	Country	City and Project	Year	Length (m)	Diameter (m)	Original cost (USD millions)	ENR Index	ENR Factor	Cost (USD) Total (millions)	Cost (USD) Per meter	Source/ Material
56	Turkey	Muğla, Marmaris	2005	406	1.20	0.75	7,446	1.15	0.86	2,116	HDPE
57	USA	Encina, CA	1964	1,370	1.20	0.35	926	9.21	3.22	2,353	G&F
58	Chile	Arauco	1997	970	1.20	3.33	7,959	1.07	3.57	3,678	HDPE
59	USA	San Elijo, CA – extension	1975	1,220	1.20	1.30	2,250	3.79	4.93	4,039	Wallis
60	Chile	Viña del Mar	1996	1,495	1.20	5.66	7,959	1.07	6.07	4,059	HDPE
61	Portugal	Leirosa, Figueira de Foz (Celbi & Soporcel)	1995	1,500	1.20	7.00	7,959	1.07	7.50	4,667	HDPE
62	USA	Aliso, CA	1976	2,620	1.20	4.80	2,950	2.89	13.88	5,296	Wallis
63	Turkey	Istanbul	1992	450	1.20	1.50	4,985	1.71	2.57	5,702	Steel
64	USA	Santa Barbara, CA	1975	2,660	1.20	4.60	2,310	3.69	16.98	6,384	Wallis
65	Portugal	Cascais	1990	1,800	1.20	12.00	4,732	1.80	21.63	6,667	HDPE
66	Hawaii	Mokapu	1974	1,547	1.20	6.20	2,577	3.31	20.52	13,263	G&F
67	USA	San Francisco, CA	1974	180	1.20	0.57	2,020	4.22	2.41	13,369	G&F
68	Turkey	Istanbul, Uskudar	1994	2@270	1.20	5.10	5,504	1.55	7.90	14,633	G&F
69	Puerto Rico	Arecibo	1976	1,070	1.20	6.80	3,400	2.51	17.06	15,940	Wallis
70	USA	EBMUD, Oakland, CA	1995	146	1.37	0.55	5,500	1.55	0.85	5,841	G&F
71	Chile	Talcahuano, Huachipato	2006	1,448	1.40	2.26	7,959	1.07	2.42	1,672	HDPE
72	Chile	Valparaiso, Loma Larga	1999	820	1.40	5.02	7,959	1.07	5.38	6,556	HDPE
73	USA	Dana Point, CA	1977	3,600	1.40	9.80	3,200	2.67	26.12	7,255	Wallis
74	USA	San Francisco, CA	1966	250	1.40	0.46	1,019	8.37	3.85	15,399	G&F
75	USA	Bellingham, WA	1973	850	1.50	0.44	1,895	4.50	1.98	2,330	G&F
76	Greece	Thessalonicki 2@1600	1998	10,520	1.60	16.00	7,959	1.07	17.14	1,630	RC
77	Romania	Constanta North	2006	5,750	1.60	22.00	7,751	1.10	24.21	4,210	GRP
78	Romania	Constanta (RO)	2006	5,750	1.60	22.00	7,751	1.10	24.21	4,210	GRP
79	France	Montpellier	2004	10,800	1.60	45.00	7,115	1.20	53.94	4,994	HDPE
80	Turkey	Istanbul, Yenikapi (Bosphorus)	1987	2@1150	1.60	7.50	4,406	1.94	14.52	6,312	Steel
81	Turkey	Istanbul, Ahirkapi	1989	2@1100	1.60	13.70	4,800	1.78	24.34	11,064	G&F
82	USA	Hampton, VA	1981	2,930	1.70	12.30	3,600	2.37	29.14	9,944	G&F

(continued)

Table A.1 Continued

	Country	City and Project	Year	Length (m)	Diameter (m)	Original cost (USD millions)	ENR Index	ENR Factor	Cost (USD) Total (millions)	Cost (USD) Per meter	Source/ Material
83	Turkey	Istanbul, Baltalimani	1994	2@270	1.70	9.50	5,504	1.55	14.72	27,258	G&F
84	Turkey	Mersin	2007	2,000	1.80	3.70	7,966	1.07	3.96	1,981	GRP
85	USA	Contra Costa County, CA	1959	520	1.80	0.17	797	10.70	1.82	3,498	G&F
86	USA	EnciNa, CA extension	1973	700	1.80	1.05	1,980	4.31	4.52	6,461	Wallis
87	USA	Encina, CA	1973	700	1.80	1.05	1,895	4.50	4.73	6,750	G&F
88	Philippines	Manila	1985	3,600	1.80	13.00	4,220	2.02	26.27	7,298	G&F
89	Puerto Rico	Ponce	1972	1,550	1.80	3.30	1,753	4.86	16.05	10,357	G&F
90	USA	Santa Cruz, CA (2)	1976	2,740	1.80	10.70	3,200	2.67	28.52	10,407	Wallis
91	USA	Santa Cruz, CA (1)	1976	2,900	1.80	12.50	3,200	2.67	33.31	11,487	Wallis
92	USA	Suffolk County, NY	1981	5,777	1.80	28.80	3,600	2.37	68.22	11,810	G&F
93	USA	Los Angeles Harbor	1995	1,585	1.80	20.00	5,500	1.55	31.01	19,565	G&F
94	Hawaii	Barbers Point	1976	3,200	1.98	12.00	2,401	3.55	42.62	13,319	G&F
95	Colombia	Cartagena	2009	3,200	2.00	28.00	8,549	1.00	27.93	8,729	HDPE
96	Hawaii	Ewa Beach	1976	3,210	2.00	11.84	3,200	2.67	31.55	9,830	Wallis
97	Hawaii	Sand Island	1976	3,816	2.10	13.60	2,401	3.55	48.31	12,659	G&F
98	Turkey	Istanbul, Kadikoy	2004	2300	2.20	12.00	7,115	1.20	14.38	6,254	Steel
99	Turkey	Istanbul, Tuzla	1995	2200	2.20	18.50	5,471	1.56	28.84	13,108	Steel
100	USA	Los Angeles County, CA	1954	3,170	2.30	2.20	628	13.58	29.88	9,424	G&F
101	USA	South San Francisco Bay	1977	11,600	2.40	34.50	3,440	2.48	85.53	7,373	Wallis
102	USA	Seattle, WA	1964	1,110	2.40	1.20	986	8.65	10.38	9,350	G&F
103	USA	EBMUD-original, CA	1949	1,820	2.40	1.20	480	17.77	21.32	11,714	Wallis
104	USA	EBMUD-extension, CA	1974	120	2.40	0.46	2,220	3.84	1.77	14,726	Wallis
105	Brazil	Rio de Janiero	1975	4,325	2.40	22.00	2,212	3.86	84.82	19,611	G&F
106	India	Bombay	1985	6,000	2.40	65.00	4,220	2.02	131.36	21,893	G&F
107	USA	Bondi, NSW	1975	3,400	2.40	28.00	2,600	3.28	91.84	27,012	Wallis
108	UK	Aberdeen	1979	1,800	2.50	13.50	3,003	2.84	38.34	21,299	Tunnel
109	Scotland	Aberdeen	1978	2,700	2.50	17.60	2,600	3.28	57.73	21,381	G&F
110	Australia	Sydney, Bondi	1986	2,300	2.50	55.00	4,300	1.98	109.08	47,426	G&F

(continued)

Table A.1 Continued

	Country	City and Project	Year	Length (m)	Diameter (m)	Original cost (USD millions)	ENR Index	ENR Factor	Cost (USD) Total (millions)	Cost (USD) Per meter	Source/Material
111	USA	San Francisco, N Point	1971	1,520	2.60	4.00	1,850	4.61	18.44	12,131	Wallis
112	USA	San Diego, CA Pt. Loma	1961	4,320	2.70	7.80	940	9.07	70.76	16,381	Wallis
113	USA	San Diego, CA	1962	4,340	2.70	10.50	872	9.78	102.69	23,661	G&F
114	USA	San Francisco, CA	1970	90	2.70	0.40	1,385	6.16	2.46	27,366	G&F
115	USA	San Diego, CA	1992	5,330	2.70	90.00	5,200	1.64	147.60	27,692	G&F
116	USA	North Head, NSW	1975	3,400	2.80	25.00	2,600	3.28	82.00	24,118	Wallis
117	USA	Orange County, CA	1969	8,350	3.00	9.00	1,269	6.72	60.48	7,243	G&F
118	USA	Los Angeles CSD (#4)	1964	3,620	3.00	4.50	1,020	8.36	37.62	10,393	Wallis
119	USA	Los Angeles County, CA	1964	3,620	3.00	4.50	986	8.65	38.92	10,752	G&F
120	USA	San Diego, CA	1995	2,652	3.05	50.00	5,500	1.55	77.53	29,234	G&F
121	USA	Malabar, NSW	1975	3,100	3.10	25.00	2,600	3.28	82.00	26,452	Wallis
122	USA	San Diego, CA	1995	1,430	3.10	36.00	5,500	1.55	55.82	39,035	G&F
123	USA	Redondo Beach, CA	1947	2,650	3.20	2.20	413	20.65	45.43	17,142	G&F
124	China	Hong Kong	1995	1,200	3.25	12.50	5,500	1.55	19.38	16,152	G&F
125	USA	El Segundo, CA	1954	2,620	3.30	2.30	629	13.56	31.18	11,902	G&F
126	USA	San Diego, South Bay	–	8,400	3.30				133.00	15,833	Tunnel
127	Australia	Sydney, Malabar	1986	3,500	3.30	55.00	4,300	1.98	109.08	31,165	G&F
128	USA	San Diego, CA	1995	5,490	3.40	90.00	5,500	1.55	139.55	25,419	G&F
129	Australia	Sydney – North Head	1986	3,400	3.50	68.00	4,300	1.98	134.86	39,665	G&F
130	USA	San Onofre, CA	1965	1,740	3.60	3.30	1,070	7.97	26.30	15,116	Wallis
131	USA	San Onofre, CA	1965	1,740	3.60	3.30	971	8.78	28.98	16,657	G&F
132	USA	Los Angeles, CA	1950	1,610	3.60	2.00	570	14.96	29.92	18,586	G&F
133	USA	Huntington Beach, CA	1957	820	3.60	1.50	724	11.78	17.67	21,547	G&F
134	USA	Los Angeles, CA	1957	8,046	3.60	20.20	724	11.78	237.94	29,572	G&F
135	Perú	Taboada	1994*	8,000	3.65	264.00	5,408	1.58	416.31	52,038	RC
136	USA	Seattle, WA	1995	168	3.66	3.20	5,500	1.55	4.96	29,534	G&F
137	Argentina	Berazategui	2008*	7,500	4.00	200.00	5,500	1.03	205.25	27,366	Tunnel
138	China	Shanghai	1984	2@1460	4.20	19.60	8,310	2.73	53.49	18,320	G&F

(*continued*)

Table A.1 Continued

	Country	City and Project	Year	Length (m)	Diameter (m)	Original cost (USD millions)	ENR Index	ENR Factor	Cost (USD) Total (millions)	Cost (USD) Per meter	Source/ Material
139	China	Shanghai	1995	1,600	4.20	23.00	5,500	1.55	35.66	22,289	G&F
140	USA	Ormoud Beach, CA	1969	1,300	4.30	3.40	1,300	6.56	22.30	17,157	Wallis
141	USA	Boston, MA	1995	7,650	4.30	151.10	5,500	1.55	234.29	30,626	G&F
142	Hong Kong	Hong Kong	1995	1,800	5.00	45.00	5,500	1.55	69.77	38,764	G&F
143	USA	San Onofre, CA	1976	6,430	5.50	65.00	3,200	2.67	173.23	26,940	Wallis
144	USA	Boston	2000	16,000	7.60	390.00	6,221	1.37	534.63	33,414	Tunnel
145	USA	Boston, MA	1995	15,200	8.10	334.20	5,500	1.55	518.19	34,092	G&F

*Proposed

Appendix A: Costs

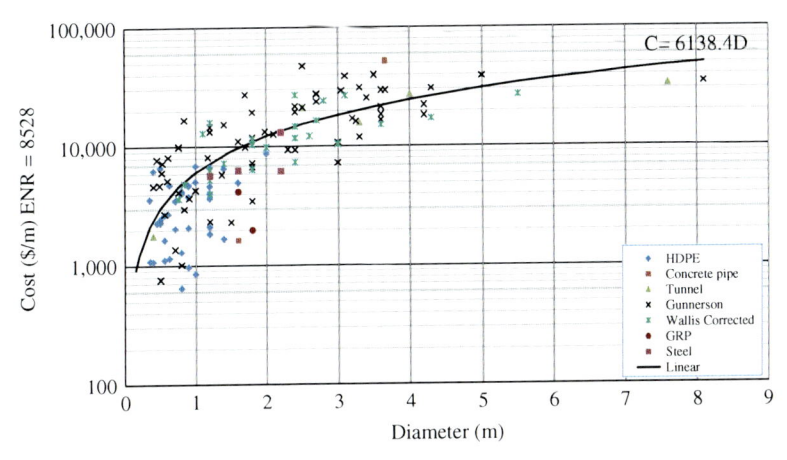

Figure A.1 Unit outfall construction costs at ENR 8528 (April 2009)

The line of best fit for a linear correlation with an intercept through zero resulted in the following equation:

$$C = 6140D$$

where C is the cost in USD per meter, and D the diameter in meters. The correlation between diameter and unit costs is quite weak, and the above equation is presented only to provide order-of-magnitude costs based on historical data. Figure A.1 implies a rapid decline in costs for diameters less than 1 m. This is due to a number of reasons including reduced mobilization costs, the use of local contractors and labor, lower equipment requirements, more pipe suppliers, direct community participation, and lower exposure to risk.

Outfalls with diameters smaller than 300 mm can often be constructed by floatation and submergence using locally available light equipment, small boats, and skilled labor. Taking advantage of this, water and sewer agencies with outfall experience sometimes construct outfalls utilizing force account methods instead of construction contracts. By doing this, the agency assumes the risk of unforeseen contingencies that are normally inherent in a contract that is let for bid. Coupled with reduced mobilization costs, this can significantly lower the cost of small diameter outfalls.

3. COST ISSUES

3.1. Overview

The following discussion is given to illustrate the factors that affect the cost of an ocean outfall. An example analysis is presented that is directed primarily to HDPE outfalls installed in an open trench under typical conditions using flotation construction.

The total outfall cost would include construction, design, field studies, operation and maintenance, and monitoring, but here we address only pipe supply and construction costs, that represent typically around 90% of the total cost. These costs (for an HDPE outfall) are mostly determined by:

- Underwater excavation, including the cost of any sheet piling needed;
- Temporary launching structure;
- Pipeline;
- Concrete ballast weights.

Other significant costs include moving equipment, boats, specialized personnel, and other resources needed for launching, towing, and flooding the outfall.

Costs are significantly affected by:

- Construction of an auxiliary structure and/or sheet piling;
- Presence of rock or corals;
- Lack of local experience;
- Water depth;
- Marine and climatic conditions;
- Lack of local resources (particularly in tourist centers).

Costs and activities related to launching depend strongly on the local coastal characteristics. Many of these costs are fixed and independent of the outfall length, so the unit cost (per installed meter) usually decreases as the outfall length increases. Cost also depends on other aspects unrelated to the project itself or to the characteristics and degree of construction difficulty (as seen in Chapter 10), such as the availability of experienced contractors and resources.

There is a high risk in outfall projects, so the professionals and supervisors involved should be highly qualified and have extensive experience. Insurance should be obtained for the equipment and resources, and the quality of all project components must also be ensured.

Appendix A: Costs

Not all countries have experience in marine outfall construction, so international tenders are usually called to allow foreign companies with the necessary expertise to participate. This can significantly affect costs as it may involve international movement of equipment and key personnel.

As a consequence of the issues discussed above, it is common in marine outfall projects that the ratio of indirect costs and general expenses to direct costs is much greater than for other types of civil engineering works.

The main components that determine project cost include:

- Preliminary studies: The costs of field surveys discussed in Chapter 5 must be included. Conceptual engineering is then carried out (Chapters 7, 8 and 9), which allows the solution to be confirmed and the cost of the project estimated and compared to alternatives. The owner can then assess the best way to finance the project. Detailed design and any environmental impact assessments are then carried out according to relevant local regulations. At the same time the marine concessions must be negotiated along with the land and easements required for onshore works. Tender bases for construction, supervision, and inspection of the project are then estimated. These activities typically take 6 to 9 months.
- Construction, including:
    (i) Works facilities;
    (ii) Pipe;
    (iii) Preparation of platform on land;
    (iv) Fabrication of reinforced concrete ballasts;
    (v) Preparation of auxiliary platform;
    (vi) Underwater excavations;
    (vii) Pipe welding and quality control;
    (viii) Mounting ballast onto pipe;
    (ix) Preparations prior to launching;
    (x) Launching, towing and flooding the outfall;
    (xi) Excavations and installation on land;
    (xii) Connections;
    (xiii) Surge chamber;
    (xiv) Removal of structures and clearing of the works site.

3.2 Direct costs

Direct costs are normally comprised of:

- Materials:
    (i) Pipes;
    (ii) Fittings;
    (iii) Concrete;
    (iv) Reinforcing steel;
    (v) Forms;
    (vi) Stainless steel bolts;
    (vii) Transport, welding and quality control of welding.
- Auxiliary structures:
    (i) Piles;
    (ii) Beams;
    (iii) Sheet piling.
- Labor:
    (i) Direct;
    (ii) Specialized and temporary.
- Equipment:
    (i) Construction (cranes, backhoes, trucks, compressors, etc.);
    (ii) Specialized (polyethylene pipe welders, removal of rock or coral);
    (iii) Temporary (boats, tugboats, communications equipment).

3.3 Indirect costs

Indirect costs are those that cannot be assigned to a particular work item, including:

- *Supervision:* Salaries and wages of senior works personnel, including engineers, constructors, administrative personnel, topographers, supervisors, risk prevention experts, and others necessary for the project, such as drivers, guards, etc.
- *Travel and board:* Travel and board for technical and administrative personnel and the labor force who are not direct or local.
- *Vehicles:* Rental, fuel, and maintenance of vehicles such as pickup trucks, trucks, and personnel transport buses.
- *Freight:* Transport of materials, pipes, metal structures, wood, reinforced steel, and other construction materials, and containers used as temporary offices and warehouses.

Appendix A: Costs

- *Works facilities:* Providing, fitting, furnishing, maintaining, dismantling and removal of temporary structures and offices. Includes warehouses, rest rooms, fences, access roads, material storage, electrical energy supply, fuel tank, lighting, etc.
- *Office:* Renting offices in the city for administration.
- *Equipment transportation:* Transporting construction equipment and machinery to the works site such as, cranes, backhoes, bulldozers, trucks, welders, compressors, and tools.
- *Insurance:* For construction risk required by the owner and that normally obtained by the contractor, such as for vehicles and machinery, and life insurance for workers.
- *Financial costs:* Taking into account that the contractor normally has to carry out works, move equipment and personnel, and supply equipment, before receiving the first payment from the client.
- *Fulfillment guarantees:* Guarantees that the client normally requests.
- *General works costs:* Communications, food and board of senior personnel, administrative personnel, equipment operators and non-local personnel, visits made by senior company personnel, representation costs, office supplies, etc.
- *Miscellaneous:* Such as municipal permits and licenses.
- *Overhead:* General expenses of the construction company's main office, which assigns part of its general expenses to be covered by the works.
- *Contingencies:* Provision for unexpected costs, such as due to poor weather conditions, the need to rent equipment such as tugboats or due to uncertainty in the exact dimensions of auxiliary structure or sheet piling, etc.
- *Profit:* The profit expected for executing the project.
- *Taxes:* As required.

3.4 Example calculation for typical HDPE outfall

The following illustrates the cost calculations involved. It is based on recent projects in Chile under normal conditions. The unit costs are summarized in Table B.2, assuming adequate local resources, and the following conditions:

Outfall diameter, OD = 710 mm
Length, L = 1,000 m
Water depth = 55 m
Length of rock zone = 150 m in the breaking wave zone
Minimum requirements of auxiliary structures, i.e. sheet piling not required.

Table B.2 Sample Outfall Cost Breakdown (April 2009)

Item	Unit	Quantity	Unit price (US$)	Total (US$)	% of total
Pipe supply	kg	93,000	2.99	278,302	12.9
Excavation of earth and rock	m^3	500	588.79	294,393	13.6
Platform and auxiliary structure	m	60	4,710.3	282,618	13.1
Land platform		1	20,948.45	20,948	1.0
Welded joints and quality control	Each	85	445.76	37,889	1.8
Fittings		1	30,707.57	30,707	1.4
Ballast and bolts	Each	200	36.66	7,331	0.3
Civil works and outfall on land		1	141,056.6	141,056	6.5
Launching of outfall		1	301,131.6	301,131	13.9
			Total direct costs:	1,394,378	
		Indirect costs @ 30% of direct costs:		418,313	19.4
	General expenses, profit and contingencies @ 25% of direct costs:			348,594	16.1
			Total (without taxes):	2,161,287	100.0

4. SUMMARY

Outfall costs vary widely depending on length, diameter, depth, construction method and materials, and local conditions. Existing data are summarized and a simple linear equation is presented to provide order-of-magnitude costs as a function of outfall diameter. It is noted that some costs are independent of outfall length, such as mobilization and demobilization of the specialized construction equipment, and construction through the surf zone can be the most expensive part. Therefore, unit costs for outfalls generally decrease as the length increases. Most HDPE outfalls at present have diameters up to about 1.4 m with costs ranging from about 1,000 to 7,000 $/m. The Cartagena, Colombia, outfall uses HDPE of 2.0 m diameter; its costs is estimated to be 8,750 $/m. Diameters greater than 2.4 m are usually concrete pipe or tunneled and generally range from 10,000 t0 20,000 $/m. Outfalls greater than 3.5 m diameter can exceed 30,000 $/m. Because of the wide variations, the numbers presented here can only be used for a preliminary estimate of the range of cost expected. It is noted that modern manufacturing techniques can produce spiral-wound HDPE pipe in diameters larger than 2.0 m, thus making HDPE a viable alternative material for even larger diameters than has previously been possible.

Appendix B
Abbreviations

ACUACAR: Aguas de Cartagena
ADCP: Acoustic Doppler current profiler
AFRI: Acute febrile respiratory illness
AIZ: Allocated impact zone
ACBM: Articulated concrete block mat
ASTM: American Society for Testing Materials
BEACH: Beaches Environmental Assessment, Closure, and Health
BOD: Biochemical oxygen demand
CCC: Criteria Continuous Concentration
CDD: Cold digestion/drying
CEPIS: El Centro Panamericano de Ingeniería Sanitaria y Ciencias del Ambiente
CEPPOL: Caribbean Environment Programme
CEPT: Chemically enhanced primary treatment
CERFS: Chemically enhanced rotating fine screens
CFD: Computational fluid dynamics

© 2010 IWA Publishing. *Marine Wastewater Outfalls and Treatment Systems.* By Philip JW Roberts, Henry J Salas, Fred M Reiff, Menahem Libhaber, Alejandro Labbe, and James C Thomson. ISBN: 9781843391890. Published by IWA Publishing, London, UK.

CL: Confidence level
CMC: Criteria Maximum Concentration
COD: Chemical Oxygen Demand
CTD: Conductivity, temperature, and depth
DAF: Dissolved air flotation
DESA: Distributed Entrainment Sink Approach
DO: Dissolved oxygen
DNS: Direct numerical simulation
DR: Dimension ratio
EEC: European Economic Community
EIA: Environmental Impact Assessment
EPBM: Earth pressure balance machine
EU: European Union
FOG: Fats, oil, and grease
GBT: Gravity belt thickening
GEF: Global Environmental Facility
GI: Gastro intestinal
GPS: Global positioning system
GRP: Glass reinforced plastic (fiberglass)
HDB: Hydrostatic design basis
HDD: Horizontal directional drilling
HDPE: High density polyethylene
ID: Inside diameter
IJS: Intermediate jacking station
ISO: International Organization for Standardization
LAC: Latin America and the Caribbean
LES: Large eddy simulation
LIF: Laser-induced fluorescence
LOAEL: Lowest-observed-adverse-effect level
LMZ: Legal mixing zone
MLLW: Mean low low water
MPN: Most probable number
MRS: Minimum required strength
MSL: Mean sea level
NAS: National Academy of Sciences
NGO: Non-Government Organizations
NIMBY: Not In My Back Yard
NOAEL: No-observed-adverse-effect level
NPDES: National Pollutant Discharge Elimination System (US)
NRC: National Research Council

Appendix B: Abbreviations

NTAC: National Technical Advisory Committee
NOAA: National Oceanic and Atmospheric Administration
OD: Outside diameter
OHHI: Oceans and Human Health Initiative
O&M: Operation and maintenance
PAE: Projected area entrainment
PE: Polyethylene
PAH: Polycyclic aromatic hydrocarbons
PAHO: Pan American Health Organization
PCB: Polychlorinated biphenyl
PDF: Probability density function
PP: Polypropylene
POM: Princeton ocean model
PVC: Polyvinyl chloride
QA/QC: Quality assurance and quality control
RANS: Reynolds-averaged Navier-Stokes
RAS: Return activated sludge
RC: Reinforced concrete
ROMS: Regional Ocean Model System
ROV: Remote operated vehicle
RWPT: Random walk particle tracking
SBR: Sequencing Batch Reactor
SDR: Standard dimension ratio
SWRCB: State Water Resources Control Board (California)
TBM: Tunnel boring machine
TDZ: Toxic dilution zone
TSS: Total suspended solids
UASB: Upflow anaerobic sludge blanket
UFW: Unaccounted for water
UNEP: United Nations Environmental Program
USD: US Dollars
USEPA: Unites States Environmental Protection Agency
UV: Ultraviolet radiation
VP: Visual Plumes
WAS: Waste Activated Sludge
WHO: World Health Organization
ZID: Zone of initial dilution

References

ACRA, A. et al. (1990). Water Disinfection by Solar Radiation; Assessment and Application. IDRC, Ottawa, p. 83.
Adams, E.E., Sahoo, D., Liro, C.R., and Zhang, X. (1994). Hydraulics of seawater purging in a tunneled wastewater outfall. *J. Hydr. Eng., ASCE*, **120**(2), 209–226.
Alder, C.A. (1973). Ecological Fantasies. Green Eagle Press, New York.
Allen, J.S. (1980). Models of wind-driven currents on the continental shelf. *Ann. Rev. Fluid Mech.*, **12**, 389–433.
APHA (1998). Standard Methods for the Examination of Water and Wastewater. 20th Edition. American Public Health Association, Washington, DC.
Arodys Leppe, Z. and Liliana Padilla, B. (1999). Emisarios submarinos de penco y tome, 5 años de vigilancia y una evaluación global de sus efectos ambientales, XIII Congreso Chileno de Ingenieria Sanitaria y Ambiental, AIDIS, Antofagasta, Chile, October 1999. (The submarine outfalls of penco and tome, 5 years of monitoring and global evaluation of their environmental effects, XIII Chilean Congress of Sanitary and Environmental Engineering, AIDIS, Anrofagasta Chile, October 1999.)
Argentina. Instituto Nacional de Ciencia y Técnica Hídricas (INCYTH) (1984). Estudio de la factibilidad de la disposición en el mar de los efluentes cloacales de la ciudad de Mar del Plata. Informe Final. Buenos Aires, Secretaría de Recursos Hídricos de Argentina.
Bahamas (2000). The Department of Statistics. *Population in Islands, Census 2000.*

© 2010 IWA Publishing. *Marine Wastewater Outfalls and Treatment Systems.* By Philip JW Roberts, Henry J Salas, Fred M Reiff, Menahem Libhaber, Alejandro Labbe, and James C Thomson. ISBN: 9781843391890. Published by IWA Publishing, London, UK.

Bartram, J. and Rees, G., (ed.) (2000). Monitoring bathing waters: A practical guide to the design and implementation of assessments and monitoring programmes. London, E & FN Spons. Published on behalf of the World Health Organization, Commission of the European Communities and US Environmental Protection Agency.

Baumgartner, D.J., Frick, W.E., and Roberts, P.J.W. (1994). Dilution Models for Effluent Discharges (Third Edition). Pacific Ecosystems Branch, ERL-N, Newport, OR. EPA/600/R-94/086, June 1994, 189pp.

Beletsky, D. and Schwab, D.J. (2001). Modeling circulation and thermal structure in Lake Michigan: Annual cycle and interannual variability. *J. Geophys. Res.*, **106**(9), 19,745–19,771.

Beletsky, D., Schwab, D.J., and McCormick, M. (2006). Modeling the 1998–2003 summer circulation and thermal structure in Lake Michigan. *J. Geophys. Res.*, **111**, 1–18.

Belize (2000). Central Statistical Office. *Census 2000*.

Bennett, J.R. and Clites, A.H. (1987). Accuracy of trajectory calculation in a finite-difference circulation model. *J. Comput. Phys.*, **68**(2), 272–282.

Bigg, P.H. (1967). Density of water in SI units over the Range 0–40 C. *Br. J. Appl. Phys.*, **8**, 521–537.

Bleninger, T. (2006). Coupled 3D Hydrodynamic Models for Submarine Outfalls: Environmental Hydraulic Design and Control of Multiport Diffusers. Thesis, Institute for Hydromechanics, University of Karlsruhe, Karlsruhe.

Blumberg, A.F. and Mellor, G.L. (1983). Diagnostic and prognostic numerical circulation studies in the South Atlantic Bight. *Journal of Geophysical Research*, **88**, 4479–4592.

Blumberg, A.F. and Mellor, G.L. (1987). A description of a three-dimensional coastal ocean circulation model. *Three Dimensional Coastal Ocean Models*.

Blumberg, A.F., Ji, Z.-G., and Ziegler, C.K. (1996). Modeling outfall plume behavior using far field circulation model. *Journal of Hydraulic Engineering*, **122**(11), 610–616.

Boehm, A.B., Sanders, B.F., and Winant, C.D. (2002). Cross-shelf transport at huntington beach. implications for the fate of sewage discharged through an offshore ocean outfall. *Environmental Science & Technology*, **36**(9), 1899–1906.

Brazil Ministerio do Interior (1976). Aguas de balneabilidade, Portoria No. 536.

Brooks, N.H. (1960). Diffusion of sewage effluent in an ocean current. *Proceedings of the First International Conference on Waste Disposal in the Marine Environment*, E.A. Pearson, (ed.), pp. 246–267. Pergamon Press, New York, 569pp.

Brooks, N.H. (1984). Dispersal of wastewater in the ocean – a cascade of processes at increasing scales. *Proc. of the Conf. Water for Resource Development*, Coeur d'Alene, Aug 14–17.

Britto, E.R. and Goncalves, F.P.B. (1979). Contribuicao ao estudo para determinacao do decaemento bacteriono. 10° Congresso Brasileiro de Engenharia Sanitária e Ambiental. Manaus, 21 a 26 Janeiro 1979.

Butler, D., May, R., and Ackers, J. (2003). Self-cleansing sewer design based on sediment transport principles. *J. Hydr. Eng., ASCE April 2003*.

Cabelli, V.J. (1979). Evaluation of recreational water quality, The USEPA approach. Chap. 14, *Biological Indicators of Water Quality*. J.A. & E.L. (eds), Chichester, Wiley.

Cabelli, V.J., Dufour, A.P., McCabe, L.J., and Levin, M.A. (1983). A marine recreational water quality criterion consistent with indicator concepts and risk analysis. *Journal of Water Pollution Control Federation*, **55**(10), 1306–1314.

References

Cabelli, V.J. (1983). Health Effects Criteria for Marine Recreational Waters. Research Triangle Park, USEPA. 98p. EPA-600/1-80-031.

Cabelli, V.J. (Undated). Epidemiology of Enteric Viral Infections.

Carhart, R.A., Policastro, A.J., Ziemer, S., Haake, S., and Dunn, W. (1981). Studies of mathematical models for characterizing plume and drift behavior from cooling towers, Vol. 2: Mathematical model for single-source (single-tower) cooling tower plume dispersion. Electric Power Research Institute, CS-1683, Vol. 2, Research Project 906-01.

Caribbean Environmental Programme (CEPPOL) & United Nations Environment Programme (UNEP). (1991). Report on the CEPPOL Seminar on Monitoring and Control of Sanitary Quality of Bathing and Shellfish-Growing Marine Waters in the Wider Caribbean. Kingston, Jamaica, 8–12 April 1991. Technical Report No. 9.

Carnelos, S.L. (2003). Urban stormwater runoff discharges to lake pontchartrain: 3-D hydrodynamic model for plume behavior and fate and transport of pathogens in the effluent. PhD Thesis, University of New Orleans, p. 474.

Carvalho, J.L.B. (2003). Modelagem e análise do lançamento de efluentes através de emissários submarinos (Modeling and Analysis of Sewage Disposal through Ocean Outfalls). Thesis, Ocean Engineering, Universidade Federal do Rio de Janeiro, COPPE, Rio de Janiero.

Carvalho, J.L.B., Feitosa, R.C., Rosman, P.C.C., and Roberts, P.J.W. (2006). A bacterial decay model for coastal outfall plumes. *Journal of Coastal Research*, **39**(Special Issue), 1524–1528.

CEDAEH (1982). Companhia Estadual de Águas e Esgotos, Divisão de Operação e Tratamento, Relatório de actividades – 1982. Río de Janeiro.

Cederwall, K. (1968). Hydraulics of marine wastewater disposal. Report No. 42, Chalmers Institue of Technology, Goteberg, Sweden.

CEPAL (1993). Estadísticas y datos básicos sobre población en América Latina y el Caribe. Comisión Económica para América Latina (CEPAL) Notas sobre la Economía y el Desarrollo, 540/541.

CEPAL (2005). Boletín Demográfico No. 75. Centro Latinoamericano y Caribeño de Demografía/División de Población (CELADE).

CEDAEH (1982). Companhia Estadual de Águas e Esgotos, Divisão de Operação e Tratamento, Relatório de actividades – 1982. Río de Janeiro.

CETESB (1999). Estudo Epidemiológico – 1999. Ocorrência de distúrbios gastrintestinais em banhistas e sua correlação com a qualidade sanitária das águas das praias, Sao Paulo, Brazil.

Cheung, V. (1991). Mixing of a round buoyant jet in a current. PhD Thesis, Dept. of Civil and Structural Engineering, University of Hong Kong, Hong Kong.

Chin, D.A. and Roberts, P.J.W. (1985). Model of dispersion in coastal waters. *J. Hydraul. Eng.*, **111**(1), 12–28.

Chin, D.A., Ding, L., and Huang, H. (1997). Ocean-outfall mixing zone delineation using doppler radar. *J. Env. Eng., ASCE*, **123**(12), 1217–1226.

Choi, K.W. and Lee, J.H.W. (2007). Distributed entrainment sink approach for modeling mixing and transport in the intermediate field. *J. Hydraul. Eng.*, **133**(7), 804–815.

Chu, V.H. (1979). L.N. Fan's data on buoyant jets in crossflow. *J. Hydr. Eng., ASCE*, **105**(HY5), 612–617.

Coburn, S.E. (1930). Survey of the pollution of rivers and lakes in the vicinity of Rochester, N.Y. *Indust. Eng. Chem.*, **22**, 1336.
Colebrook, C.F. (1939). Turbulent flow in pipes, with particular reference to the transition region between the smooth and rough pipe laws. *Journal of the Institution of Civil Engineers*, **11**, 133–156.
Colombia. Ministerio de Salud. (1979). Disposiciones Sanitarias sobre Aguas – Artículo 69 Ley 05.
Colwell, R.R., Orlob, G.T., and Schubel, J.R. (1996). Mamala bay study management. Report MB-1, Mamala Bay Study Commission, Honolulu, Hawaii.
Committee on Bathing Beach Contamination of the Public Health Laboratory Service. (1959). Sewage contamination of coastal bathing waters in England and Wales. A bacteriological and epidemiological study. *J. Hyg.*, **57**, 435.
Committee on Water Quality Criteria, National Academy of Sciences, National Academy of Engineering (1972). *Water Quality Criteria*. Washington, DC, EPA-R3-73-033.
Committee on Evaluation of the Safety of Fishery Products. (1991). Sea food safety. Forid E. Ahmed, (ed.), Food and Nutritional Board. Institute of Medicine. National Academy Press, Washington, DC.
Connolly, J.P., Blumberg, A.F., and Quadrini, J.D. (1999). Modeling fate of pathogenic organisms in coastal waters of Oahu, Hawaii. *J. Env. Eng., ASCE*, **125**(5), 398–406.
Cooper, C.K. (1988). Parametric models of hurricane-generated winds, waves, and currents in deep water. *Proc. 24th Annual Technology Conf.*, Houston, TX.
Cooper-Smith, G. (2001). Chemically enhanced primary treatment: The UK experience. *Water*, **21**, 10.
Costa Rica (2004). Supreme Decree No. 32133-S, December 7, 2004.
Council of the European Union (2004). Directive 2004 concerning the management of bathing water quality. Interinstitutional File: 2002/0254.
Couriel, E.D. (1993). Near field physical modeling of malabar deepwater outfall. *11th Australasian Conference on Coastal and Ocean Engineering*, Barton, ACT, pp. 85–89.
Csanady, G.T. (1979). What drives the waters of the continental shelves. *Oceanus*, **22**(3), 28–35.
Csanady, G.T. (1983a). Dispersal by randomly varying currents. *J. Fluid Mech.*, **132**, 375–394.
Csanady, G.T. (1983b). Advection, diffusion, and particle settling. E.P. Myers and E.T. Harding, (eds.), MIT, Cambridge, Mass., pp. 177–246.
Cuba. Ministerio de Salud. (1986). Higiene comunal, lugares de baño en costas y en masas de aguas interiores, requisitos higiénicos sanitarios. pp. 93–97. La Habana, Cuba.
Davidson, M.J., Papps, D.A., and Wood, I.R. (1990). The behavior of merging buoyant jets. *Recent Research Advances in the Fluid Mechanics of Turbulent Jets and Plumes*, Viana do Castelo, Portugal, June 1993.
Daviero, G.J. and Roberts, P.J.W. (2006). Marine wastewater discharges from multiport diffusers III: Stratified stationary water. *Journal of Hydraulic Engineering*, **132**(4), 404–410.
Davis, D.A. and Ciani, J.B. (1976). Technical report: Wave forces on submerged pipelines. Civil Engineering Laboratory, Naval Construction Battalion Center, Port Hueneme, California.

References

Davis, L.R. (1999). *Fundamentals of Environmental Discharge Modeling.* CRC Press, Boca Raton, FL.

Davis, L.R., Davis, A., and Frick, W. (2004). Computational fluid dynamic application to diffuser mixing zone analysis – case studies. *3rd International Conference on Marine Wastewater Disposal MWWD 2004*, Catania, Italy, Sept 27–October 2, 2004.

Dimou, K. (1992). 3-D Hybrid Eulerian-Lagrangian/Particle tracking Model for simulating mass transport in coastal water bodies. Thesis, MIT Thesis, Massachusetts Institute of Technology.

Ding, L., List, E.J., Hicks, N., Hannoun, I.A., and Steele, A. (2008). Joint water pollution control plant outfall modeling. *MWWD 2008*, Cavtat, Croatia, October 27–31, 2008.

DMSA (1971). Estudio Integral para determinar el estado de contaminación de las playas del litoral central. Dirección de Malariología y Saneamiento Ambiental, Ministerio de Sanidad y Asistencia Social. Caracas.

Doneker, R.L. and Jirka, G.H. (1999). Discussion of mixing of inclined dense jets' by Roberts, P.J.W., Ferrier, A., and Daviero, G. (1997). *J. Hydraul. Eng.*, **125**(3), 317–319.

Duer, M. (2000). Use of elastomeric Duckbill valves for long-term hydraulic and dilution efficiency of marine diffusers. Marine Wastewter Discharges MWWD2000, Genova, Italy, 28 Nov–1 Dec 2000.

Dufour, A.P. (1984). Health effects criteria for fresh recreational waters. EPA-600/1-84-004, U.S. Environmental Protection Agency, Cincinnati, USEPA, 33p.

Dufour, A.P. and Ballentine, P. (1986). Ambient Water Quality Criteria for Bacteria – 1986 (Bacteriological ambient water quality criteria for marine and fresh recreational waters). Washington, D.C. USEPA. 18p. EPA A440/5-84-002.

Ecuador. Ministerio de Salud Pública (1987). Instituto Ecuatoriano de Obras Sanitarias, Proyecto de normas reglamentarias para la aplicación de la Ley.

Editorial América, S.A. (1983). Almanaque Mundial 1984. Panamá.

El Salvador (2000). Dirección general de estadística y censos. *Proyecciones de Población de el Salvador 1995–2025.*

ENCIBRA (1969). Engineering-Science Inc. Marine sewage disposal for Rio de Janeiro, City of Rio de Janeiro, Brazil.

EEC (1976). Council directive of 8 December 1975 concerning the quality of bathing water. *Official Journal of the European Communities*, **19**, L31.

EU (2006). Directive 2006/7/EC of the European Parliament and of the Council of 15 February 2006 concerning the management of bathing water quality and repealing Directive 76/160/EEC.L 64–37 Official Journal of the European Union, Strasbourg, 15 February 2006.

Fan, L.N. (1967). Turbulent buoyant jets into stratified or flowing ambient fluids. Report No. KH-R-15, W.M. Keck Lab. of Hydraulics and Water Resources, California Institute of Technology, Pasadena, California.

Fan, L.N. and Brooks, N.H. (1969). Numerical solutions of turbulent jet problems. Technical Report KH-R-18, W.M. Keck laboratory of hydraulics and water resources, California Institute of Technology, Pasadena, California.

Fischer, H.B., List, E.J., Koh, R.C.Y., Imberger, J., and Brooks, N.H. (1979). *Mixing in Inland and Coastal Waters.* Academic Press, New York.

Fleisher, J.M. (1991). A reanalysis of data supporting US federal bacterial water quality criteria governing marine recreational waters. *Research Journal, Water Pollution Control Federation*, **63**, 259.

Fleisher, J.M., Kay, D., Salmon, R.L., Jones, F., Wyer, M.D., and Godfree, A.F. (1996). Marine waters contaminated with domestic sewage: Nonenteric illnesses associated with bather exposure in the United Kingdom. *American Journal of Public Health*, **86**(9), 1228–1234.

Food and Drug Administration (FDA) (1989) Revision. Sanitation of shellfish growing areas. National shellfish sanitation program, manual of operations Part I. Center for Safety and Applied Nutrition. Division of Cooperative Program, Shellfish Sanitation Branch, Washington, DC.

French, J.A., Johnson, W.A., Mills, J.A., and Marsh, G.S. (1993). Protecting the boca raton outfall before and after Hurricane Andrew. *Hydraulic Engineering (1993)*, San Francisco, California, pp. 543–548 July 25–30, 1993.

Frick, W.E. (1984). Non-empirical closure of the plume equations. *Atmospheric Environment*, **18**(4), 653–662.

Frick, W.E., Baumgartner, D.J., and Fox, C.G. (1994). Improved prediction of bending plumes. *Journal of Hydraulic Research*, **32**(6), 935–950.

Frick, W.E., Khangaonkar, T., Whitman, R., Dufour, A., Yang, Z., and Korinek, G. (2003). Predicting the movement and bacteria concentrations of river plumein lake Michigan using hydrodynamic modeling. Great Lakes Beach Association, *2nd Annual Meeting*, 21–22 October, 2003, Muskegon, Michigan.

Frick, W.E., Roberts, P.J.W., Davis, L.R., Keyes, J., Baumgartner, D.J., and George, K.P. (2003). Dilution models for effluent discharges, 4th Edition (Visual Plumes). EPA/600/R-03/025. US EPA, Athens, Georgia, 131pp.

Frick, W.E. (2004). Visual plumes mixing zone modeling software. *Environmental Modeling and Software*, **19**, 645–654.

Frick, W.E. (2005). Models for submarine outfall – validation and prediction uncertainties. Keynote Address. HELECO '05, Athens, Greece, 3–6 February 2005.

Garber, W.F. (1956). Bacteriological standards for bathing waters. *Sewage Industr. Waters*, **28**, 795.

García Agudo, E. (1991). CETESB Superintendent. Personal communication.

Goda, Y. (2003). Revisiting Wilson's formulas for wind-wave prediction. *Journal of Waterway, Port, Coastal, and Ocean Engineering, ASCE*, **129**(2), 93–95.

Grace, R.A. (1978). *Marine Outfall Systems.* Prentice-Hall. Englewood Cliffs, 600pp.

Grace, R.A. (2005). Marine outfall performance. I : Introduction and flow restoration. *Journal of Performance of Constructed Facilities, ASCE*, 347–358.

Grace, R.A. (2005). Marine outfall performance. II: Stabilization and case studies. *Journal of Performance of Constructed Facilities, ASCE*, 359–369.

Grace, R.A. (2009). *Marine outfall construction.* ASCE Press, Reston, Virginia.

Grigg, R.W. and Dollar, S.J. (1995). Environmental protection misapplied: alleged versus documented impacts of a deep ocean sewage outfall in Hawaii. *Ambio*, **24**(2), 125–128.

Gunnerson, C.G. and French, J.A. (1996). *Wastewater Management for Coastal Cities*, Springer.

Hamilton, P., Singer, J., and Waddell, E. (1995). Mamala bay study: Ocean current measurements. Science Applications International Corporation, Raleigh, NC, July 15, 1996.

Harleman, D.R.F. and Murcott, S. (1998). Low cost nutrient removal demonstration study. *Tech. Rep.*, World Bank UNDP, BRA/90/010.

References

Harleman, D.R.F. and Murcott, S. (2001). An innovative approach to urban wastewater treatment in the developing world. *Water*, **21**, 44–48.

Haugland, R.A., Siefring, S.C., Wymer, L.J., Kristen P. Brenner, K.P., and Dufour, A.P. (2005). Comparison of enterococcus measurements in freshwater at two recreational beaches by quantitative polymerase chain reaction and membrane filter culture analysis. *Water Research*, **39**, 559–568.

Helliwell, P.R. and Webber, N.B. (1973). The use of the solent model for an investigation into sewer outfall location. *Mathematical and Hydraulic Modeling of Estuarine Pollution*, Water Pollution Research Laboratory, 18th and 19th April 1972.

Helmer, R. Hespanhol, I., and Silva, L.J. (1991). Public health criteria for the aquatic environment: Recent WHO guidelines and their application. *Water Science and Technology*, **24**(2), 35–42.

Hydroscience, Inc. (1977). Guanabara bay water quality study, Brazil. Pan American Health Organization/FEEMA.

Huber, H., Tanik, A., and Gercek, M. (1995). Case studies on preliminary treatment facilities at marine outfalls. *Wat. Sci. Tech.*, **32**(2).

Huber (2003). The Huber Group, Personal communication.

Hulstrom, F. (1939), Transportation of detritus in water in Trask, P.D., Recent marine sediments. American Association of Petroleum Geologists, Tulsa, OK, pp. 5–21.

Hwang, R.R. and Chiang, T.P. (1995). Numerical simulation of vertical forced plume in a crossflow of stably stratified fluid. *J. Fluids Eng.*, **117**, 696–705.

Hwang, R.R., Chiang, T.P., and Yang, W.C. (1995). Effect of ambient stratification on buoyant jets in crossflow. *Journal of Engineering Mechanics*, **121**, 865–872.

Instituto Nacional de Ciencia y Tecnica Hidricas (INCYTH) (1984). Estudio de la factibilidad de la disposición en el mar de los efluentes cloacales de la ciudad de Mar del Plata. Argentina. 3 volúmenes.

Isaacson, M.S., Koh, R.C.Y., and Brooks, N.H. (1983). Plume dilution for diffusers with multiple risers. *J. Hydr. Eng., ASCE*, **109**(2), 199–220.

Jamaica (2001). Statistical Institute of Jamaica. *Census 2001*.

Janson, L.E. (2005). Plastic pipes for water supply and sewage disposal. 4th Edition, Borealis, Stockholm, Sweden.

Jirka, G.H. and Harleman, D.R.F. (1979). Stability and mixing of a vertical plane buoyant jet in confined depth. *J. Fluid Mech.*, **94**(2), 275–304.

Jirka, G. (2004). Integral model for turbulent buoyant jets in unbounded stratified flows. Part I: Single round jet. *Environmental Fluid Mechanics*, **4**(1), 1–56.

Jirka, G. (2006). Integral model for turbulent buoyant jets in unbounded stratified flows. Part II: Plane jet dynamics resulting from multiport diffuser jets. *Environmental Fluid Mechanics*, **6**, 43–100.

Japan. Environmental Agency (1981). Environmental laws and regulations in Japan (III) water.

Jones, F. and Kay, D. (1990) Recreational water quality: the relationship between epidemiological studies and recreational activities in water. Report of a biological standards seminar, Middlesex Polytechnic, 9 February 1990.

Kay, D., Wyer, M., McDonald, A., and Woods, N. (1990). The application of water-quality standards to UK bathing waters. *Journal of the Institution of Water and Environmental Management*, **4**(5), 436–441.

Kay, D., Fleisher, J.M., Salmon, R.L., Wyer, M.D., Godfree, A.F., Zelenauch-Jacquotte, Z., and Shore, R. (1994). Predicting Likelihood of gastroenteritis from sea bathing; results from randomized exposure. *Lancet*, **344**(8927), 905–909.

Kinsman, B. (1965). Wind waves. Prentice-Hall, Englewood Cliffs, New Jersey.

Koh, R.C.Y. (1988). Shoreline impact from ocean waste discharges. *J. Hydr. Eng., ASCE*, **114**(4), 361–376.

Koochesfahani, M.M. and Dimotakis, P.E. (1985). Laser-induced fluorescence measurements of mixed fluid concentration in a liquid plane shear layer. *AIAA J.*, **23**(11), 1700–1707.

Koppl, S. and Formmann, C. (2004). Eco-efficient wasteater treatment of river and sea outfalls, MWWD 2004 *3rd International Conference on Marine Wastewater Disposal and Marine Environment*, Catania, Italy, September 27 October 2, 2004.

Lam, K.M., Lee, W.Y., Chan, C.H.C., and Lee, J.H.W. (2006). Global behaviors of a round buoyant jet in a counterflow. *J. Hydraul. Eng.*, **132**(6), 589–604.

Law, A.W.-K., Lee, C.C., and Qi, Y. (2002). CFD modeling of a multi-port diffuser in an oblique current. *MWWD2002*, Istanbul, Turkey, September 16–20, 2002.

Lee, J.H.W. and Jirka, G.H. (1981). Vertical round buoyant jet in shallow water. *J. Hydr. Div., ASCE*, **107**(HY12), 1651–1675.

Lee, J.H.W. and Neville-Jones, P. (1987). Initial dilution of horizontal jet in crossflow. *J. Hydr. Eng., ASCE*, **113**(5), 615–629.

Lee, J.H.W., Cheung, Y.K., and Cheung, V. (1987). Mathematical modelling of a round buoyant jet in a current: An assessment. *Proceedings of International Symposium on River Pollution Control and Management*, Shanghai, China, October 1987.

Lee, J.H.W., Karandikar, J., and Horton, P. (1998). Hydraulics of "Duckbill" elastomer check valves. *J. Hydraul. Eng.*, **124**(4), 394–405.

Lee, J.H.W., Wilkinson, D.L., and Wood, I.R. (2001). On the head-discharge relation of a Duckbill elastomer check valve. *J. Hydr. Res.*, **39**(6).

Lee, J.H.W., Guo, Z.-R., and Yau, T.W.C. (2001). Numerical and experimental study of unsteady salt water purging in Hong Kong sea outfall model. *J. Hydr. Res.*, **39**(1), 83–91.

Lee, J.H.W. and Chu, V.H. (2003). *Turbulent Jets and Plumes: A Lagrangian Approach*. Kluwer Academic Publishers, Boston.

List, E.J., Gartrell, G., and Winant, C.D. (1990). Diffusion and dispersion in coastal waters. *J. Hydr. Eng., ASCE*, **116**(10), 1158–1179.

Liu, W.C., Kuo, J.T., Young, C.C., and Wu, M.C. (2007). Evaluation of marine outfall with three-dimensional hydrodynamic and water quality modeling. *Environmental Modeling & Assessment*, **12**(3), 201–211.

Ludwig, R.G. (1976). The planning and design of ocean disposal systems. Prepared for the Pan American Health Organization Symposium on Wastewater Treatment and Disposal Technology. Buenos Aires.

Ludwig, R.G. (1983). Marine outfall planning and design. ENCIBRA S.A., Sao Paulo, Brazil.

Ludwig, R.G. (1988). Environmental impact assessment, siting and design of submarine outfalls. EIA Guidance Document, MARC Report No. 43. Monitoring and Assessment Research Centre, King's College, University of London and World Health Organization.

Mellor, G.L. and Yamada, T. (1982). Development of a turbulence closure-model for geophysical fluid problems. *Reviews of Geophysics*, **20**(4), 851–875.
Mendez-Diaz, M.M. and Jirka, G.H. (1996). Buoyant plumes from multiport diffusers in deep coflowing water. *J. Hydr. Eng., ASCE*, **122**(8), 428–435.
Mexico. Secretaría de desarrollo urbano y ecología (SEDUE). (1983). Breviario Jurídico Ecológico. Subsecretaria de Ecología.
Millero, F.J., Chen, C.-T., Bradshaw, A., and Schleicher, K. (1980). A new high pressure equation of state for seawater. *Deep-Sea Research*, **27A**, 255–264.
Millero, F.J. and Poisson, A. (1981). International one-atmosphere equation of state for seawater. *Deep-Sea Research*, **28A**, 625–629.
Moore, B. (1959). Sewage contamination of coastal bathing waters in England and Wales: A bacterial and epidemiological study. *J. Hyg.*, **57**, 435.
Moore, B. (1975). The case against microbial standards for bathing beaches. In *Discharge of Sewage from Sea Outfalls*. A.L.H. Gamerson, (ed.), London, Pergamon Press, 103p.
Morton, B.R., Taylor, G.I., and Turner, J.S. (1956). Turbulent gravitational convection from maintained and instantaneous sources. *Proceedings of the Royal Society of London*, **A234**, 1–23.
Murcott, S., Dunn, A., and Harleman, D.R.F. (1996). Chemically enhanced wastewater treatment for agricultural reuse in Mexico. International Association of Water Quality, Biennal Conference, Singapore.
Muellenhoff, W.P., Soldate, A.M., Jr., Baumgartner, D.J., Schuldt, M.D., Davis, L.R., and Frick, W.E. (1985). Initial mixing characteristics of municipal ocean outfall discharges: Volume 1. Procedures and Applications. EPA/600/3-85/073a. (November 1985).
Myers, E.P. and Harding, E.T. (1983). Ocean disposal of municipal wastewater: Impacts on the coastal environment, MIT Sea Grant College Program, Cambridge, Mass.
MWRA (2004). Massachusetts Water Resources Authority effluent outfall ambient monitoring plan Revision 1, March 2004. Boston: Massachusetts Water Resources Authority. Report ENQUAD ms-092. 65p. http://www.mwra.com/harbor/enquad/pdf/ms-092.pdf
National Technical Advisory Committee (NTAC) (1968). Water quality criteria. Washington, DC, Federal Water Pollution Control Administration, 7p.
Nekouee, N., Roberts, P.J.W., McCormick, M.J., and Schwab, D.J. (2007). 3d Numerical Prediction of the Grand River Plume. *IAGLR 50th Annual Conference on Great Lakes Research*, University Park, Penn., May 28–June 1, 2007.
Netherlands Antilles (2001). Central bureau of statistics. *Population and Housing Census 2001*.
Neville-Jones, P.J.D. and Chitty, A.D. (1996). Sea outfalls – construction, inspection, and repair. An engineering guide. CIRIA Report 159, London, UK.
Nicaragua (2000). Instituto Nacional de Estadística y Censos. Dirección de Estadística Sociodemográfica. *Estimaciones y Proyecciones de la Población Total por Año Calendario, según Departamento y Municipio*.
NRC (1993). *Managing Wastewater in Coastal Urban Areas*. National Research Council. Committee on Wastewater Management for Coastal Urban Areas, National Academy Press, Washington, DC.
Ocean Surveys (1999a). Sosua and sosua alternate alignments oceanographic studies for the design of submarine outfalls. OSI Report No. 98ES109.2.

Ocean Surveys (1999b). Puerto plata alignment oceanographic studies for the design of submarine outfalls. OSI Report No. 98ES109.3.
Ocean Surveys (1999c). Barahona and samana alignments. oceanographic studies for the design of submarine outfalls. OSI Report No. 98ES109.4.
Pasquill, F. and Smith, F.B. (1983). *Atmospheric Diffusion.* E. Horwood, (ed.), New York, Chichester, West Sussex.
Peru Ministerio de Salud (1983). Modificaciones a los Artículos 81° y 82° Reglamento de los Títulos I, II y III de la Ley General de Aguas. Decreto Supremo N° 007-83-SA. Perú.
Pearson, E.A. (1971). Guidelines for conduction of bacterial dissapearance rate (T-90) studies for marine outfall design. Unpublished paper, April.
Petrenko, A.A., Jones, B.H., Dickey, T.D., and Hamilton, P. (2001). Internal tide effects on a sewage plume at sand Island, Hawaii. *Cont. Shelf Res.*, **20**, 1–13.
Petrofisa Plásticos, S.A. Cañerías Submarinas en P.R.F.V. Buenos Aires, Argentina.
PipeLife (2002). Technical catalogue for submarine installations of polyethylene Pipes. Pipelife Norge AS, December 2002.
Pruss, A. (1998). Review of epidemiological studies on health effects from exposure to recreational water. *International Journal of Epidemiology*, **27**, 1–9.
Puerto Rico (2000). US Census Bureau. *Census 2000.*
Puerto Rico (1983) Junta de Calidad Ambiental. *Reglamento de Estándares de Calidad de Agua*, 28 de febrero de 1983.
Quetin, B. and de Rouville, M. (1986). Submarine sewer outfalls – a design manual. *Marine Pollution Bulletin*, **17**(4), 133–183.
RAS (2000), Technical norm of the water and sanitation sector in colombia (Resolution 1096 of November 17, 2000 del Reglamento de Agua y Saneamiento), Article 180.
Rawn, A.M. and Palmer, H.K. (1929). Pre-determining the extent of a sewage field in sea water. *Trans. ASCE,* Paper No. 1750, 1036–1081.
Rawn, A.M., Bowerman, F.R., and Brooks, N.H. (1960). Diffusers for disposal of sewage in seawater. *Proceedings of the American Society of Civil Engineers, Journal of the Sanitary Engineering Division*, **86**, 65–105.
Roberts, P.J.W. (1977). Dispersion of buoyant wastewater discharged from outfall diffusers of finite length. Report No. KH-R-35, W.M. Keck Laboratory of Hydraulics and Water Resources, California Institute of Technology.
Roberts, P.J.W. (1979). Line plume and ocean outfall dispersion. *J. Hydr. Div., ASCE*, **105**(HY4), 313–330.
Roberts, P.J.W. (1979). Two-dimensional flow field of multiport diffuser. *J. Hydr. Div., ASCE*, **105**(HY5), 605–611.
Roberts, P.J.W., Snyder, W.H., and Baumgartner, D.J. (1989a). Ocean outfalls. I: Submerged wastefield formation. *J. Hydr. Eng., ASCE*, **115**(1), 1–25.
Roberts, P.J.W., Snyder, W.H., and Baumgartner, D.J. (1989b). Ocean outfalls. II: Spatial evolution of submerged wastefield. *J. Hydr. Eng., ASCE*, **115**(1), 26–48.
Roberts, P.J.W., Snyder, W.H., and Baumgartner, D.J. (1989c). Ocean outfalls. III: Effect of diffuser design on submerged wastefield. *J. Hydr. Eng., ASCE*, **115**(1), 49–70.
Roberts, P.J.W. and Williams, N. (1992). Modeling of ocean outfall discharges. *Wat. Sci. Tech.*, **9**, 155–164.
Roberts, P.J.W. and Snyder, W.H. (1993a). Hydraulic model study for the boston outfall. I: Riser configuration. *J. Hydr. Eng., ASCE*, **119**(9), 970–987.

References

Roberts, P.J.W. and Snyder, W.H. (1993b). Hydraulic model study for the boston outfall. II: Environmental performance. *J. Hydr. Eng., ASCE*, **119**(9), 988–1002.

Roberts, P.J.W. and Webster, D.R. (2002). Turbulent diffusion. In *Environmental Fluid Mechanics – Theories and Applications*, H. Shen, (ed.), ASCE, Reston, Va., p. 467.

Roberts, P.J.W. (1999a). Modeling the Mamala Bay Plumes. I: Near Field. *J. Hydr. Eng., ASCE*, **125**(6), 564–573.

Roberts, P.J.W. (1999b). Modeling the Mamala Bay Plumes. II: Far Field. *J. Hydr. Eng., ASCE*, **125**(6), 574–583.

Roberts, P.J.W. (1999c). Closure to Discussion of mixing of inclined dense jets' by Roberts, P. J. W., Ferrier, A., and Daviero, G. (1997). *J. Hydraul. Eng.*, **125**(3), 317–319.

Roberts, P.J.W., Hunt, C.D., and Mickelson, M.J. (2002). Field and model studies of the boston outfall. *Second International Conference on Marine Waste Water Discharges, MWWD2002*, Istanbul, Turkey, September 16–20, 2002.

Roberts, P.J.W. (2005). Modeling of wind effects on bacterial transport for the cartagena ocean outfall. Prepared for the World Bank, Washington, DC, Atlanta, 29 July 2005.

Roberts, P.J.W. and Villegas, B. (2006). Three-dimensional modeling of bacterial transport for the cartagena ocean outfall. Prepared for Aguas de Cartagena, Cartagena, Colombia, Atlanta, 20 April 2006.

Saint Lucia (2000). Goverment Statistics Department. Central Statistical Office. *Census*.

Salas, H. and Robinson, K. (2006). Summary Report, Experts Consultation Development of a Protocol for Epidemiological Investigations in Recreational Bathing Waters, Mexico City, Mexico, 28–30 November 2005. CEPIS/BS/SDE/PAHO, Lima, Peru.

Salas, H. and Robinson, K. (2006). Protocol for epidemiological investigations in recreational bathing waters in latin America and the Caribbean. CEPIS/BS/SDE/PAHO Lima, Peru.

Saliba, L.J. and Helmer, R. (1990). Health risks associated with pollution of coastal bathing waters. *World Health Statistics*, **43**(3), 177–187, World Health Organization.

Rosenburg, M.L. (1976). Shigellosis from swimming. *J. Am. Med Assoc.*, **236**, 1849.

Saint Lucia (2000). Goverment Statistics Department. Central Statistical Office. *Census Population 2000*.

Schatzmann, M. (1979). An integral model of plume rise. *Atmos. Env.*, **13**(5), 721–731.

Scott, W.J. (1951). Sanitary study of shore bathing waters. *Bull. Hyg.*, **33**, 351.

Schwab, D.J. and Bedford, K.W. (1994). Initial implementation of the great lakes forecasting system: A real-time system for predicting lake circulation and thermal structure. *Water Pollut. Res. J. Can.*, **29**, 203–220.

Simons, G.W., Hilscher, R., Ferguson, H.F., and Gage, S. de M. (1922). Report of the committee of bathing places. *Amer. J. Pub. Health*, **12**(1):121–123.

Signell, R.P., Jenter, H.L., and Blumberg, A.F. (2000). Predicting the physical effects of relocating Boston's sewage outfall. *Est., Coastal and Shelf Sc.*, **50**, 59–72.

Sotiropoulos, F. (2005). Introduction to statistical turbulence modeling for hydraulic engineering flows. *Computational Fluid Dynamics*, P.D. Bates, S.N. Lane, and R.I. Ferguson, (eds.), John Wiley & Sons, Ltd.

Spiegel, E.A. and Veronis, G. (1960). On the Boussinesq approximation for a compressible fluid. *Astrophys. J.*, **131**, 442–447.

Sternberg, R.W. (1972), Predicting initial motion and bed-load transport of sediment particles in the shallow marine environment. *Shelf Sediment Transport Processes and Patterns*. D.J.P. Swift, D.B. Duane, and O.H. Pilkey, (eds.), Dowden, Hutchinson, and Ross, Inc., Stroudsburg, PA, pp. 61–82.

Stevenson, A.H. (1953). Studies of bathing water quality and health. *J. Am. Public Health Assoc.*, **43**, 529.

Streeter, H.W. (1951). Bacterial quality objectives for the Ohio river: A guide for the evaluation of sanitary conditions of waters used for potable supplies and recreational uses. Cincinnati, Ohio River Valley Water Sanitation Commission.

Suh, S.-W. (2006). A hybrid approach to particle tracking and eulerian-lagrangian models in the simulation of coastal dispersion. *Environmental Modelling & Software*, **21**(2), 234–242.

Suriname (2004). General Bureau of Statistics in Suriname. *Census 2004*.

Swanson, C. and Isaji, T. (2008). Use of a Lagrangian particle model to simulate the far field transport and fate of effluent discharged from the proposed lacsd outfall. *MWWD 2008*, Cavtat, Croatia, October 27–31, 2008.

SWRCB (2005). Water quality control plan, ocean waters of California. State Water Resources Control Board, California.

Tang, H.S., Paik, J., Sotiropoulos, F., and Khangaonkar, T. (2008). Three-dimensional numerical modeling of initial mixing of thermal discharges at real-life configurations. *J. Hydraul. Eng.*, **134**(9), 1210–1224.

Teeter, A.M. and Baumgartner, D.J. (1979). Prediction of initial mixing for municipal ocean discharges. CERL Pub. 043, 90pp. U.S. Environmental Protection Agency Environmental Research Laboratory, Corvallis, Oregon.

Tian, X. and Roberts, P.J.W. (2003). A 3D LIF system for turbulent buoyant jet flows. *Experiments in Fluids*, **35**, 636–647.

Tian, X., Roberts, P.J.W., and Daviero, G.J. (2004a). Marine wastewater discharges from multiport diffusers I: Unstratified stationary water. *Journal of Hydraulic Engineering*, **130**(12), 1137–1146.

Tian, X., Roberts, P.J.W., and Daviero, G.J. (2004b). Marine wastewater discharges from multiport diffusers II: Unstratified flowing water. *Journal of Hydraulic Engineering*, **130**(12), 1147–1155.

Tian, X., Roberts, P.J.W., and Daviero, G.J. (2006). Marine wastewater discharges from multiport diffusers IV: Stratified flowing water. *J. Hydraulic Engineering*, **132**(4), 411–419.

Trinidad and Tobago (2000). Central Statistical Office. *2000 Housing and Population Census*.

Turner, J.S. (1986). Turbulent Entrainment: The development of the entrainment assumption, and its application to geophysics. *Journal of Fluid Mechanics*, **173**, 431–471.

Union Tribune (2002). Treated sewage waiver signed, by Terry Rodgers, Union Tribune, San Diego, September 14, 2002, (also in California Coastal Coalition).

United Nations (1991). Demographic Yearbook (1991). Department of Economic and Social Development, Statistical Division.

United Nations (2003). Word urbanization prospects: The 2003 revision population database. Department of Economic and Social Affair/Population Division.

United Nations (2004). Statistics Division. *Population Density and Urbanization*.

United Nations Environment Programme (UNEP)/World Health Organization (WHO) (1983). Assessment of the state of microbial pollution of the mediterranean sea and proposed measures. Document UNEP/WG. 91/6, 1983.

United Nations Environment Programme (UNEP) (1985). Report of the fourth ordinary meeting of the contracting parties to the convention for the protection of the mediterranean sea against pollution and its related protocols. Genoa, 9–13 September 1985, Document UNEP/IG. 56/5, 1985.

USEPA (1976). Office of Water Planning and Standards. Quality Criteria for Water. Washington, DC, USEPA, 537p. EPA-440/9/76-023.

USEPA (1984). Water quality criteria. Request for comments. *Federal Register*, **49**(102).

USEPA (1985). Test methods for escherichia coli andenterococci in water by the membrane filter procedure. U.S. Department of Commerce. NTIS. 30p. EPAS-600/4-85/076.

USEPA (1986). Bacteriological ambient water quality criteria availability. *Federal Register*, **51**(45), 8012.

USEPA (1986). Bacteriological ambient water quality criteria for marine and fresh recreational waters. EPA 440/5-84-002. U.S. Environmental Protection Agency, Office of Research and Development, Cincinnati, OH.

USEPA (1991). Technical support document for water quality-based toxics control. EPA 505/2-90-001, U.S. EPA Office of Water, Washington, DC.

USEPA (1994). Amended Section 301(h) Technical Support Document. Oceans and Coastal Protection Div. (4504F), Office of Wetlands, Oceans and Watersheds. EPA 842-B-94-007, September 1994.

USEPA (1997). Method 1600: Membrane filter test method for enterococci in water, United States Environmental Protection Agency, Office of Water, EPA-821-R-97-004, May 1997.

Venezuela (1978). Reglamento parcial No. 4 de la ley orgánica del ambiente sobre clasificación de las aguas.

Wade, T.J., Calderon, R.L., Sams, E., Beach, M., Brenner, K.P., Williams A.H., and Dufour A.P. (2006). Rapidly measured indicators of recreational water quality are predictive of swimming-associated gastrointestinal illness. *Environmental Health Perspectives*, **114**(1), January 2006.

Wade, T.J., Calderon, R.L., Brenner, K.P., Sams, E., Michael Beach, Haugland, R., Wymer, L., and Dufour, A.P. (2008). High sensitivity of children to swimming-associated gastrointestinal illness results using a rapid assay of recreational water quality. *Epidemiology*, **19**(3), May 2008.

WHO (1975). Guide and criteria for recreational quality of beaches and coastal waters. Bilthoven, 28 Oct.–1 Nov. 1974.

WHO (1977). Health criteria and epidemiological studies related to coastal water pollution. Athens, 1–4 March 1977.

WHO (1978). First report on coastal water quality monitoring of recreational and shellfish areas (MED VII). WHO United Nations Environment Programme (UNEP). WHO/EURO Document ICE/RCE 206(8), Copenhagen.

WHO (1986). Correlation between coastal water quality and health effects: Report of a joint WHO/UNEP meeting. Follonica, 21–25 October 1985. Copenhagen, WHO Regional Office for Europe, WHO/EURO document ICP/CEH 001 M06.

WHO (1998). Guidelines for safe recreational-water environments. Volume 1: Coastal and fresh-waters. Draft for consultation. Geneva: WHO; 1998. WHO/EOS/98.14. 208pp.

WHO (1999) Health-based monitoring of recreational waters: the feasibility of a new approach (the Annapolis Protocol). Geneva, World Health Organization.

WHO (2003). Guidelines for safe recreational water environments. Volume 1, Coastal and fresh waters. World Health Organization. http://www.who.int/water_sanitation_health/bathing/srwe1/en/ Geneva, Switzerland, p. 80.

Winant, C.D. (1980). Coastal circulation and wind-induced currents. *Ann. Rev. Fluid Mech.*, **12**, 271–302.

Winiarski, L.D. and Frick, W.E. (1976). Cooling tower plume model. USEPA Ecological Research Series, EPA-600/3-76-100, USEPA, Corvallis, Oregon.

Winslow, C.E.A. and Moxon, D. (1928). Bacterial pollution of bathing beaches in new haven harbor. *Am. J. Hyg.*, **8**, 299.

Wood, P.C. (Undated). Public health aspects of shellfish from polluted waters. Chapter 13.

Wood, I.R., Bell, R.G., and Wilkinson, D.L. (1993). *Ocean Disposal of Wastewater.* World Scientific, Singapore.

World Bank (2003). World Development Report (WDR).

WRC (1990). Design guide for marine treatment schemes. Vols I to IV. Water Research Center (WRc) Report No. UM 1009, Swindon, UK.

Wright, S.J. (1977). Mean Behavior of Buoyant Jets in a Crossflow. *J. Hydr. Div., ASCE*, **103**(5), 499–513.

Wright, S.J. (1984). Buoyant jets in density-stratified crossflow. *J. Hydr. Eng., ASCE*, **110**(5), 643–656.

Wong, D.R. and Wright, S.J. (1988). Submerged turbulent buoyant jets in stagnant linearly stratified fluids. *J. Hydr. Res.*, **26**(2), 199–223.

Zhang, X.Y. (1995). Ocean outfall modeling – interfacing near and far field models with particle tracking method. Thesis, Civil Engineering, MIT, Cambridge, MA.

Zhang, X.Y. and Adams, E.E. (1999). Prediction of near field plume characteristics using a far field circulation model. *J. Hydraul. Eng.*, **125**(3), 233–241.

Zimmerman, J.T.F. (1986). The tidal whirlpool: A review of horizontal dispersion by tidal and residual currents. *Netherlands J. Sea Res.*, **20**, 133–156.

Index

A
Aberdeen, case study 356–358
ACBM *see* articulated concrete block mat
acceleration, under passing waves 233–236
acoustic Doppler current profiler (ADCP) 140
  data, vessel mounted 142
  Hawaii 141
activated sludge process 174–175
ACUACAR 416–419, 421, 443–444, 446
acute febrile respiratory illness (AFRI) 36, 38–41
air
  elimination from pipes 276–279
  fill ratio 324
Airy (linear) analyses, waves 234–236
AIZ *see* allocated impact zone
allocated impact zone (AIZ) 19
American Public Health Association study 26–33

American Society for Testing Materials (ASTM) standards 248
  bending 274
  glass reinforced plastics (GRP) pipes 216
  high density polyethylene (HDPE) outfalls 250–253, 255
anchored pipes 269
anodes 220–221
articulated concrete block mats (ACBM) 266–268
assembly 311–312
ASTM *see* American Society for Testing Materials
Auger type anchors 262–263

B
bacteria
  American Public Health Association study 26–33

bottle method 157–158
California Ocean plan 21–22
contaminants 26–47
decay data 156–158
density 30, 32
lowest-observed-adverse-effect level (LOAEL) 4
measurement 157–158
monitoring 437–438
no-observed-adverse-effect level (NOAEL) 40–41
Santa Marta Bay 437–438
standards 21–22, 33–42
swimmers 30, 32
wastewater 409
ballast
  attachment 318
  costs 460
  friction coefficients 260
Bandra outfall 361
barges 332–334
bathing water criteria 34–35, 43–44
bathymetry
  charts 153–154
  data 151–153
  Mamala Bay 124
  Santa Marta Bay 435
beaches 410–411
Beaches Environmental Assessment, Closure, and Health (BEACH) 33
bending, pipe stress 271–274
Biarritz 372–373
biochemical oxygen demand (BOD)
  case studies 430–432
  pollutants 16–17, 20
  treatment 172–173, 180–182, 185, 191–195
  treatment costs 200–201
  wastewater 161, 168–170, 406, 409
biological characteristics, California Ocean plan 24
biological sampling 407–408
BOD see biochemical oxygen demand
bolts, costs 460
Boston 132–133
  case study 358
  plumes 134

bottle method 157–158
bottom pull method 327–330, 328–329
box model 91
buoys 399
butt fusion
  joints 319
  welding 312–315

C

California Ocean plan 20–25
  water quality limits 23
Caribbean Environment Programme (CEPPOL) 34
Cartagena 128–132
  case study 414–428
  coliforms 131
  hydrodynamic model 130
  monitoring 410
  preliminary treatment 189
  public relations program 442–446
  sanitation sector 416–418
  social aspects 427–428
  wastewater disposal 421–427
  World Bank support 418–421
case studies
  Aberdeen 356–358
  Biarritz 372
  biochemical oxygen demand (BOD) 430–432
  Boston outfall 358
  Cartagena 414–428
  Castro Urdiales 380–381
  Concepcion 428–434
  Europipe 369–370
  Horden 370–371
  Isle of Man 381–383
  Mumbai outfall 360–361
  numerical modeling 123–132
  Rockaway Beach 383–384
  Santa Marta 434–438
  South Bay Ocean 359–360
Castro Urdiales
  case study 380–381
  horizontal directional drilling (HDD) 381
CCC see criteria continuous concentration
CDD see cold digestion/drying

centerline dilution
  diffusion 83
  jets 64
centrifuge thickening 177–178
CEPIS *see* El Centro Panamericano de Ingenierýa Sanitaria y Ciencias del Ambiente
CEPPOL *see* Caribbean Environment Programme
CEPT *see* chemically enhanced primary treatment
CERFS *see* chemically enhanced rotating fine screens
CFD *see* computational fluid dynamics
check valves
  diffusers 290–293
  duckbill 283, 291–292
chemical oxygen demand (COD) 169, 189–190, 195, 432
chemical sampling 406–407
chemically enhanced primary treatment (CEPT) 181, 192–193
  costs 200–201, 204, 301
  and filtration 193–194
  removal capabilities 301
  ultraviolet radiation (UV) 193–194
chemically enhanced rotating fine screens (CERFS) 195–196
  costs 301
  and filtration 196–197
  removal capabilities 301
  solids separation 195–196
civil works 460
clamshell bucket 338
clogging 276
CMC *see* criteria maximum concentration
cnoidal analyses 234–236
coastal wastewater discharge 52, 170
coastal waters
  density stratification 56–57
  hydrodynamics 54
  motion of 55–56
  wastewater discharge 52, 170
  winds 57–59
COD *see* chemical oxygen demand
cold digestion/drying (CDD) lagoons 179

coliforms 409
  Cartagena 131
  Columbia 168
compound limits 23
computational fluid dynamics (CFD) 96
  direct numerical simulation (DNS) 101–102
  models 101–103, 135
Concepcion 428–434
concrete ballast 315–318
  pipe stabilization 257–262
  Santa Marta Bay 436
  weights 258
concrete pipes 207, 210–213
  joints 211
  tunnel boring machine (TBM) 365
conductivity, temperature, and depth (CTD) 145–146
connections 208–209
construction offices 30
contaminants
  bacteria 26–47
  box model 91
CORMIX models 96, 108–109, 122, 279
corrosion protection
  ductile iron 220
  maintenance 394–396
  pipes 219–221
costs 447–460
  chemically enhanced primary treatment (CEPT) 200–201, 204, 301
  chemically enhanced rotating fine screens (CERFS) 301
  direct 458
  Engineering News Record (ENR) index 449–454, 449–455
  glass reinforced plastic (GRP) 448–455
  high density polyethylene (HDPE) 459–460
  indirect 458–459
  maintenance 400
  treatment processes 181–182
coupled particle tracking model 121
criteria continuous concentration (CCC) 25–26
criteria maximum concentration (CMC) 25

CTD *see* conductivity, temperature, and depth
currents
   data 138–144
   drag coefficient 239, 242–243
   inertia coefficient 244
   lift coefficient 239, 241–242, 245
   oceanic 224–226, 236–240
   outfall construction 302–303
   pipe stress 269–270
   polar scatter 143
   Reynold's number 241
cutting removal 367
   tunnels 354

D

DAF *see* dissolved air flotation
data
   bacterial decay 156–158
   bathymetry 151–153
   currents 138–144
   density stratification 145–149
   geophysics 153–155
   meteorology 149–151
   tides 149
   water quality 155
   waves 149
Delft3D model 111, 115, 122, 129–131
density stratification 145–149
   coastal waters 56–57
   conductivity, temperature and depth (CTD) profiling 146–147
   plumes 57
deposits 392
DESA *see* distributed entrainment sink approach
design
   diffusers 281–284
   wave 226–228
developing countries
   effluent standards 168–169
   wastewater management guidelines 164–167
diameters 255–256
diffusers 281–284
   check valves 290–293
   clogging 276

   configuration of 279–281
   deposits 392
   design 279–293
   far field dilution 73
   hydraulic calculations 286–290
   installation 344–346
   maintenance 396–397
   near field dilution 73
   outfall construction 344–346
   plumes 280
   port configurations 284–286
diffusion
   centerline dilution 83
   turbulent mixing 82
dimension (diameter) ratio (DR) 251, 253–256, 270, 273–274, 324
   standard (SDR) 379
direct numerical simulation (DNS) 101–102
direct pipe installation 388
discharge
   California Ocean plan 21
   flow 398–399
   fluxes 67
   maintenance 398–399
   multiport diffuser 67
   pipe diameters 254
   post-discharge monitoring 406–408
   pre-discharge monitoring 405–406
disinfection 176–177
   ultraviolet radiation (UV) 176–177
dissolved air flotation (DAF) 173
dissolved oxygen (DO) 115–116, 170, 430–431
distributed entrainment sink approach (DESA) 109, 121
DNS *see* direct numerical simulation
DO *see* dissolved oxygen
DR *see* dimension ratio
drag
   anchored pipes 269
   coefficient 238–239, 242–243
dragline bucket 337
drilling 376–384
drive shafts 356
driven type anchors 262
drogues 145
duckbill check valves 283, 291–292

Index   483

ductile iron pipes 207, 214–216
  corrosion protection 220
  joints 214
dyes 398

E
ear infections (EI) exposure
  risk for acute febrile respiratory
    infection (AFRI) 39
  risk for gastrointestinal (GI)
    infection 39
earth pressure balance
  machine (EPBM) 353–354, 360, 367
  tunnel boring machine (TBM) 354
echo sounders 153–154
ECOM model 110–111, 120
EEC *see* European Economic Community
effluent standards 167–169
  California Ocean plan 21
  developing countries 168–169
  industrialized countries 167
EIA *see* Environmental Impact
    Assessment
El Centro Panamericano de Ingenierýa
    Sanitaria y Ciencias del
    Ambiente (CEPIS) 9
elastic modulus 270
ELCOM model 111
Engineering News Record (ENR) index
    449–455
enterococci
  swimmers 30
  wastewater 409
entrainment models 96–101
environmental impact assessment (EIA),
    non-government organizations
    (NGOs) 427
environmental impact assessment (EIA)
    155, 427–428, 440, 442–444
EPBM *see* earth pressure balance machine
epidemiology, WHO guidelines 37
EU *see* European Union
Eulerian models 96, 115–116
  wave views 233
European Economic Community (EEC),
    bathing water criteria 34–35,
    43–44

European Union (EU), water quality
    criteria 34
Europipe 369–370
excavation costs 460
expansion type rock anchors 262

F
far field model 109–114
  diffusers 73
  ECOM model 110–111, 120
  Regional Ocean Model System
    110–111
  statistical 84–89
  transport 85
  wastewater mixing 54, 81–90
fats, oil, and grease (FOG) 170
Fenton's Fourier series analyses, waves
    234–236
fiberglass *see* glass reinforced plastic
fish
  lipid content 407
  marine monitoring 407
  mercury 407
  moisture content 407
  polychlorinated biphenols (PCBs) 407
fishing 399
fittings 460
flange plates 291
flexibility of pipes 210
floating crane barges 333–334
flotation
  outfall construction 308–322
  pipe stress 270–271
flow
  drag coefficient 238
  duckbill check valves 292
  flushing velocities 395
  multiport diffusers 71, 77
  Reynolds number 238
  surge chambers 278
  wastewater 409
flowing currents
  jets 76
  merging plumes 70–73, 76–80
  single plume 68–70
flushing
  flow velocities 395
  wastewater mixing 54, 91–92

FOG *see* fats, oil, and grease
forces, high density polyethylene (HDPE) outfalls 260
friction coefficients 260

G
gantries 317
GBT *see* gravity belt thickening
GEF *see* Global Environmental Facility
geophysics data 153–155
geotechnical investigations 350–351
glass reinforced plastic (GRP) 205–208
    American Society for Testing Materials (ASTM) standards 216
    costs 448–455
    International Organization for Standardization (ISO) 216
    pipe connections 208–212, 217
    pipes 207, 216–217, 216–219, 304, 333, 357, 398
Global Environmental Facility (GEF) 167
global positioning system (GPS) 144, 151, 302, 321–322, 330, 333, 407
global warming 202–203
global water supply 162–163
government, and wastewater management 166–167
GPS *see* global positioning system
gravitational spreading
    multiport diffusers 80
    parallel currents 80
gravity belt thickening (GBT) 177–178
grease 409
grouted type rock anchors 262
GRP *see* glass reinforced plastic
guidelines
    developing countries 164–167
    microbial water quality 46
    shellfish 46
    wastewater management 164–167

H
Hawaii 141
HDB *see* hydrostatic design basis
HDD *see* horizontal directional drilling
HDPE *see* high density polyethylene
head loss 254

high density polyethylene (HDPE) pipe
    pullback 382
    towing 383
    pipes, ocean containers 305
    see also polyethylene
high density polyethylene (HDPE) pipes 218–219
    ASTM standards 250
    concrete ballast weights 258
    damage to 399–400
    diffusers, tapered 283
    forces on 260
    ISO standards 249
    maintenance 399–400
    pressure ratings 253
    stresses on 268–274
    transport 305–306
hole opener assembly 382
Horden outfall
    case study 370–371
    microtunneling 371
horizontal buoyant jets 60–66
horizontal directional drilling (HDD) 348–349, 356
    Castro Urdiales 381
    direct pipe 388–389
    marine outfalls 376–379, 381
    pipe thrusters 384
    river crossing 377
horizontal ports 286
hydraulics
    diffusers 286–290
    polyethylene pipes 274–279
hydrodynamic models 109–114
    Cartagena 130
    coastal waters 54
    Delft3D model 111, 115, 122, 129–131
    Ipanema 128–129
hydrostatic design basis (HDB) 251–253, 274
hydrostatic pressure 245–246

I
IJS *see* intermediate jacking station
indigenous species 47

Index

industrialized countries, and effluent standards 167
inertia coefficient 244
installation
  diffusers 344
  mechanical anchors 344
  outfall construction 307–335
  submarine outfalls 215
intake installation 379–380
intermediate jacking station (IJS) 365, 371
International Organization for Standardization (ISO)
  bacteria 34, 405
  glass reinforced plastic (GRP) pipes 216
  high density polyethylene (HDPE) outfalls 249
  minimum required strength (MRS) 274
  polyethylene pipes 248–253, 255
Ipanema 128–129
Isle of Man 381–383
ISO *see* International Organization for Standardization

J

jack-up platform 380
jacking stations 365–366, 368
  intermediate (IJS) 365, 371
jet nozzles 342
  cutter head 340–341
  suction dredge 339
  suction pipe 340
jet sled 342
jets
  centerline dilution 64
  flowing 76
  laser-induced fluorescence (LIF) images 61, 65, 69
  multiport diffuser 66, 70
  stationary 61, 64–66, 68
  stratified current 74–75, 74–76
  unstratified current 70
joints
  concrete pipes 211–212
  costs 460
  ductile iron pipes 214
  PVC pipes 219
  testing 319

L

La Plata treatment plant 186
LAC *see* Latin America and the Caribbean
lagoon thickening 177–178
Lagrangian models 96, 116–119
  plumes 99
  wave views 233
Lake Michigan
  nesting model 114
  particle tracking model 117–118
large eddy simulation (LES) 102
laser-induced fluorescence (LIF) 60–61, 65–67, 69, 71–72, 79
Latin America and the Caribbean (LAC) 8–10, 48–49
launching 319–321
lay barges 332–333
leak detection 398
legal mixing zone (LMZ) 19
length-scale models 96
LES *see* large eddy simulation
LIF *see* laser-induced fluorescence
lift coefficient 239, 241–242, 245
linear (Airy) analyses 234–236
linings 354–356
lipids 407
LMZ *see* legal mixing zone
LOAEL *see* lowest-observed-adverse-effect level
long-term flushing 54, 91–92
lowest-observed-adverse-effect level (LOAEL) 40–41

M

maintenance 391–392
  activity list 401
  buoys 399
  clean pipe bore 392–394
  corrosion protection systems 394–396
  costs 400
  diffusers 396–397
  discharge flow recording 398–399
  fishing 399
  high density polyethylene (HDPE) 399–400
  logs 399
  PIG 396, 398, 401

remote operated vehicle (ROV) 396
seabed erosion 400
total headloss recording 399
visual inspection 397–398
Mamala Bay 123–127
marine organisms
  compound limits 23
  submarine pipes 239
marine outfall installation 379–380
  drilling 376–384
  microtunneling 364–375
  tunneling 351–375
marine waters, and illness in swimmers 36
mathematical models 95–135
  computational fluid dynamics (CFD) 101–103
  CORMIX 108–109
  coupling 119–123
  distributed entrainment sink approach (DESA) 109, 121
  entrainment 96–101
  Eulerian 96, 115–116
  hydrodynamic 109–114
  Lagrangian 96, 116–119
  length-scale 96
  nesting technique 113–114
  NRFIELD 107–108
  turbulence 96
  VISJET 109
  visual plumes 103–107
mean low low water (MLLW) 149
mean sea level (MSL) 362–363, 374–375, 385–387
mechanical anchors
  installation 344
  outfall construction 344
  stabilization 262–263
mercury 407
merging plumes
  flowing currents 70–73
  flowing stratified current 76–80
  stationary stratified current 75–76
meteorology
  data 149–151
  stations 150
microbial water quality
  guidelines 46

indigenous species 47
recreational waters 44–45
shellfish 46
standards 47
World Health Organization (WHO)
  guidelines 40–41
microtunneling 374–375, 385–387
  Horden outfall 371
Mike model 111
minimum required strength (MRS) 249–252
  International Organization for Standardization (ISO) 274
mixing zone
  concept of 18–20
  models 96–109
  outfalls 19–20
MLLW *see* mean low low water
mobile jack-up platforms 330–331
moisture content 407
monitoring 403–404
  beaches 410–411
  Cartagena 410
  marine fish 407
  outfalls 405–410
  parameters 404–405
  program of 412
  wastewater 408–409
  WHO schedule 412
MRS *see* minimum required strength
MSL *see* mean sea level
multiple buoyant jets 66–68
multiport diffusers
  discharge fluxes 67
  flow 71
  gravitational spreading 80
  near field dilution 78
  parallel currents 80
  plumes 66, 70–72
  rise height 79
  stratified currents 77–79
  unstratified current 72
Mumbai outfall 360–361

N

National Pollutant Discharge Elimination System (US) (NPDES) 18, 103

National Research Council (NRC) 16, 162
near field dilution
  deep outfalls 295
  diffusers 73
  multiport diffusers 78
  plumes 73
  stratified currents 78
  unstratified current 73
near field models 96–109
  plumes 67, 81
  schematic 125
  wastewater mixing 54, 59–81
nesting models 114
nesting techniques 113–114
NGOs see non-government organizations
no-observed-adverse-effect level (NOAEL) 40–41
non-government organizations (NGOs)
  environmental impact assessment (EIA) 427
  public assessment 439–440, 443, 445
NPDES see National Pollutant Discharge Elimination System (US)
NRC see National Research Council
NRFIELD models 107–108
numerical modeling 123–132
nutrients 409

O

O&M see operation and maintenance
ocean containers 305
oceanic force
  currents 224–226, 236–240
  hydrostatic pressure 245–246
  turbulent mixing 82–84
  waves 226–236, 240–245
oil 409
open clamshell bucket 338
operation and maintenance (O&M) costs 160, 173, 175–176, 178–180, 182, 188, 191, 193–195, 197, 200, 203, 400
outfall construction
  currents 302–303
  installation methods 307–335
  offices 303
  pipe supply 304–306

public relations 307
regulations 301
sea conditions 302–303
seabed conditions 301–302
storage yards 306–307
surf zone conditions 303
tides 302–303
waves 302–303
winds 302–303
outfalls
  Boston 133
  coastal waters 52
  costs 447–460
  deep 293–296
  launching, costs 460
  loading on 271
  microtunneling 374–375, 385–387
  mixing zone 19–20
  monitoring 405–410
  plumes 295
  shallow 296–298
  wastewater discharge 52
oxidation ditch process 174
oxygen, dissolved oxygen (DO) 115–116, 170, 430–431

P

PAE see projected area entrainment
PAH see polycyclic aromatic hydrocarbons
PAHO see Pan American Health Organization
Pan American Health Organization (PAHO) 445
parallel currents 80
parameters
  monitoring 404–405
  wastewater 409
particle tracking models
  coupled 121
  Lake Michigan 117–118
  random walk particle tracking (RWPT) 116–117
PCB see polychlorinated biphenyl
PDF see probability density function
PE see polyethylene
Penco outfalls 429, 431

pH 409
physical characteristics 22
physical models 133–134
physical sampling 406–407
pier assembly 311
PIG 393–394
  maintenance 396, 398, 401
pilot bore 376–377
pipe bore 392–394
pipe diameters
  discharge 254
  head loss 254
  velocity curves 254
pipe materials 205–208
  characteristics 207
  costs 207
pipe stress
  bending 271–274
  currents 269–270
  flotation 270–271
  waves 269–270
pipe supply
  costs 460
  outfall construction 304–306
pipe thrusters 384
pipes
  air elimination 276–279
  concrete 207, 210–213
  concrete ballast stabilization 257–262
  connections 208–209
  corrosion protection 219–221
  dimension (diameter) ratio (DR) 251, 253–256, 270, 273–274, 324, 379
  ductile iron 207, 214–216
  flexibility 210
  glass reinforced plastic (GRP) 207, 216–217
  high density polyethylene (HDPE) 218–219
  internal pressure 209
  mechanical anchor stabilization 262–263
  polyethylene (PE) 205, 207
  polypropylene (PP) 205, 207
  polyvinyl chloride (PVC) 207, 217–219

resistance 209–210
self-cleaning velocities 275–276
steel 207, 213–214
trench installation 263–266
platforms 460
plumes
  Boston outfall 134
  centerline dilution 64
  currents 124
  deep outfalls 295
  density stratification 57
  diffusers 280
  entrainment 97
  flow 71
  flowing 76
  Lagrangian model 99
  laser-induced fluorescence (LIF) 60–61, 65–67, 69, 71–72, 79
  multiport diffusers 66, 70–72
  near field dilution 73
  near field properties 67, 81
  round 100
  stationary 61, 64–66, 68, 81
  stratified 74–76
  temperature 124
  unstratified current 68, 70, 72
  visual plumes (VP) 103–106
polar scatter 143
pollutants
  biochemical oxygen demand (BOD) 16–17, 20
  coastal wastewater discharge 170
  wastewater 161, 169–170
polychlorinated biphenyl (PCB) 5
  fish 407
  wastewater 408–409
polycyclic aromatic hydrocarbons (PAH) 16, 170
polyethylene (PE) pipes 205, 207
  diameter 253–256
  diameters 255–256
  hydraulics 274–279
  reference standards 248–253
  stabilization of 256–268
  stresses 268–274
  see also high density polyethylene (HDPE) pipes

Index

polyethylene (PE) resins
  cell classification limits 252
  elastic modulus 270
polypropylene (PP) 205, 208, 304
polyvinyl chloride (PVC) 205
  characteristics 207
  internal pressure 209
  pipe connections 208
  pipes 207, 217–219
  resistance of 209
POM see Princeton ocean model
port configurations 284–286
positioning 321–322
PP see polypropylene (PP)
pre-reaming 378
preliminary treatment
  Cartagena 189
  solid waste 197–199
pressure
  high density polyethylene (HDPE)
    pipes 253
  pipes 209
pressure balance
  earth pressure balance machine (EPBM)
    353–354, 360, 367
  rock cutting head 352
  tunnel boring machine (TBM) 353
Princeton ocean model (POM) 110–112, 114
  POMGL 114, 117
probability density function (PDF) 39
programs 412
projected area entrainment (PAE) 100
public assessment 439–440, 443, 445
public relations
  outfall construction 307
  program 439–446
pull-back 378–379
pulling head 320
PVC see polyvinyl chloride

Q

quality assurance and quality control (QA/QC) 137, 405, 408

R

radioactivity 24
rail systems 310
random walk particle tracking (RWPT) 116–117
RANS see Reynolds-averaged Navier-Stokes
RAS see return activated sludge
RC see reinforced concrete
reception shafts 356
recreational waters
  microbial water quality 40–41, 44–45
  WHO guidelines 40–41
reference standards 248–253
Regional Ocean Model System (ROMS) 110–111
regulations 301
reinforced concrete (RC) 362, 374–375, 451
remote operated vehicle (ROV) 152, 392
  deep outfalls 293
  diffuser maintenance 396
  inspections 293, 392, 397, 401
  maintenance 396
  sea bed 392
removal capabilities
  CERFS 301
  chemical enhanced primary treatment (CEPT) 301
  treatment processes 181
resistance 209–210
return activated sludge (RAS) 168–169, 174
Reynold's number
  currents 241
  flow 238
Reynold's-averaged Navier-Stokes (RANS) 102, 111
rigs
  jack-up platform 380
  layouts 378
Rio de Janiero 127–128
rise height 79
riser (MWRA) 133
risks of illness
  acute febrile respiratory illness (AFRI) 39
  ear infection (EI) exposure 39
  gastrointestinal illness (GI) 39
  swimmers 36

river crossing 377
rock cutting head 352
Rockaway Beach 383–384
ROMS *see* Regional Ocean Model System
rotating fine screen system 184–188, 190
ROV *see* remote operated vehicle
RWPT *see* random walk particle tracking

S

salinity profiles 148
sample monitoring 408–410
Sand Island 126–127
   visitation frequencies 89
sanitation sector, Cartagena 416–418
Santa Marta Bay 434–438
   bathymetry 435
SBR *see* sequencing batch reactor
screens, volume 199
SDR *see* standard dimension ratio
sea conditions 302–303
seabed
   conditions 301–302
   erosion 400
   excavation 336–342
   imaging 155
   trenches, backfilling 265, 267
   tunnel boring machine (TBM) 369
secondary treatment 190
sediments 264–265
self-cleaning pipes 275–277
sequencing batch reactor (SBR) 174–176
shafts 368–369
shellfish
   microbial water quality guidelines 46
   standards 45–47
shields 351–353, 367
single plume
   flowing currents 68–70
   stationary stratified flow 73–75
sludge
   dewatering 178–179
   gravity belt thickening (GBT) 177–178

management 199–200
   stabilization 178
   thickening 177–178
slurry 353
social aspects 427–428
solid waste treatment 197–199
South Bay outfall 39
   case study 359–360
stabilization by ultraviolet radiation (UV) 251
standard dimension ratio (SDR) 379
standards
   bacterial water quality 35–45
   indigenous species 47
   microbial water quality 47
   shellfish 45–47
   toxicity 25–26
State Water Resources Control Board (California) (SWRCB) 21, 44
stationary current
   horizontal buoyant jet 60–66
   multiple buoyant jet 66–68
stationary plumes 81
stationary stratified current
   merging plumes 75–76
   single plume 73–75
steel pipes 207, 213–214
storage yards 306–307
straight diffusers 280
stratified currents
   jets 74–76
   multiport diffusers 77–79
   near field dilution 77–78
   plumes 74–76
   rise height 79
stresses 268–274
submarine outfalls
   installation 215
   marine organisms 239
   pipes 210–219, 239
submersion
   air fill ratio 324
   outfall construction 322–327
subsea excavation 342–344
surf zone conditions 303
surface drifters 144

surge chambers, flow 278
swimmers
  bacteria density 30, 32
  enterococcus density 30
  risks of illness, marine waters 36
SWRCB *see* State Water Resources Control Board (California)

T
T-diffusers 280
tapered high density polyethylene (HDPE) diffusers 283
TBM *see* tunnel boring machine
TDZ *see* toxic dilution zone
Teledyne Benthos SIS-1625 155
Telemac model 111
temperature profiles 148
testing 319
tides
  data 149
  outfall construction 302–303
Tome outfalls 429–430
total headloss recording 399
total suspended solids (TSS) 161
  Columbia 168
  monitoring 409, 430–431
  treatment 181, 191–195
  treatment costs 200–201
  wastewater 170, 409
towing
  high density polyethylene (HDPE) pipe 383
  outfall construction 321–322
toxic dilution zone (TDZ) 20
toxicity standards 25–26
transport 305–306
trapezoidal ballast weight 259
treatment
  biochemical oxygen demand (BOD) 172–173, 180–182, 185, 191–195, 200–201
  costs 181–182, 200–201
  nutrient removal 176
  physico-chemical 190–197
  plants 186

preliminary 172, 183–190, 197–199
primary 172–173
removal 181
secondary 173–176
solid waste 197–199
tertiary 176
trenches
  backfilling 265, 267
  pipe installation 263–266
trenchless techniques
  advantages of 30, 348–349
  geotechnical investigations 350–351
  outfall construction 348–388
trestles 334–335
trickling filters 174
TSS *see* total suspended solids
tunnel boring machine (TBM) 351–353, 367
  concrete pipes 365
  recovery, seabed 369
tunnels
  cutting removal 354
  linings 354–356, 367
  marine installation 351–375
  outfalls 362–363
  segments 355
turbidity 409
turbulence
  mixing 82
  models 96

U
UASB *see* upflow anaerobic sludge blanket
UFW *see* unaccounted for water
ultraviolet radiation (UV) 173
  CERFS 196
  chemically enhanced primary treatment (CEPT) 193–194
  disinfection 176–177
  stabilization 251
unaccounted for water (UFW) 419
UNEP *see* United Nations Environmental Program
United Nations Environmental Program (UNEP) 33–34, 36–37, 44, 46

United States Environmental Protection
    Agency (USEPA) 25
  bacterial contaminants 26, 28–33, 35,
    43–46, 177, 405, 411
unstratified current
  jets 70
  multiport diffusers 72
  near field dilution 73
  plumes 68, 70, 72
upflow anaerobic sludge blanket (UASB)
    176
  reactors 182
USEPA see United States Environmental
    Protection Agency
UV see ultraviolet radiation

V
velocity
  curves 254
  self-cleaning pipes 275–277
  under passing waves 233–236
vertical ports 286
visitation frequencies, Sand Island 89
VISJET models 109
visual inspection, maintenance 397–398
visual plumes (VP) 103–106, 103–107
  inputs 106
  output 107
vortex grit chambers 188–190
VP see visual plumes

W
walls 211–212
waste activated sludge (WAS) 174
wastewater
  bacteria 409
  biochemical oxygen demand (BOD)
    406, 409
  coliforms 409
  enterococci 409
  flow 409
  grease 409
  monitoring 408–409
  nutrients 409
  oil 409
  parameters 409
  pH 409

  pollutants 161, 169–170
  total suspended solids (TSS) 409
  turbidity 409
wastewater discharge
  coastal waters 52
  in-situ bacterial measurement 157
wastewater disposal 163–164
  Cartagena 421–427
wastewater management
  biochemical oxygen demand (BOD)
    161, 168–170
  guidelines, developing countries
    164–167
wastewater mixing
  far field 54, 81–90
  long-term flushing 54, 91–92
  near field 54, 59–81
wastewater sampling 408
wastewater treatment
  global warming 202–203
  principals of 171–180
  processes 172–177
  technology 180–201
water
  coastal water motion 55–56
  waves 231–233
water quality
  criteria 33–34
  data 155
  general issues 16–18
  limits, California Ocean plan 23
  models 115–119
  objectives, California Ocean plan 21
  recommendations 47–49
  standards for bacteria 35–45
  standards for shellfish 45–47
  standards for toxics 25–26
  World Health Organization standards
    33–42
water utilities 165–166
waves 302–303
  acceleration under passing 233–236
  Airy (linear) analyses 234–236
  analysis of 229, 235
  cnoidal analyses 234–236
  data 149
  deep water transformation 231–233

Fenton's Fourier series analyses 234–236
induced oceanic force 240–245
oceanic force 226–236
pipe stress 269–270
significant wave period 231
sinusoidal 233
velocity under passing 233–236
wind-driven waves, prediction of 228–231
WHO *see* World Health Organization
wind-driven waves 228–231
winds 302–303
coastal waters 57–59
working platforms 309–312

World Bank support, Cartagena 418–421
World Health Organization (WHO)
epidemiology guidelines 37
monitoring schedules 412
recreational water guidelines 40–41
standards for bacteria 33–42
world population growth forecast 164
Worli outfall 361

Y
Y diffusers 280

Z
zone of initial dilution (ZID) 20